Conservation and Management of Tropical Rainforests

An Integrated Approach to Sustainability

Eberhard F. Bruenig

Chair of World Forestry
University of Hamburg
Germany

CAB INTERNATIONAL

Canopy of Mixed Dipterocarp forests in the Mulu area (see Figs 1.16 and 1.17) with great diversity of crown and canopy architecture and spatial variety of gaps in the A and B layers. Reproduced courtesy of Peter Wee.

the
UNIVERSITY
of
GREENWICH

Contents

Preface

Tropical rainforests have fascinated northern scientists, explorers and El Dorado seekers over the past two millenia. The Chinese geographer Li Daw Yuan in the fifth century AD in *Commentaries on the Book of Rivers* described tropical rainforests of 'aromatic lauraceous trees and dense, dark and lush forest with lianes and rich in bird life' in which roamed herds of elephants and rhinoceroses (cit. in Wang, 1961). In the typical Chinese tradition he saw diversity, richness and variety of forms and species of the forests as an expression of the universal harmony. A.v. Humboldt (1847) similarly saw the luscious and varied Amazonian tropical rainforest in the Casiquiare–Rio Negro region expressing harmony and vitality. He postulated the existence of unifying fundamental principles of nature which shape vegetation and landscapes. Forestry professor A. Moeller, who studied the species-rich and diverse Mata Atlantica rainforest in Brazil towards the end of the nineteenth century, recognized the important roles of mutualism, symbiosis and biodiversity. His later *Dauerwaldidee* (Moeller, 1922, 1929) introduced the concept of biodiversity as a universally valid tool of sustainable forest management (Bier, 1933). Measures to protect endangered species and tropical forests were already thought necessary in the mid-nineteenth century (Schlich, 1889; Beccari, 1904). Since the 1920s, commercial greed, political misdemeanour and epidemic increase of the human species have begun locally to threaten tropical rainforests. Even in the 1920s sustainability was recognized as the priority problem for world forestry in the twentieth century, first by scientists (Heske, 1931a,b), later by practitioners and lastly by the public. From the beginning it was realized that the problem was more social and political than technical or scientific.

During the past 200 years natural and anthropological research has amassed a wealth of scientific knowledge. Native subsistence farmers and forest dwellers possess practical experience which has been supplemented during the past 100 years by commercially orientated, largely expatriate agriculturists and foresters. The stock of knowledge and experience is still much less than the ecological and economic importance of the tropical rainforest deserves and demands, but it is sufficient to improve present practices and to design feasible low-risk and adequately sustainable forest management and integrated land-use systems. However, tropical deforestation continued to accelerate and was exacerbated in the 1960s by a timber boom, which exceeded natural resource potentials during the 1970s and 1980s. Growing public awareness and concern spawned both welcome and unwelcome consequences. Poorly qualified dilettantes claimed competence to pass judgement on forestry science and practice. In the 1980s it became fashionable to ignore existing knowledge and experience, to pretend that we know nothing about the tropical rainforest, to reinvent the wheel in tropical ecology and in forestry science and practice, and to alienate the perception of native life, customs and aspirations. Confusion and confrontation led to acrimonious antagonism that thwarted progress towards the recovery of sustainability. On the other hand, the increasing interest in rainforests exposed a degree of overuse and abuse of resources that threatened the long-term social and political stability and sustainability of life support systems in many tropical countries. In the field of forestry, the obstacle to sustainability lies not in the nature of the tropical rainforest but in human ethics and our genetically programmed behavioural tendencies. Our biological heritage and fashionable ideologies have lured all humankind into a state in which 'the present chaotic and anachronistic commercial, social and intellectual sectors, and the antiquated and self-centred sovereign-state system, cannot deal effectively or equitably with the problems facing mankind' (Garnett quoted by Myint Zan, 1994).

My motivation and purpose in writing this book are to:

- highlight that there is a large body of knowledge, experience and tradition available to ecology and forestry in the tropical rainforest which is poorly applied;
- correct the more notorious scientific myths and public misunderstandings of the ecology of the rainforest and of the cultural systems involved;
- describe the principles of integrated conservation and management, which have achieved reasonable levels of practicability and promise to approach sustainability in spite of the complex dynamics and uncertainties of forestry;
- identify relevant unifying basic laws that regulate the processes in the forest structure–function system and which are fundamental to ecosystem viability and stability (Grossman and Watt, 1992) in forests of all biomes;
- demonstrate the complexity of the dynamic interactions within and between natural and cultural ecosystems, making each case unique in spite of the universality of basic laws and principles;

- discuss features of the natural forest and socio-cultural ecosystems which can be understood as adaptive mechanisms and can be mimicked in the design of self-sustainable forests that are viable, robust and tolerant, capable of coping flexibly with uncertainty;
- define principles of feasible tactics and strategies for forest management and conservation, applying holistic approaches, compromising between fundamental universality and specific uniqueness, especially for Sarawak.

The scheme is based on the hypothesis that there are fundamental natural principles *sensu* Humboldt (1847) which apply to forest ecosystems everywhere, including the rainforest, but that each ecosystem is a unique combination of past and present structural, dynamic and process features. I shall argue for the rationale to mimic nature and utilize inherent ecosystem dynamics and indeterminism to improve self-sustainability and economic viability. I shall also argue that the failures to stop the currently rampant overuse, misuse and abuse of the rainforest and to return to a state of sustainability do not lie in the nature of the rainforest ecosystem, lack of knowledge or experience but in the inadequate states of the human mind and the devaluation of traditional culture. Social and political behaviour have become more and more chaotic, anachronistic, antiquated and have created dangerously non-sustainable conditions in many parts of the world. In the tropical rainforest biome, as elsewhere, commercial, social and political forces continue to cause, condone and conspiratorily support overuse, misuse and abuse of the natural and human resources. Under the present conditions of degraded culture, society as a whole and its intellectual exponents have not the power to halt the global plunder (Bruenig, 1981, 1989d,e). Mitigation requires national and international consensus, holistic and comprehensive strategies and internationally coordinated programmes including the realm of human culture. Throughout the treatise I apply the traditional holistic concept of sustainability that has evolved in multiple-use forestry. This age-old concept combines natural and political ecology, as defined by Bryant (1992), and integrates the active processes and linkages within and between natural and cultural levels of the ecosystem hierarchy of the rainforest biome (Grossmann, 1979; Bruenig *et al.*, 1986). The same human motivations and social forces that presently drive towards biological holocaust and cultural collapse, if suitably harnessed and domesticated, could help to design and implement holistic strategies towards sustainability. Basic human motivations, unconstrained by culture, lead to social ills and evils that endanger sustainability of development of natural and human resources globally. Without reinstating traditional ethics and moral values and a profound change of human attitudes, the plundering of the globe and of the tropical rainforest will go on to the bitter end.

The important subjects of system analysis, system model simulation and data and information systems are only briefly touched on because these will be treated in another book (Bruenig *et al.*, 1994). Communal forestry is not treated as a separate subject in this treatise, in spite of it having been my favourite subject as a student and practitioner, because the principles, strategies and problems are not fundamentally different from those described in the text.

Most of the research examples and general information are from South-East Asia and Amazonia, where I have undertaken research and practical work in forest management since 1954. Sarawak is used throughout the book to demonstrate the universal problems and show how solutions can be approached. I gave this preference to Sarawak because my work there gave me a unique insight into the forests and culture in a rainforest country with a long and distinguished history of rainforest conservation, research and management and integration of forestry with other sectors. It is also an example for the very early introduction of community-orientated forestry using the principle of sustainability, originally for forest preservation and the utilization of non-timber forest products and mangrove wood and later also of timber, followed by a period of competitive over-exploitation that in recent years was countered by the political determination to turn the tide.

Acknowledgements

Ecological research and technology development and application in research and practice have been supported by the Sarawak Government 1954–1963 and 1988 to present; by the German Research Foundation (DFG), Bonn, since 1965 to present; by IBM, Hamburg, and Siemens, München and Singapore, for data processing; by the German Federal Ministry for Research and Technology, Bonn, for ecosystem research in China; by the German Agency for Technical Cooperation (GTZ), Eschborn for system-model simulation. Progress in research and practice was made possible by the enthusiasm, high self-discipline and strong motivation of technical and scientific collaborators in the field and office. There are many to whom I owe thanks, but particularly to my Sarawak collaborators of the 1950s, especially Joseph Yong, Bojeng bin Sitam, Ilias bin Piae and Ignatius Chua. The cooperation, participation, understanding, indulgence and friendship of the people of Kampong Bako under Tua Kampong Suhaidi, of the Iban in the upper Batang Ai under Penguluh Ngali and of some of the early timber licencees provided invaluable information and testing grounds without which the scientific research and practical development work could not have succeeded.

The text has been word-processed by Mrs Anja Köhler, Mrs Thea Schweizer, Mrs Siegrid Holz, Mme Lim Poi Joo and Miss Julia Philip. Mr Hermann Wohltorf, Mme Rosalind Lai and Miss Tan Pek Eng drew diagrams and maps. Mrs Christina Waitkus contributed photographical work, Mrs Frauke Ruhnke, Sonja Csomos and Ramona Saw Yeng Meng rendered valuable editorial assistance. Dr Thomas Smaltschinski and Mr William Then prepared computer graphics.

I dedicate this book to all who contributed in so many ways to its information base and to its preparation, but particularly to the people of Sarawak who gave me the inspiration and motivation to sustain the often tough and tedious work against discouraging odds. Sarawakians are blessed with the privilege to live in and make use of one of the most wondrous natural and cultural systems on earth, which it will not be easy to sustain in a world of unsustainable change. I would be happy if this book makes a small contribution to their efforts to sustain the unique world of Sarawak.

Cover photograph: The shifting 3-dimensional mosaic of phasic crown association units creates a vertical and horizontal heterogeneity of process intensities and habitat features. In the sclerophyllous mesophyll A-storey *Shorea elliptica* Burck and *Dipterocarpus sarawakensis* V. Sl. form a translucent rough surface layer. The B/C storeys with *Dipterocarpus confertus* V. Sl. and *Koompassia malacensis* Maing. ex Taub., *Dialium wallachii* (Baker) Prain form a heterogeneous species-rich layer over the variable D-storey at the ground. Semengoh Arboretum, Sarawak, 22 July 1995.

We gratefully acknowledge the financial assistance provided by Syarikat Samling Timber SDN BHD towards enabling the colour printing of this cover.

E.F. Bruenig
Kuching, May 1995

Figures and Tables

Figures

Tables

Diversity at β-level in Bornean Mixed Dipterocarp (MDF) forest: riparian fringes, almost flat alluvials, undulating MDF, herring-bone patterned low hill MDF, hill MDF on long slopes and ridges, montane forests and woodland (courtesy of Sarawak Government, 1962).

Acronyms, Abbreviations and Symbols

A	age; area
a	year (*annum*)
a.a.c.	annual allowable cut (m.a.i. ± st. dev.)
a.g.	above ground
BM	biomass per hectare
BNP	biological net production/productivity
Btg	batang, main river
CIFOR	Centre for International Forestry Research
d	diameter of a tree trunk at standardized height of measurement above ground (1.3 m or equivalent)
ET	evapotranspiration (suffix o = actual, p = potential)
FAO	Food and Agricultural Organization of the UN
F.R.	Forest Reserve (gazetted, part of PFE)
FRIM	Forest Research Institute Malaysia
g	cross-sectional area of a tree trunk at height of d
G	sum of g per ha
GATT	General Agreement on Trade and Tariffs (see WTO)
GNP	Gross National Product
GTZ	German Association for Technical Cooperation
h	height of a tree from ground to certain points of measurement
h/d	h/d ratio, indicator of tree slenderness and sturdiness
ha	hectare ($10{,}000 \ m^2$)
HP	Humus podzol, with prefix S = shallow, M = medium, D = deep
i	increment of d, h or v of an individual tree
I	increment of $\bar{d}$, $\bar{h}$ or V per ha
IUBS	International Union of Biological Sciences under ICSU (International Council of Scientific Unions)

IUCN	The World Conservation Union (originally International Union of Conservation of Nature)
IUFRO	International Union of Forest Research Organizations
IPCC	Intergovernment Panel on Climate Change
ITW	Initiative Tropenwald
KF	Kerangas forest, analogue to evergreen Caatinga
KrF	Kerapah forest, wet peat facies of KF
LEI	Lembaga Ecolabel Indonesia
LSU	Landform–soil unit in site mapping
MAB	Man and the Biosphere Programme of Unesco of the UN
m.a.i.	mean annual increment of $\bar{d}$, $\bar{h}$, G or V over a period (planning period, felling cycle, rotation)
MDF	Mixed Dipterocarp forest, zonal formation in Malesia
MMUS	Modified Malaysian Uniform System (higher pg d-limit)
MPSF	Mixed Peatswamp forest, P.C.1
MUS	Malaysian Uniform System (low pg d-limit)
NCR	Native customary right
NGO	Non-government organization
N.P.	National Park
NPP	Net primary production/productivity (apparent gross photosynthesis–respiration)
NTFP	Non-timber forest product (syn. minor f.p., auxiliary f.p.)
P	precipitation (rainfall, dew and fogdrip)
P.C.	Phasic community, especially in PSF
PEP	Potential economic production/productivity (which can become actual yield)
P.F.	Protected Forest (gazetted, part of PFE)
PFE	Permanent Forest Estate (gazetted)
pg	poison-girdling (trees in TSI)
PM	phytomass, plant biomass
PSF	Peatswamp forest
Q	Solar radiation balance
RIL	Reduced impact logging, corrective of SL
RME	reliable minimum estimate, mean of a statistical population of the growing stock minus its standard deviation, st. dev. (Dawkins, 1958)
RMP	Regional Management Plan
RP	Research plot
S	sungai, small river
SAG	Species association group (ASSG in Fig. 1.10)
SIS	Selection Improvement System, SMS in the Philippines
SL	Selective logging
SMS	Selection Silviculture Management System, syn. selection management system
sp., spp.	species, singular and plural
SS	selection silviculture

t	tonne (in Figures also used for time)
TP	trial plot
TPA	Totally Protected Area (gazetted)
TPI	Tebang Pileh Indonesia, Indonesian SMS
TSI	timber stand improvement
v	volume of a single tree in terms of BNP or PEP
V	volume as above of the tree growing stock on an area, usually 1 ha
WTO	World Trade Organization
z_0	aerodynamic roughness parameter, key indicator of the complexity of forest canopy structure

'The reflective, observing mind perceives nature as unity in diversity, unity of bonding the manifold diversities of form and mixture to the essential wholeness of natural phenomena and natural forces of life. The most important result of sensible, well conceived physical research, therefore, is to recognize the unity within diversity.'

A.v. Humboldt, 1847 (transl. by the author)

'Any form of utilization and concession to use forests must only extend up to those limits within which the forest's natural capacity for regeneration and their economic state and value are sustained ... constrained within the limits of physical non-damaging compatibility.'

Hundeshagen, 1828 (transl. by the author)

'This know also, that in the last days perilous times will come. For men shall be lovers of their own selves, covetous, boasters, proud, blasphemers ...'.

Paul in II Timothy 3: 1–2

The Tropical Rainforest Ecosystem 1

1.1 Tropical Rainforest Biome

The rainforest biome around the equator is home of the zonal forest formation class of the dense, tall, evergreen, wet forest (Champion, 1936) or the predominantly evergreen superhumid to humid, or according to Ashton (1995) 'fire-sensitive aseasonal evergreen', tropical lowland forest (Baumgartner and Bruenig, 1978). Moisture availability is adequate on average zonal soils and sites in the lowlands for the existence of mesophil evergreen tree species that can endure sporadic periods of drought stress. The latest and most reliable data on the areal extent and the state of the tropical forest are available in the series of three volumes *Conservation Atlas of Tropical Forests* prepared and published by IUCN (The World Conservation Union) since 1990–1991 (Collins *et al.*, 1991; Sayer *et al.*, 1992). The biome includes a wide range of geology, tectonics, evolutionary history, past and present climate, landform, exposure, atmospheric chemistry, soil and vegetation (Tables 1.1 and 1.2). There are profound differences between geographic regions. At the largest scale, there are climatic, geological and geomorphological, floristic and faunistic differences between Asia, Africa and America. At the medium scale, there are differences between climatically similar biogeographic regions, such as Borneo, Peninsular Malaysia and Sumatra within Malesia (Fig. 1.1), which are caused by differences in history of evolution, migration of species, geology, geomorphology, size of land surface and proportions of the various soil and site types in the landscape. Differences in past and present interactions between vegetation, animals and humans add to the extreme heterogeneity in the tropical rainforest biome, which is continued at landscape and forest ecosystem levels and will be discussed later.

Primeval tropical rainforest covered more than 90% of the biome's

1

Table 1.1. Climatic parameters in the humid tropics.

Zonal location	Equatorial belt and tropical areas with constant flow of moist air-masses outside this belt	Subequatorial to outer tropics with influence of trades, monsoons and monsoon-like alternating winds
Climate type	Tropical perhumid (wet), isotherm, non-seasonal to weakly seasonal, diurnal variation > annual variation	Tropical humid (moist), isotherm, seasonal with predominantly summer rainfall
Amount of annual isolation on the ground (kW cm^{-2})	South-East Asia 9.8–11.2 Congo basin 8.4–9.1 Amazon basin 7.0–8.4	~ 11.2
T_m(°C) mean annual	28	25
Annual march of seasonal variation	3	15–20
Diurnal variation	9	20
Wind	Predominance of tropical low pressure trough, low velocities, except very high in convectional bursts and squalls, local tornadoes; frontal winds moderate, high velocities rare	High velocities during summer (typhoon, hurricane, cyclone), low during dry season, strong effect of tropical convergence zone. Local frontal storms toward the end of dry season (*habub* in Africa)
Relative humidity (mean %)	95/100 night, 60/70 daytime, occasionally 40 at noon, little seasonal variation, in episodic dry periods < 40%	90/100 wet season, 60/80 dry season
Potential evaporation: actual precipitation ratio, E_p/P_o	< 1	< 1
Annual precipitation (mm), range of means and min./max.	average = 3000–4000 minimum = 50 (T_m + 12) = 2000 maximum > 10,000	average = 1300–3000 minimum = 1000 maximum = very variable

Adapted from UNESCO (1978).

land surface before the advent of humankind. This forested area has been drastically modified and reduced since humankind learnt to master the biotechnological problem of cutting the tall tropical evergreen forest for slash-and-burn cultivation in the very wet equatorial climate. There is possibly hardly any tropical rainforest in the world that has not been influenced and modified in some way by humankind (Dilmy, 1965) (Table

Table 1.2. Vegetation formations in the humid tropics in the two climates in Table 1.1.

	Megatherm–hydrophilous	Megatherm–tropophytic
Characteristic habitus of zonal formation:	Hygromorph–mesomorph, sclerophyll–mesophyll	Tropomorph
Zonal	Superhumid to humid, ombrophilous, predominantly evergreen equatorial wet forest and semi-evergreen wet tropical forest	Humid to subhumid, semi-deciduous and deciduous tropical forest
Edaphic formations	Littoral forest Mangrove forest and woodland Freshwater swamp forest Swamp grassland Riparian forest Peatswamp forest (with phasic communities) Simple sclerophyll, xeromorph forests on ultra-basics, podzols, skeletal and related soils	Littoral forest Mangrove forest and woodland Freshwater swamp forest Swamp grassland Riparian forest (often relic gallery) Evergreen forest (on moister, deep and well-drained soils) Sclerophyll forest (sandy terraces, podzols, and skeletal soils)
Physiographic formations	Submontane forest (wet) Montane forest (wet to moist) Alto-montane forest (moist) Alto-montane moss forest (wet, misty) Alto-montane woodland and scrub (moist)	Similar to the perhumid climate zone, except for species composition, conifers increase in southern and northern hemisphere, bamboo species become more frequent in the northern hemisphere
Degraded formations	Secondary mesophyll forest and woodland Secondary microphyll sclerophyll forest and woodland Pine forest to pine savannah Grassland (*Imperata cylindrica*) Karst-woodland Sclerophytic savannah	Secondary forest Savannah Pine forest, pine woodland or pine savannah Karst-woodland Sclerophytic xeromorphic savannah

Adapted from UNESCO (1978).

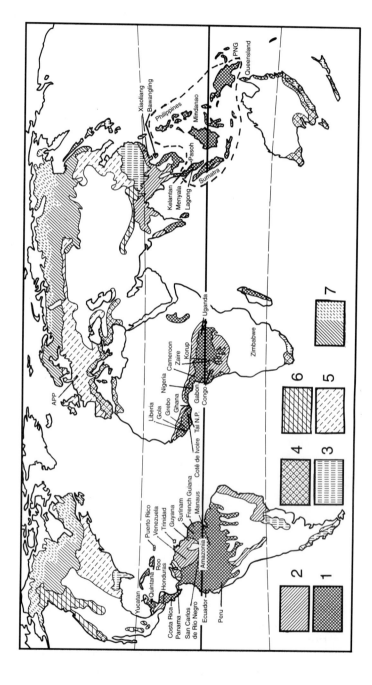

Fig. 1.1. Locations outside Borneo mentioned in the text. The broken line delimits the Malesian Floral Region, which is characterized by a high species richness and high dominance of *Dipterocarpaceae* and a generally high plant species diversity with very high values of α- and β-diversity. Borneo is probably the regional centre of tree-species richness and α-diversity and, the Sabal area, RP 146, has the highest recorded values (Fig. 1.16; Weiscke, 1982; Droste, 1995). The Malesian region is faunistically divided between east and west by the various versions of the Wallace line (not shown). APP is Auermühle where the goal-orientated production programmes were developed (Bruenig, 1995). The location of FRIM, Kepong, Selangor, on the map is west of Pasoh and south of Lagong. The zonal formations of the potential natural vegetation are: 1, superhumid to humid evergreen and semi-evergreen tropical forest; 2, humid to subhumid semi-deciduous and deciduous tropical forest; 3–7, non-tropical. (Adapted from Bruenig, 1987c.)

Table 1.3. The changes of forest areas and world population taking an optimistic and a pessimistic view.

Category of forest	Area (million km^2)				
	1965	1975	1985	2000	2050
Closed natural forest (optimistic)					
Equatorial tropical predominantly evergreen, wet to moist (perhumid/humid)	5.5	5.0	4.4	4.0	3.0
Tropical seasonal predominantly rain-green, deciduous, wet to moist (subhumid)	7.5	6.5	6.0	5.5	4.5
Open natural forest (optimistic)					
Tropical, seasonal rain-green and evergreen (alluvial, montane), moist to dry (semi-arid)	7.5	6.5	6.0	5.5	4.5
Sum of tropical natural forest					
Optimistic	20.5	18.0	16.4	15.0	12.0
Pessimistic	20.5	17.0	15.0	12.5	8.0
Tree plantations in tropics					
Forestry	0.04	0.05	0.08	0.16	?
Agriculture	0.20	0.25	0.27	0.28	?
Total forest of the earth					
Optimistic	38.2	36.0	34.5	33.7	30.5
Pessimistic	38.2	35.5	33.0	29.0	24.0
World population (1000 millions)					
Optimistic	–	4.0	5.0	5.8	7.0
Pessimistic	–	4.2	5.3	6.6	11.0

From: Ist eine Klimaänderung unausweichlich? – Der Raubbau an den Wäldern ist bedrohlich. *Die Umschau* 85 (1985) 3, 153–155.

1.3). Commercial forestry has modified (logged) tropical rainforest but the area is unreliably recorded – probably about one-third of the existing rainforest has been logged, and the decline of growing stock and in area is continuing (Bruenig, 1981, 1989b–e; Amelung and Diehl, 1992). Rainforest destruction is overwhelming, with more than 90% due to deforestation, mostly caused by agricultural expansion. Forestry and logging modify the growing stock, rarely causing deforestation directly but possibly paving the way for it (Bruenig, 1989e; Amelung and Diehl, 1992). There are regional differences in the contributions of traditional migratory (shifting) agriculture (prevailing

Table 1.4. Area in 1850 and 1985 of tropical forest formations and changes by land uses in million hectares and forest area logged in Latin America (including Surinam and Guyana) (Houghton *et al.*, 1991); and in Sarawak, Malaysia, for 1840 (estimated from probable population densities) and 1990 (Forest Department Sarawak, undated brochure, 14th Commonwealth Forestry Conference, Kuala Lumpur, 1994) (see also Table 5.1).

Tropical forest formation class	1850	1985	Change in area					Total change	Logged forest
			Crops	Pasture	Fallow	Degraded	Plantation		
Evergreen	226	212	7	3	0	4	0	−14 (6%)	0
Seasonal	616	445	48	54	30	32	7	−171 (28%)	21
Open	380	211	27	101	7	34	0	−169 (44%)	0
Sum	1222	868	82	158	37	70	7	−354 (29%)	21
	1840	1990							
Evergreen in Sarawak (total land area 12.3)	12.1	8.7	0.5	0	3[a]	0.2	0.2	−3.7 (30%)	~ 4.5

[a] Secondary woodland, only partly active fallow in practised shag (shifting rotational agriculture).

in Asia), farming and husbandry (prevailing in tropical America, Table 1.4) to deforestation.

1.2 Rainforest Macroclimate

Throughout the Quarternary, temperature and precipitation fluctuated. The rainforest flora, fauna and microbes had to endure long (millenia) and medium (centuries) periods during which the temperature and moisture were considerably higher or lower than that of present (UNESCO, 1978; Flenley, 1988; Heany, 1991; Taylor, 1992; Verstappen, 1994). The large-scale atmospheric circulation systems that determine the present macroclimate in addition fluctuate on the scale of years to decades. In the Americas, it is predominantly the Hadley cell circulation. In Africa, South Asia and Malesia and Australo-Pacific region, the very variable monsoonal circulation is superimposed on the Hadley cell pattern. A comprehensive description of tropical climates and climate change is given by Hendersen-Sellers and Robinson (1988). The climate is neither as uniform as climate diagrams suggest (Fig. 1.2 and Table 1.1) nor is it unequivocably favourable for plants. Stress factors include the great intensity of radiant heat influx, extreme midday moisture saturation pressure deficits, high proportion of high-intensity rainfall

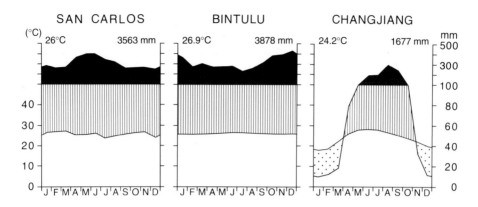

Fig. 1.2. Climate diagram of the apparently everwet equatorial climate (examples San Carlos de Rio Negro, Amazonia, Venezuela, and Bintulu, Sarawak) and of the seasonal tropical climate with one dry season at the outer margin of the tropics (example Changjiang, Hainan) (*World Climate Atlas* of Walter and Lieth, 1960–1967; data for San Carlos supplied by J. Heuveldop). San Carlos de Rio Negro, Territorio Amazonas, Venezuela, and Bintulu are typical for the 'everwet' tropical rainforest climate. Both locations are subject to occasional, episodic drought. The seasonal tropical climate of Changjiang permits predominantly evergreen mesophyll forest only on alluvial lowland and on montane sites. Black represents logarithmic scale above 100 mm per month, supposed to denote 'everwet'. Stippled below the temperature line denotes 'dry period'.

events, occurrence of more or less unseasonal and unpredictable episodically very severe and prolonged periods of drought, and the high frequency of heavy lightning strikes, storms, squalls, tornadoes and aerial micro-bursts of high velocity.

The occurrence of prolonged and physio-ecologically effective drought conditions were suspected by Schulz (1960) in Surinam and proved for Sarawak by Bruenig (1966, 1969a, 1971a), for the Amazonian Caatinga near San Carlos de Rio Negro by Heuveldop (1978; Bruenig *et al.*, 1979) and confirmed for Borneo by Baillie (1972, 1976), Whitmore (1984, p. 59), Wirawan (1987) and Woods (1987). The El Niño droughts in Malesia in recent decades demonstrate the seriousness of the ecological and economic effects of prolonged drought in the 'ever-wet' tropical rainforest climate. In contrast, prolonged periods of supersaturation stress the vegetation by heavy leaching, high rates of surface-water runoff and erosion, and low radiation. A high incidence of lightning gaps and windthrow can be observed on aerial photographs of the uniform canopy of Peatswamp forests in Sarawak (Figs 1.3 and 1.4; Bruenig, 1973b) and there is evidence of strong fluctuations of the incidence of gap formation between years. Severe and extensive storm damage in tropical rainforest is well documented (Browne, 1949; Anderson, 1961a; Whitmore, 1974, 1975a; Basnet *et al.*, 1992). Additional climatic heterogeneity is caused by topographical features. Large plumes of clouds originate regularly at coastal mountains, stretch far inland and seed rainfall along their path (Fig. 1.5). The spatial patchiness of atmospheric and soil moisture, caused

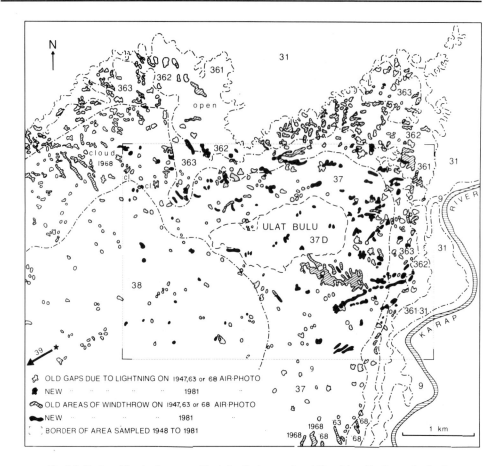

Fig. 1.3. Zoning of the phasic communities in the Peatswamp forest. Type 9 is cultivation on the riparian fringe, rice fields and secondary forest fallow. Then follow the forest phasic community PC1 or forest type 31, PC2: 361–363, PC3: 371–373, PC4: 3.8, outside the map area further southwest follow PC5 and 6.37D: 110-ha gap caused by lethal defoliation of alan bunga (*Shorea albida*) by an unidentified caterpillar *ulat bulu* before 1948 (Anderson, 1961b). The map has been drawn from a sequence of aerial photographs (1947, 1963, 1968, 1981) showing effects of disturbance events caused by lightning, windthrow and *ulat bulu*. Incidence and severity of the disturbance are related to differences in the canopy roughness, stature of the trees, tree-species richness and diversity. Note the concentration of gaps in the aerodynamically rough and tall canopy of 363 and adjacent 37, and the decline of gap size with crown size in 37 and 38. Karap river is a tributary of Batang Baram, Sarawak. For further explanation of forest types see Section 6.5 and Bruenig (1969b).

by physiographic heterogeneity at small to medium scale, complicates land-use and forestry planning.

An illustration of the effects of the great spatial heterogeneity of climatic conditions, and accordingly of effects at very large regional scale, is the intensity of discharge of water and sediments from land surfaces into the seas (Figs 1.6 and 1.7). The extremely high discharge of water and

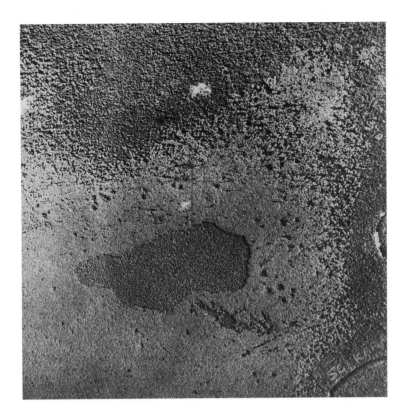

Fig. 1.4. Diversity of vegetation and of disturbances at β-level in Peatswamp forest in the same area as shown in Fig. 1.3 (status in 1981) (courtesy Sarawak Government, 1982).

sediments per square kilometre land surface in the South-East Asian Archipelago is related to the very high intensities of rainfall events and high relief energy. These extremely high discharge rates are indicators of the fragility and vulnerability of the land surface, which impose severe limitations on sustainable land and forest use in the Malesian region. This, and the tendency to abnormal atmospheric circulation patterns in the region, are serious but usually overlooked problems and are obstacles to sustainable forestry, environmental protection and land use generally. The consequences of the exacting rainforest climate for sustainable forestry management and silviculture are as follows.

1. Heavy leaching of minerals and organic matter from the soil and heavy rates of surface and within-soil erosion.
2. No dry-season recovery of the nutrient regime as in deciduous tropical forests.
3. High temperatures and consequently high speed and intensification of

Fig. 1.5. Diversity of the physiography of the landscape affects heterogeneity of meso-climate: plume of clouds are regularly formed by the peninsular Bako N.P. and, travelling far inland, affects the heterogeneity of the rainfall distribution pattern. Gunung Santubong on the left has only formed the noon cloud cap on this day, but often also is the source of very compact plumes. The plain between Bako and Santubong is a mosaic of mangrove, beach forest and Kerangas forest on terraces, mostly strongly modified or converted. The Santubong hill (810 m a.s.l) and the Bako N.P. carry a mixture of very diverse MDF and KF forests and secondary vegetation.

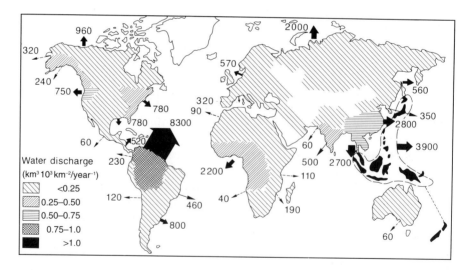

Fig. 1.6. Mean annual total discharge of fluvial waters from the catchment areas (regions) to the oceans (arrowed numbers) and the average rates of annual discharge (runoff) of water in cubic kilometres per 1000 km^2 land surface (hatching). The Amazon catchment has the highest total water yield, but the Malesian Archipelago has the highest runoff per 1000 km^2 land surface. (From Milliman and Meade, 1983.)

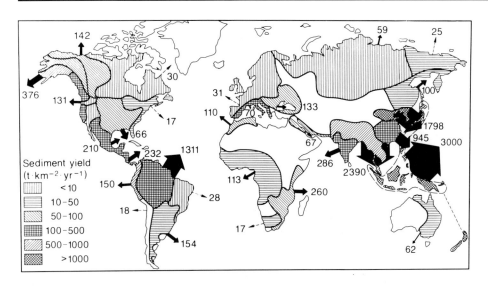

Fig. 1.7. Mean annual fluvial sediment flows from regional drainage areas into the oceans (arrowed numbers) in million tonnes and the average rates of annual sediment yields in tonnes per square kilometre land surface (hatching). The Malesian Archipelago has the highest water runoff rate and by far the highest sediment yield per unit land surface (km^2). (From Milliman and Meade, 1983.)

chemical processes on the plant surface, in the plants and in the soil.

4. Intense solar radiation, high temperature, high humidity and occasionally strong vapour saturation deficit, individually and in combination, stress and strain the plants and favour photochemical and biochemical reactions with natural and anthropogenic pollutants.

5. Prolonged drought periods create fire risks (Goldammer and Seibert, 1989) and cause mortality among seedlings and saplings (Becker and Wong, 1993); in severe cases growth is checked and there is mortality among trees.

6. High rates of damage from wind and lightning.

In conclusion, the equatorial tropical climate is exacting and puts severe limitations on forest utilization, sustainable management and conservation. The year-round hot weather favours outbreaks of pests and diseases, especially in monocultures. The combination of constantly high temperatures and the often very high rainfall intensities that leach and erode the soils make unsustainable any harvesting and management system that does not maintain high levels of complexity and the protective functions of the canopy. The occasional high, episodically extreme, wind speeds blow off leaves (Woell, 1989) and crowns (Anderson, 1961a), or throw trees or tracts of forests (Browne, 1949). Native trees are physiologically and morphologically adapted to compromise between the need for high transpiration (cooling, nutrient pumping) and the avoidance of

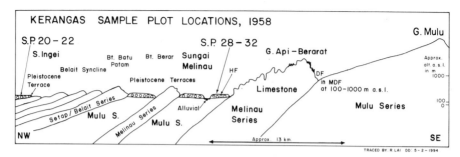

Fig. 1.8. Diversity of geologic substrate along a 13-km transect from Gunung Mulu in north-eastern Sarawak westwards to Sungai Ingei, Ulu Belait, in south-western Brunei, which is an example of the heterogeneity of parent material and topography at landscape scale, in this case formed by a strong uplift of a Miocene and Pliocene Tertiary synclinal trough in which rather varied argillaceous and arenaceous materials had been deposited. Older materials, including limestone, broke through the surface. New heavy erosion and sedimentation in the trough and unlifting of the edges created several levels of Pleistocene and Holocene deposits. This geologically and geomorphologically diverse and active landscape supports a great variety of very different tropical soils. SP 20–22 and 28–32 indicate the location of kerangas sample plots of 1959 (Bruenig, 1966/68). The plot HF of Proctor *et al.* (1983) and Newbery and Proctor (1984) is adjacent to RP 28 and 29. DF is their MDF plot.

overheating, drought and storm damage (Gates, 1965, 1968; Bruenig, 1966, 1969a,b, 1971c), but extreme conditions can overstrain even native tree species.

1.3 Rainforest Soils

Rainforest soils are extremely heterogeneous and their taxonomy is confusing. The soils in rainforest can vary profoundly at a scale of a few square metres to square hectometres according to the spatial variation of the geological substrate, geomorphology, climate and the long-term influences of vegetation (Figs 1.8–1.10). A pantropical synopsis of tropical soils and their complicated nomenclature in the various systems of classification is given by Schmidt-Lorenz (1986) and Longman and Jenik (1987). Burham gives a brief but useful regional review for South-East Asia (Burham in Whitmore, 1975a). The forest soils of Sarawak have been intensively pedologically studied, classified, assessed and mapped since the 1950s. A useful overview of the soils from an agricultural and forestry point of view is given by Andriesse (1962). A description of the great variety of soils, including a critical contribution to the laterite discussion, and identification of distinctive features that are important for the growth of tree species and communities, is given in the now classic reconnaissance report of the soils of the Brazilian Amazon region by Sombroek (1966). He states that:

any differences in forest characteristics within an area of uniform
climatic conditions and non-existent or uniform anthropogenic
influences, are therefore apt to show a correlation with lower
category soil differences. Among these are moisture holding
capacity, the total available amounts of the various plant nutrients,
and the penetration possibilities for roots.

This also holds true in Bornean upland forests and are the primary
features used in forest soil mapping for forest management planning in
Germany. Uncertainties, however, remain, mainly caused by human
ignorance and by the openness of the system of interacting factors of soil,
geomorphology, climate, flora, fauna, vegetation and physio-ecology
(Bruenig, 1970a; Ashton and Bruenig, 1975; Weiscke, 1982; Newbery, 1985,
1991; Newbery *et al.*, 1986; Ashton and Hall, 1992).
    In the 1950s, Dutch soil scientists surveyed forest and agricultural soils
for the Departments of Agriculture and Forestry Sarawak and produced
a basic classification scheme (Andriesse, 1962; Dames, 1962) on which the
currently used classification could be built (Tie, 1982). The diagnostic field
and laboratory parameters used for upland soils are: soil organic matter,
drainage, Munsell colour, presence of cemented spodic, oxic, argillic or
cambic horizons, soil texture, nature of cation exchange complex, pH,
parent material. Peats are classified by depth, substrate, vegetation,
ground water table, ash content. These parameters accord broadly to the
soil properties that Sombroek used to differentiate between types, and are
key indicators of sustainable productivity: moisture-holding capacity (soil
organic matter, SOM), rootability (soil texture, horizons), nutrients (Munsell colour, cation exchange capacity (c.e.c.) and pH). For forestry planning,
soil and site data have been incorporated in the forest inventory
procedure in Sarawak since the mid-1950s (Bruenig, 1961b, 1963, 1965a).
This information, as elsewhere in the world, was subsequently hardly
applied in forestry and agricultural planning and development (Ashton
and Bruenig, 1975). The very strong timber orientation of the Food and
Agriculture Organization (FAO) forest inventory in Sarawak in the 1970s
(Bruenig, 1976) and the lack of a comprehensive integrated multisectoral
spatial information system (Bruenig, 1992) blocked further advancement
towards sustainability and facilitated the mushrooming of unregulated
timber exploitation.
    The typical 'zonal' mesophytic tropical rainforests grow on red–
yellow, heavy textured clays to moderately textured, more or less sandy
clay loams. Structure, water-holding capacity, depth of rootability and
content and distribution of SOM vary widely. Their combination determines fragility of the soil and sets the limits for sustainable forest
modification and manipulation. Mixed Dipterocarp forest (MDF) soils
(cambisols, acrisols, ferralsols, udult ultisols) in Sabah have generally
Mg^{2+} and K^+ as dominant cations and, on average, an effective cation
exchange capacity (e.c.e.c.) of 80% of c.e.c. In contrast, the sandy and very

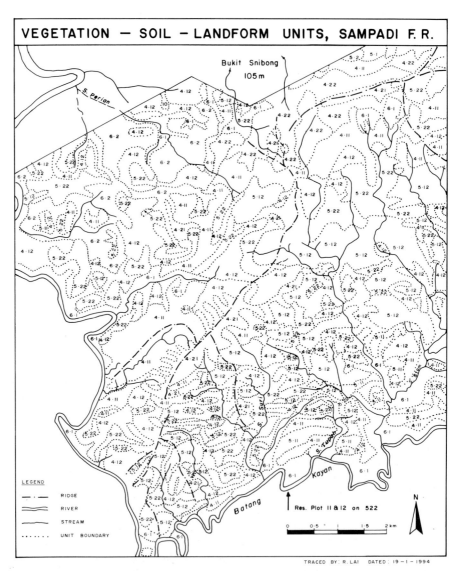

VEGETATION — SOIL — LANDFORM UNITS, SAMPADI F. R.

TRACED BY: R. LAI   DATED: 19−1−1994

acid (pH < 3–4.5) kerangas and padang soils have Al^{2+} and 30% e.c.e.c. (humult ultisol/ferralsol) or Fe^{3+} and 10% e.c.e.c. (arenosol/podzol) (Lagan and Glauner, 1994). The very acid sandy humult ultisols, podzols and oligotrophic peat soils (Anderson, 1961a) are absolute forest soils (Bruenig, 1969c). Therefore, they will be an important component of the eventual permanent forest estate, at least in some parts of the biome, such as Borneo, Guyana–Surinam and Amazonia. These soils are inherently infertile and extremely fragile. This excludes agriculture and constrains forestry severely (Bruenig, 1966, 1969c; Klinge, 1969). In Borneo these

Fig. 1.9. (*opposite*) The intricate medium-scale (1–250 ha) mosaic pattern of site-related heterogeneity or β-diversity in tropical rainforest in an area of approximately 55 km^2 in Sampadi Forest Reserve, West Sarawak. Each unit contains a finer fabric of heterogeneity of site conditions (soil and land form) at a spatial scale of square metres to hectometres (source: forest inventory and soil survey report 1954–5 by T.W. Dames and E.F. Bruenig, Forest Department, Forest Department Sarawak file 524.7 and map no. 194 of 1956). The mapped forest types (FT) are: 4. Mixed Dipterocarp forest on plateau sandstone formation of upper Cretaceous to Eocene age, subdivided by landform (undulating, cuesta, steepness), forest stature, dominant tree species and characteristic non-tree indicator plant species (*Weiserpflanzen*) which also expresses soil type. 5. Kerangas forest on dip-slopes of the plateau sandstone formation and on Pleistocene terraces, subdivided by landform/geological substrate, forest stature, dominant and indicator species, on the various kerangas soil types (Fig. 1.19). 6. Riparian (alluvial) forest types, subdivided into river-bank flood plain (6.1) and interior freshwater swamp forest (6.2). 10. Secondary vegetation from encroachment by shifting cultivation. Kerangas sample plots SP 11 and 12 had been located in FT 522 between Sungai Suri and S. Tupah. The area is now deforested by illegal slash-and-burn cultivation and plantation establishment (Sungai = river, stream).

---

infertile upland soils are mainly podzolic arenosol, podzolic stagnic gleyic clay or excessively drained sandy podzol (spodosol) and the related kerangas peatbogs or kerapah (Bruenig, 1966, 1989c).

The kerangas and kerapah soils and the analogous Amazonian and Guyana–Surinam caatinga soils (Klinge, 1969) differ in some essential ecologically important features from the clay–loam and clay soils (ultisols, oxisols, acrisols, ferralsols) under MDF (Dames, 1962). The main features of forest-management relevance (Bruenig, 1966, 1974; Bruenig and Schmidt-Lorenz, 1985) are: extremely low nutrient content and high acidity; extremely unbalanced, drought-prone water regime; thick layer of litter over fibrous, very acid surface raw humus or peat rich in non-hydrolysable polyphenols (tannins) and allelopathic compounds (Bruenig, 1966; Bruenig and Sander, 1983); shallow to deep eluvial albic soil horizon often overlying a reddish-brown, hardened, illuvial humic or humic-iron pan; discharge of tea-coloured strongly humic 'black-water'. The peatswamp soils of Sarawak were described by Anderson (1961a, 1964, 1983) who commented on the diversity of their geomorphic features. The surface may be flat or dome-shaped, the substrate varied in topography and nature (Fig. 1.11). Anderson suspected that the almost permanent waterlogging at the periphery and periodic or episodic drought in the centre were the major determinants of the vegetation types rather than the nutrient contents of the peat. The organo-chemical nature of the peat may influence abundance and distribution patterns of individual species, such as Ramin (*Gonystylus bancanus* (Miq.) Kurz), in the various phasic communities *sensu* Anderson (1961a) more than nutrients (Bruenig and Sander, 1983). The same applies in principle to Kerangas forests (Bruenig, 1966; Riswan, 1991; Section 1.7).

The interaction of moisture, aeration, nutrients, SOM (living and dead) and biological soil conditions determine the soil's capacity to support plant growth. The productivity of the plant determines the amount of food available to the soil organisms. The soil organisms

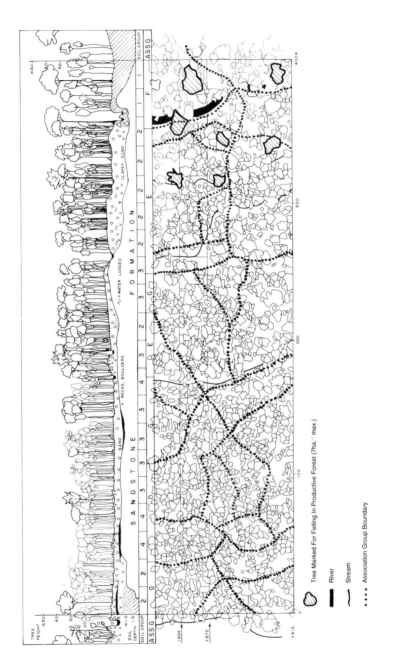

Tree Marked For Felling In Productive Forest (7ha.⁻¹ max.)

River

Stream

Association Group Boundary

Fig. 1.10. (*opposite*) Pattern at small spatial scale (order of metres to 100 m² = small-gap size) of terrain and soil type and canopy structure along a 20-m wide E–W transect in Sabal Forest Reserve, Sarawak, 1963. The rooting systems and depths on the various soil types correspond to conditions shown in Fig. 1.14. The profile of forest and soil (top) runs between Y 675 and 695 from X 0 to 400 (in metres). The mixed dipterocarp forest (X 360–400, soil group 1) is part of the area that was selectively logged in 1978 and had the highest tree-species richness in 1990 recorded in Sarawak, described in Sections 1.8, 2.8 and 6.3; Tables 1.6 and 6.2. Soil groups are: (1) deep sandy–clayey loam, ultisol, (2) medium deep sand over clayey sand, podzol, (3) medium deep sand over hard Bh (Ortstein) or sandstone, podzol, (4) shallow sand, podzol. D.E.F.G: association groups of trees > 30 cm diameter, 30 nearest neighbour FANTASM-B divisive classification (Bruenig *et al.*, 1979; Weiscke, 1982). Dotted line: boundary between association groups. Canopy projection: crowns of A- and upper B-layer trees. Thick-lined crowns: merchantable trees marked for felling in selection silviculture management in 1.5 ha production forest (MDF and *Agathis* KF on soil groups 1 and 2).

maintain crumb structure, porosity and thereby aeration, rootability, infiltrability, absorptivity and water-holding capacity of the soil, and contribute to the conservation and cycling of nutrients. Their importance, especially of mycorrhiza (Mikola, 1980), in tropical rainforests was realized by Moeller (1922, 1929) over a century ago during ecological research in Brazilian Atlantic rainforest. His understanding of mutualism and symbiosis helped to design sustainable silvicultural management systems, especially for the restoration of degraded lands and forests, in Germany (Moeller, 1922, 1929; Bruenig, 1984a, 1989a) and also a century later in tropical rainforests of China (Yang and Insam, 1991). While there are ample data available on annual fine-litter fall (on average 5–9 t ha^{-1}), there is little quantitative evidence on the production and turnover rates of coarse litter. Rates of decay of trunks vary widely in relation to species, soil type and contact with the ground. In Sabal Forest Reserve, Research Plot (RP) 146, in Sarawak, Bindang (*Agathis borneensis*) and Meranti (*Shorea* spp.), cross-cut logs and whole trunks of trees felled in 1978, but not extracted, were strongly decomposed in 1990 if they were in direct contact with the ground (sandy-loamy humult ultisol and sandy podzol). Where the same trunk spanned a hollow, the wood was still firm. Browne (1949) reports that chengal (*Neobalanocarpus heimii* (King) Ashton) logs were still being extracted from the storm-thrown MDF in Kelantan, Peninsular Malaysia, almost 70 years after the calamity. Trunks of Belian (*Eusideroxylon zwageri* Teijsm. & Binn.) trees felled by the grandfather, who thereby established a customary property right, were still being utilized in the forest by the grandchild. The high spatial and temporal variations of decay and mortality (Section 1.11) make assessments of coarse-litter production difficult.

The organic matter of standing dead and dying trees, of decayed wood in hollow living trees, of fine and coarse litter and the various compartments of SOM (Fig. 1.12) have important and manifold ecological functions in forests: as sources of nutrients for the roots of living trees and as food and nutrients for soil organisms; as aggregate-stabilizing substances in the soil improving soil stability, crumb cohesion, porosity and aeration,

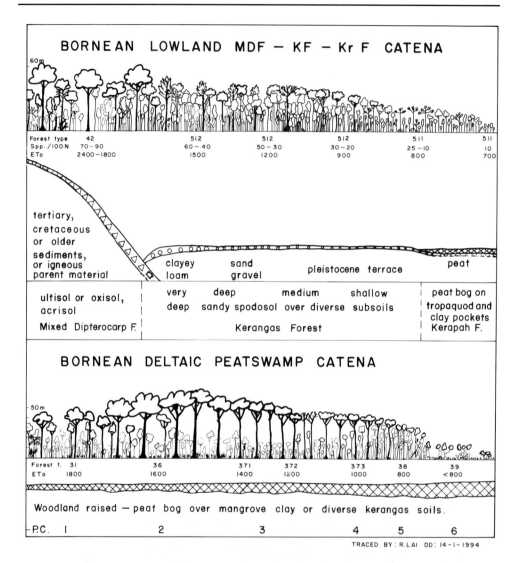

Fig. 1.11. The change of canopy structure with site in Borneo. Forest types according to Bruenig (1969b) in the upland (almost flat or undulating lowland to upper hillforest on dry land) forest catena (see Figs 1.23 and 1.24) and Anderson (1961a, 1983) in the peatswamp catena. SPP/100N: number of tree species > 1 cm diameter in 100 random individuals. ET_0: Evapotranspiration rate. Top: mixed dipterocarp forest (type 42), *Agathis*-bearing kerangas (type 513), typical kerangas (type 512), poor kerangas (511), kerapah or related very poor types (510). Bottom: deltaic peatswamp forest, mixed ramin-bearing peatswamp forest (type 31 and PC1), *Shorea albida* phasic communities alan (PC2), alan bunga (PC3), padang alan (PC4), mixed (PC5) and open padang (PC6). The crowns have been pictorially depicted by a continuous outline. In reality the mostly cauliflower- or broccoli-shaped crowns are an open compound of subcrowns with more or less bunched leafage that transmit light particularly well if the sun is high around noon, but intercept more in the morning and afternoon.

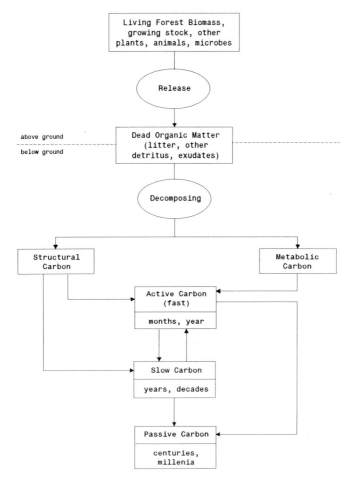

Fig. 1.12. Inputs of litter and exudates above and below ground that are directly decomposed and polymerized into persistent humous matter or pass through a chain of biological metabolism before finally entering the three sectors of the functional soil carbon pool. The system is not closed and decomposition and metabolic processes and the soil carbon pools discharge carbon in various forms into the living biomass and the environment (atmosphere, hydrosphere, lithosphere below pedosphere). (Adapted from Fig. 5, TSBF report 1992.)

and thereby rootability; and as an absorption and exchange complex for water, mineral nutrients and trace elements, but also as a sink for phosphorus. The patchy deposition of coarse litter, especially trunks and crown-wood, and the heterogeneous chemical nature and differing decomposition rates of coarse and fine litter will create heterogeneity of chemical and physical soil conditions. These, and the diverse root exudates, will affect the activities of soil microbes and invertebrates, and influence the germination of seeds, growth of roots, establishment of seedlings and the gap dynamics above (Facelli and Pickett, 1991). The importance of microbial biomass in the soil lies in the formation and glueing of mineral soil–humus aggregates and in its role as sink, store and source of nutrients in the soil. The nutrient-preserving functions of live and dead SOM are particularly important in the hot and wet rainforest climate with high decomposition rates and heavy soil flushing by percolating water (Insam, 1990; Yang and Insam, 1991). The usually high SOM content (Scharpenseel and Pfeiffer, 1990; Sombroek, 1990; Kehlenbeck, 1993) accords with its important role in rainforest soils. This contrasts with the popular conviction that rainforest soils contain little humus and the recent statement by Kira (1995) that 'the accumulation of dead organic matter as the A_o-layer and soil humus decreases toward warmer habitats'. Microbes, fungi, invertebrates and other soil organisms will penetrate deep into the soil as long as habitat and food are attractive. Deep, porous and well-aerated soils contain humus and are biologically active to several metres depth. Living and dead SOM are the decisive, filtering interface between the forest, the hydrosphere and, indirectly, the atmosphere. SOM is the crucial element in creating and maintaining crumb structure, porosity and rootability. Deep rooting is vital to sustain prolonged drought periods, and possibly to retain nutrients. Next to the canopy (Section 1.6), SOM is the key element for ecosystem self-sustainability and sustainability of forest management in the tropical rainforest.

In most rainforest soil types the soil-aggregate structure is fragile and susceptible to breakdown and compaction. The SOM–mineral (clay/silt/sand) aggregates collapse easily under stresses from exposure, rain-splash or impact. Recovery depends fully on biological activity (roots and organisms). The benefit of soil freezing, which speeds recovery in cool and cold temperate soils, is absent. Restoration of porosity, crumb structure and rootability after collapse in rainforest soils is, therefore, a slow, long-term and complex process (Droste, 1995). The climate-related tendency to rapid decomposition of organic matter, the sensitivity of the clay and loam soils to physical impact, and the susceptibility of soils on sloping ground to erosion and internal soil slumping impose limitations on the design of low-risk and high self-sustainability forest management systems (Jentsch, 1933/1934).

1.4 Large- and Medium-scale Dynamic Changes

Plate tectonics, geomorphological changes and climatic fluctuations during the Tertiary and Quarternary have affected the floral, faunal and vegetational evolution of the rainforests. Tree pollen spectra of the equatorial tropical rainforest zone show evidence of climate-related changes of the vegetation (Livingstone and Hammen, 1978; Flenley, 1979, 1992; Horley and Flenley, 1987; Newsome and Flenley, 1988; Heaney, 1991; Taylor, 1992). A general picture emerges of fairly rapid changes of vegetation associated with increasing or decreasing temperature and, independently, precipitation, and of altitudinal and lateral shifting, shrinking and expansion of the areas inhabited by the tropical rainforest and its subdivisions. Tectonic movements, erosion and sedimentation have created a typical diverse array of site and soil conditions that continue to evolve – accordingly the forest vegetation is diverse and dynamic (Section 1.5). The tropical rainforest has survived climatic and tectonic changes effectively by evasion (shifting) and adaptation. Figure 1.13 gives a generalized schematic illustration of the distribution of broad forest types in relation to landform created by tectonic uplifting or subsidence in Sarawak. The relationship between structurally distinct

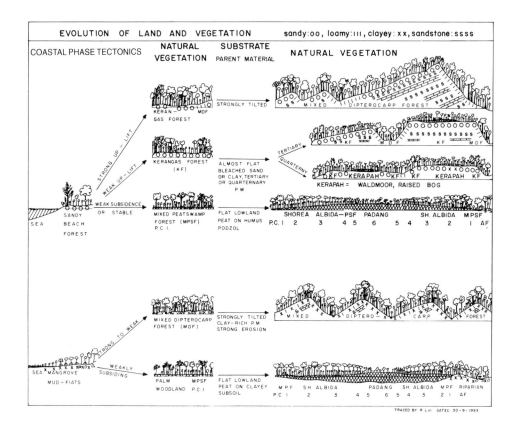

Fig. 1.13. Simplified and generalized illustration of the primary succession of vegetation and soil formation in tropical rainforest in Borneo starting with littoral deposition and proceeding in relation to tectonic movement (tilted uplifting, flat horizontal uplifting, stable, subsidence) and nature of substrate (arenaceous or argillaceous). MDF, mixed dipterocarp forest (the zonal forest formation); KF, kerangas forest (edaphic formation on well to excessively drained soils); Kerapah, water-logged facies of kerangas forest (woodland bog); MPF, mixed peatswamp forest (perimeter of peatswamp); *Sh.albida*, alan (*Shorea albida*) – phasic communities towards the peatland centre if a raised peatswamp dome is formed (Anderson, 1961a, 1964).

upland forest types in the Amazonian rainforest is very similar, except for the absence of large tracts of oligotrophic Peatswamp forests.

Pollen spectra of various deltaic Peatswamp forests in Sarawak (Anderson and Muller, 1975) showed rather similar primary–successional changes of floristic structure that correspond to the sequence of the phasic communities on the present peatswamp surface. However, this view has been questioned recently by Yamada (1995; Section 6.5). In contrast, exploratory pollen analysis of Kerapah peatbogs on terraces of Holocene (Dalam Forest Reserve near Miri) and Pleistocene (Gunung Mulu National Park) origin and on a Tertiary high plateau (Merurong Plateau at 730–1125 m a.s.l.) showed no regular pattern, but there were differences between the locations and erratic changes of species composition in the profile, including disappearance and reappearance of species that are difficult to explain. The indication is that the present pattern of species distribution and the presence or absence of regeneration are poor indicators of the direction of stand dynamics. So many site and chance factors interact in the Kerangas/Kerapah forests in such a diverse manner that no consistent picture of species distribution, association and succession manifests itself in the pollen diagrams and in the living vegetation (Bruenig, 1966, 1974; Proctor *et al.*, 1983a; Newbery and Proctor, 1984; Newbery, 1991). In the Peatswamp forests and some Kerapah types, the long-term dominance of one actively peat-forming, competitive tree species over large tracts may create natural monocultures that maintain themselves over many thousands of years. The single-species dominance of the very site-tolerant and aggressively competitive *Shorea albida* Sym. in the Peatswamp forests in Sarawak is the best-documented example (Anderson, 1961a). Gap formation caused by lightning and windthrow and large-scale catastrophic collapse caused by an insect pest (so-called *ulat bulu*, hairy caterpillar) are very common (Figs 1.3 and 1.4; Section 6.7).

Fire is, in contrast to previous views, being recognized as an ecological factor in primeval rainforests (Goldammer and Seibert, 1989). On the easily drying Bornean kerangas and Amazonian caatinga soils, episodic droughts make the forest susceptible to spontaneous or human-induced fires. Examples are the fires in Sabah and Sarawak around 1880–1890, 1930–1933, 1958, 1983 and 1991. Vegetation recovers rapidly after an initial fire by coppicing and seed germination. Nutrient losses cease within a year or two, as the studies in San Carlos (Jordan, 1989) and Kalimantan (Riswan and Kartaniwata, 1991) showed. Repeated fires consume the SOM and reduce the ecosystem eventually to an arrested state of degradation such as the white-sand savannah described from Surinam (Heyligers, 1963), the Bana in Amazonia, or the various types of Padang in Malesia (e.g. Bako National Park; Bruenig, 1961a, 1965b). Equally fire-prone is limestone vegetation, e.g. the sclerophyll, xeromorph woods on Gunung Api (Fire Mountain) in Mulu National Park, Sarawak (Bruenig, 1966). Rainforest on deep clay soils, slopes and valley bottoms, and riverine freshwater swamp and alluvial fringe forests do not dry out easily. If any fires occur, they are usually light and quickly followed by regrowth of the original type of forest. Shifts of climate towards greater aridity or stronger irregularity may cause fire risk where the forest is fragmented by expanding selective logging, shifting cultivation and agriculture. Since the 1940s to 1960s, there has been an increase in the severity and frequency of drought events and the number of associated damaging fires in the rainforests of Ghana (Orgle *et al.*, 1994). Another complex of factors possibly effecting variation and change are natural and anthropogenic atmospheric pollutants and nutrients, which enter the ecosystem in form of trace gases, aerosols and solid substances, affecting the productivity and health of the rainforest ecosystem.

As a result of the natural life cycle of trees and the various perturbations and disturbances, the taxonomic structure of the rainforest is not static but variable at small to medium spatial and temporal scale; catastrophic collapse may be caused by the synergistic interaction of several internal and exogenous perturbing factors; species spectra, species richness and diversity are not constant but change in response to short- to long-term disturbances; locally, plant and animal species may become extinct, as indicated for trees in Kerangas pollen profiles, but may reappear; constant changes of the floristic and spatial structure of the ecosystem seem to improve its resilience, adaptability and flexibility, while simple natural 'monocultural' forests are more vulnerable to persistent damage. Knowledge of the history of changes helps to under-stand the present capacity for change. The comparison of the present vegetation with pollen profiles in Kerangas peats and Peatswamp forests have proved to be useful indicators of the forest dynamics (Anderson, 1961a; Wood, 1965; Bruenig, 1966; Muller, 1972; Anderson and Muller, 1975). Alan (*S. albida*) may invade into Peatswamp forest phasic community PC1 and initiate succession to PC2, 3 and 4. Kerangas forest species may successfully invade MDF, survive and initiate raw humus formation and greying of surface soil. The raw humus-forming *Casuarina nobilis* Whitm., planted in 1939/1940 on a patch of sandy clay–loam ultisol/acrisol in 6th Mile Forest Reserve near Kuching, had produced several centimetres of raw humus and caused considerable bleaching of the surface soil to 5–10 cm depth by 1954–1955. In Sampadi Forest Reserve a single large *C. nobilis* was found in MDF in Kerangas RP 12 (Fig. 1.9). Possibly a surviving gap colonizer, the tree had similarly accumulated a wide disc of thick raw humus–root mat and initiated topsoil greying (Bruenig, 1966, 1974, pp. 20 and 65). On the plateau of the present Bako National Park, MDF was cleared in the mid-nineteenth century by the local communities to grow the tannin-rich gambir, *Uncaria gambir* Roxb., for the production of catechu, cutch and betel leaves (Burkill, 1935, vol. II, pp. 2198–2204). The sandy clay–loam ultisol/acrisol degraded quickly, cultivation was aban-doned and a secondary medium-deep humic podzol under a typical tall Kerangas forest had developed by 1955. By this time, MDF species began to invade but did not look very healthy (Kerangas forest RP 17; Bruenig, 1966, 1974). The examples show the risk to sustainability if the natural dynamics of the forest and the species–site interactions are not well considered. Newbery and Proctor (1984) hypothesize that the highly dynamic nature of the rainforest makes it likely that soil properties change in the course of gap formation and regeneration. Revitalization of the soil biology, mineralization of moder and formation of mull during the natural-regeneration gap-phase is well documented in beech and spruce forests in northwest Germany. It would seem plausible to assume that similar effects happen in tropical rainforest, particularly in large gaps into which pioneer (early secondary forest) species invade. The general features of long-term soil and vegetation dynamics in patches, initiated by gap formation of different spatial scales and temporal frequencies, in tropical rainforest are very similar to those described in semi-deciduous tropical forests (e.g. Taylor *et al.*, 1995) and basically similar to the processes known and traditionally used in forestry practice of naturalistic silviculture in mixed temperate forests.

1.5 Rooting Sphere

The main, but not exclusive (Fig. 1.14), habitat (rhizosphere) of the roots of trees and other plants is the soil including the litter layer (pedosphere) and decaying trees. Arboreal epiphytes and habitually soil-rooting epiphytes, such as strangler figs, have roots in the canopy. Some tree species may develop epibiotic roots from the trunk or root upward from the soil to supplement deficiencies in the edaphic water and nutrient supply. In rainforest, as in temperate forest, roots penetrate into the rotting heart of live or dead standing trees, into the decomposing heart and inner bark of fallen tree trunks, or grow downward or upward among lichens and mosses on bark surfaces to scavenge nutrients and water. As in temperate tree species, the rooting systems of tropical rainforest trees differ between species, but also vary within a species according to soil and light conditions. The variety of shape and depth of rooting systems is as great as the variety of crown forms (Fig. 1.15). Contrary to a common misconception, tropical rainforest trees are not characteristically 'shallow-rooted'. Similar to temperate forest trees, the root systems may be shallow or deep, spreading or compact, a combination of shallow surface roots, medium-deep spreading roots and deep tap roots, or sinker roots from wide-spreading horizontal roots. Soedjito (1988) describes the root system of 40 saplings (diameter 3.0–4.5 cm, height not reported but probably 3.5–6 m) of 40 species in lower montane MDF near Long Barang, East Kalimantan (Fig. 1.16). The root systems are medium-deep and wide-spreading (e.g. *Macaranga* spp., *Ficus tamarans*, all secondary-forest species) or a combination of shallow surface roots, intense medium-deep rooting and deep-reaching tap roots (*Shorea parvifolia, Agathis dammara, Baccaurea*

Fig. 1.14. Diversity of rooting: the roots of the stilted Kerangas tree *Casuarina nobilis* Whitm. embrace and grow upward on a *Calophyllum sclerophyllum* Vesque. Sabal F.R., Sarawak.

macrocarpa, Nauclea subdeta). The tap root depths vary from 35 cm (sub-canopy species) to 55–60 cm (*Shorea* and *Agathis*). For saplings of this size and growing in shade, this cannot be considered shallow-rooted.

In the rainforest soils of the CERP–MAB Ecosystem Study Project in Bawangling the carbon content, base saturation and c.e.c. are relatively high in the Ah horizon and decline markedly with soil depth but, depending on soil types, carbon contents and mineral nutrients can still

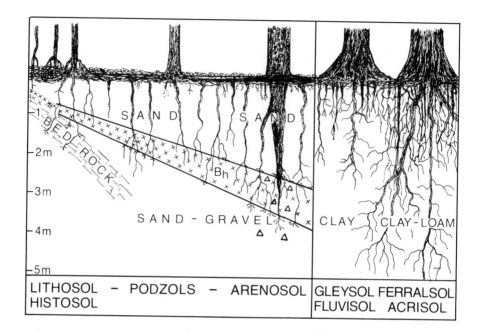

Fig. 1.15. Structure and depth of the root systems in relation to tree species and soil. Surface feeder roots often form dense litter/root mats (lithosol, histosol, shallow to giant podzol, and transition to arenosol or humult ultisol). The big, deep reaching tap root is of *Agathis borneensis* Warb, which on giant podzol may reach deeper than 5 m. Strong tap and sinker roots occur in all sufficiently friable soils especially in more sandy humult utlisol or ferralsol/acrisol, and less on udult ultisol clay and gleysol. Dense feeder roots develop also at greater depth if stimulated by water and nutrients. Sinker and feeder roots also develop along structural clefts in blocky clay soil, pass through soft parts of the B_h in podzols and penetrate in rock cracks deeply into the substrate. Wide-spreading lateral roots stabilize the tree by pulling against swaying wind forces, especially effectively if joined to buttresses (e.g. *Shorea albida*) and grafted to roots of other trees. Variation between and within species and between sites is great. (Compiled from profile descriptions, windthrows and observations along road cuts by the author.)

be high below 2 m (Kehlenbeck, 1993, pp. 84–103). This would be an incentive for roots to penetrate into these soil layers. Rooting in 17 profiles of 13 major soil types was observed right to the bottom of the 1.5–2 m deep pits. Deep fresh road cuts outside the research area showed roots penetrating as deep as 10 m in friable, porous soils in Dipterocarp-bearing forest and lower montane oak–conifer forest. In Sarawak, over 200 soil profiles in the 57 Kerangas research plots showed a close correlation of rooting depth with soil texture and drainage. Rooting depth ranged from 25 cm in shallow humus podzol and gleysol to over 5 m in deep humus podzol (DHP) (Bruenig, 1966, 1974, pp. 75–76, table 6; Fig. 1.15, Table 1.5). In the sandy soils of Kerangas tree roots are more strongly concentrated in the Ah and Ae horizons than in sandy humult ultisol soils of the MDF. Few roots penetrate into the B and C horizons of Kerangas soils except in

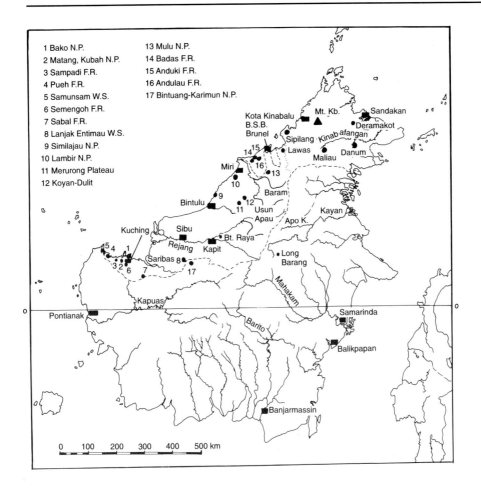

Fig. 1.16. Locations in Sarawak, Sabah, Brunei Darussalam and Kalimantan that are mentioned in the text. Sabal is Sabal Forest Reserve containing the > 600 ha RP 146, which has the highest species richness per hectare recorded in Borneo (Droste, 1995).

DHP and sandy RYP (red–yellow podzolic, or humult ultisol/acrisol) soils. Strongly tap-rooted species, such as *A. borneensis* Warb., reach as deep as 3–5 m into the B and, more rarely, into the C or D horizon. These roots are probably important generally for nutrient catching, but especially for the water supply during episodic droughts. On the same site, the stilt-rooted, extremely sclerophyllous and xeromorph *C. nobilis* Whitm., in mixture with the microphyllous *A. borneensis*, has a much shallower but wide-spreading root system. In MDF, roots penetrate even deeper and can still be observed at 7–10 m depth in road cuts. Rainforest trees, as trees in other climates, grow their roots where they find the conditions to do so. Wherever the soil and the weathering parent material are physically

Table 1.5. Water availability, wilting point and rooting depth in different soil types have been estimated for typical soil texture and soil organic matter content. Transpiration rates accord with those in Section 1.7, and decline with gradual drying of soils. Water stress is assumed to begin when the soil water potential can no longer be overcome to extract sufficient amounts of water to maintain adequately leaf activity and temperature. The estimated length of rainless periods required to create water stress in natural site-adapted forest is a useful guide for matching of tree species, stand structure, silvicultural system and site conditions in Mixed Dipterocarp forest (MDF) on humult and udult ultisol/acrisol, in Kerangas forest (KF) on shallow, medium and deep humus podzol and in Kerapah forest (KrF) on stagnic gleysol overlain by a thin peat layer. Values are recalculations of tables 3 and 4, Bruenig (1971a), with new data on the structure and depth of the root systems (for rooting depth see Fig. 1.15, for forest structure see Fig. 1.11).

Forest	Soil	Field capacity (%)	Wilting point (%)	Rooting depth		Plant available water			Dry days to persist water stress
				A (cm)	B/C (cm)	In soil (cm)	Biomass (cm)	Sum (mm)	
MDF	Humult ultisol/acrisol, sandy loam	24	6	50	600	63	2	650	>100
MDF	Udult ultisol/ferralsol clay	46	26	30	400	40	2.5	425	>70
KF	DHP sand/gravel	20	1	30	300	20	2.5	225	50
KF	MHP sand/gravel	24	2	25	120	15	2	170	33
KF	MHP sand/sandstone	26	2	20	60	10	1	110	25
KF	SHP sand/sandstone	26	2	15	30	5	0.5	55	15
KrF	PGL gleyic clay	32	20	20	60	7	0.7	77	20

penetrable and chemically and hydrologically attractive, roots will grow. Germinating seedlings rapidly develop a vigorous tap root with a brush-like collar of nutrient-feeding roots in the F-layer and SOM-rich surface soil layer. These 10–30 cm long tap roots often die during prolonged periods of water saturation and regrow when the water level subsides.

Thick and dense root mats, intricately mixed with decaying litter and raw humus, can be found on almost any soil type. In MDF on ultisol/acrisol they are small, patchy and usually confined to the base of decaying large trees. They cover the surface in extremely oligotrophic peatswamps and on humic podzols, on lithosols, on limestone karst or quartzitic sandstone, and also on some very dense or rocky oxisol/ferralsol soils. Nutrients in the dead organic matter are efficiently recycled and losses to the hydrosphere are very small (Jordan *et al.*, 1980; Jordan, 1985). Within the root mat and the Ah horizon of Kerangas soils, Schmidt-Lorenz and von Buch found abundant ectomycorrhiza, leguminous and non-leguminous (*Podocarpus* spp., *Falcatifolium* spp.) nodules and animal faecal pellets (reported in Bruenig, 1966). The proportion of root biomass in the total living tree biomass generally seems to increase with decreasing fertility/productivity of the soil in tropical rainforest (Klinge, cit. in Kurz, 1983). The same is well documented for forests in Germany (e.g. Vavoulidou-Theodorou and Babel, 1987). In tropical rainforest the root biomass as a proportion of total biomass is about 10–15% on heavy clayey soils well supplied with water, 15–20% on average loamy-clay oxisols and clay-loam ultisols. It increases with increasing sandiness and unfavour-ability of soil conditions and reaches 50–60% on very adverse sites such as the bana in San Carlos de Rio Negro. A possible explanation is that the roots function as a reservoir of nutrients and water (Klinge cit. in Kurz, 1983). The heavy weight of the intricately bonded fine-root mineral soil/SOM complex, together with a low height/diameter ratio, makes the trees wind resistant.

The dogmatic myths of shallow-rootedness and lack of litter and humus in tropical rainforests have obscured the ecological importance of the diversity of structure of rooting systems and of the importance of an adaptable and deep-reaching rooting system for nutrient preservation and emergency water supply in tropical rainforest soils. They have also led to the thesis that almost all nutrients are stored in the above-ground living biomass. But, as Whitmore (1990) points out, in many tropical rainforests most nutrients are fairly evenly divided between above- and below-ground live and dead biomass. This may even apply to extremely oligotrophic conditions if the root mass and SOM content are relatively large. It is therefore wrong to assume that all fertility rests in the living biomass and in the very top soil horizon (Ah), that it cannot be restored once the phytomass and topsoil have been disturbed, and then to deduce that there is no possibility of sustainable forest management if it involves removal of tree biomass. In reality, not only are the physical, chemical and biological conditions of the humus layer and surface soil amenable and crucial to sustainable forest management, but so are the structure, SOM, nutrients and rootability of the deeper soil layers. Forest management in natural forests therefore should retain the spectrum of natural tree species with site-adapted diverse rooting systems and litter production. In planted forests, mixing tree species with diverse litter quality and rooting habits should secure full utilization of the rootable soil volume and adequate SOM quality and quantity.

1.6 Canopy

Potential sources of stress in tropical rainforests are chronic scarcity of nutrient reserves in the intensely weathered parent material and diurnally recurrent and unpredictable episodic heat load and water deficiencies in the leaves. The frequency and severity of episodic climatic extremes vary strongly between localities. Their effects on energy and water balance vary according to soil and site types (Bruenig, 1966, 1971a). If this is so, there should be differences in the physiognomy of the forest canopy between sites that could be explained as risk avoidance or risk reduction with respect to nutrient loss and damage from heat, drought or supersaturation. Canopy differences between sites could indicate feasible strategies of coping with climatic and nutrient stress (Bruenig, 1970a, 1984c, 1986b, 1991b). The classic, theoretical and experimental work of

Fig. 1.17. Canopy of Mixed Dipterocarp forest in the Mulu area from above (see frontispiece colour photo of the same area). The great richness and diversity of form and species in the canopy enables rapid reaction ('chaotic swinger' type) and reduces the risk of damage from biotic and abiotic factors. (Reproduced courtesy of Peter Wee.)

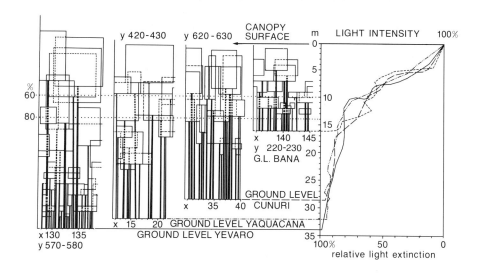

Fig. 1.18. Stand profile along a gradient of declining quality for plant and microbial life and growth of chemical, physical and hydrological soil conditions as shown in Fig. 1.11, top. The 'structural tree species' of the A-layer and the upper part of the B-layer monopolize and regulate the major processes in the forest ecosystem of primary production, evapotranspiration and energy partitioning. Up to 80% of the incoming direct and diffuse light is intercepted in the canopy of the structural tree species (above the approximate 80% levels). The stand profiles are computer prints from field data of the International Amazon Ecosystem Study at San Carlos de Rio Negro. The crown rectangles represent measured or visually estimated crown length and average diameter. Extinction curves from Heuveldop (1978). Yevaro, *Eperua purpurea* Benth.; yaquacana, *Eperua leucantha* Benth; cunuri, *Micandra sprucei* Benth., which are characteristic species differentiating and dominating the respective forest type (association group according to Bruenig *et al.*, 1978).

Denmead (1964), Gates (1965, 1966, 1968), Kung (1961), Lemon *et al.* (1969), Monteith (1963) and Reifsnyder (1967) provided information for interpreting the ecological significance of the differences in rooting, of the physiognomy of the canopy and of the morphology, chemistry and physiology of leaves in tropical rainforest in Kerangas in Borneo and in evergreen Caatinga in Amazonia (Bruenig 1966, 1969a, 1970a, 1971a, 1976; Bruenig *et al.*, 1978, 1979, 1978). These forests, growing under ecologically more unfavourable conditions than the 'zonal' forests, are forced to adapt more rigorously to stresses and thus can be more easily observed. Regulatory mechanisms are more obvious and can be studied more easily than in the 'zonal' forests. The general conclusions (Bruenig, 1966, 1970a; Bruenig *et al.*, 1979) are that the canopy of the tropical rainforest is naturally adapted and can be silviculturally manipulated in a manner which reduces the effects of site-specific stresses and the risks of damage, but at the same time permits the trees to exploit fully the more favourable conditions during average situations. The aerodynamic roughness of the canopy surface (macrostructure of canopy) affects the interception of radiant energy, air moisture, air pollutants, regulates free and forced

convection, wind interception and turbulence. The aerodynamic roughness and shape of the typical 'cauliflower' crown, the bunched or more
diffuse leaf distribution, and leaf orientation determine ventilation,
energy interception, radiative load distribution, transmission of light and
of sensible and non-sensible (latent) heat. The leaf morphology and
chemistry (xeromorphic sclerophylly) affects transpiration rates, heat and
drought resistance (Fig 1.11), accessibility of organic matter and nutrients
to consumers and decomposers, and the humus-forming properties of the
litter (Fig. 1.9; Section 1.7). The structural similarity of tree crowns in the
A and B canopy layers of tall, species-rich complex zonal rainforest and
the whole canopy of the lower, simpler edaphic climax forests on podzols
and related unfavourable soils is paralled by the similarity of light
extinction (Figs. 1.18 and 1.24). The corresponding brightness at ground
level of the poorer types of Kerangas and Caatinga forests explains the
relative ease of regeneration. The main problem of regeneration is not
light but episodic water supply deficiency, which can cause drought
strain and mortality among regenerating trees (Becker and Wong, 1993;
Dalling and Tanner, 1995) and in severe cases also among upper-canopy
trees (Woods, 1987) (Section 1.7).

1.7 Hydrology and Nutrients

Forests affect atmospheric and edaphic moisture and the hydrological
conditions of the site by the stature (height), structure (roughness) and leaf
physiognomy of the canopy, the structure of the root system, and the
physiological features of the absorption and utilization of output of water
by the trees. The observed changes of canopy and roots along ecoclines
(Figs 1.11 and 1.15) suggest that the natural primeval rainforest is structurally adapted to the climatic, edaphic and atmospheric conditions of the
site, especially to episodically occurring extreme events. Figure 1.11
illustrates the general trend of change along the two catenas MDF–
Kerangas forest (KF)–Kerapah forest (KrF). This catena is structurally
analogous to the Amazonian catena Mixed Upland forest (TF)–Caatinga
forest (CF)–Bana woodland (Bruenig *et al.*, 1979), the corresponding
ecocline lowland–montane, oxisol–podzol in Bawangling, Hainan (Bruenig, 1986a; Bruenig *et al.*, 1986), and the Peatswamp forest sequence of the
phasic communities PC 1–6 according to Anderson (1961a). With increasing adversity of site, especially high risk of alternating drought and water
saturation, the aerodynamic roughness of the canopy decreases, while
sclerophylly and xeromorphy increase. In Kerangas and Caatinga forests,
crowns are smaller and the leaves more sclerophyllous, coriaceous and
smaller than in sclerophyll/mesophyll MDF and TF. However, as adaptation to site can be achieved by different combinations of the various
adaptive features, variability and overlap are wide (Bruenig, 1970a).
Classification of forest and woodland vegetation in the MDF–KF and

TF–CF complexes by leaf size spectra alone is not generally feasible (Newbery, 1991) in spite of obvious and visible physiognomic trends (Huang and Bruenig, 1987). This diversity of adaptive features offers a range of opportunities for designing site-specifically adapted canopies with certain microclimatic and physio-ecological characteristics for sustainable naturalistic silvicultural management in artificially established (planted or otherwise) forests and for water catchment management (Bruenig and Sander, 1983).

The soil in tropical rainforests has a typically high infiltrability. The humus-rich, porous soil surface absorbs rainfall very rapidly. Overland water flow occurs only during extremely heavy and intense downpours or after the soil has been saturated, but then it can be substantial and cause erosion and flash-floods. Heuveldop (1978; Fig. 1.19) in San Carlos and Waidi *et al.* (1992) in Danum Valley showed that, for an annual non-seasonal rainfall of 3600 mm (100%), about 12–17% was interception loss, 80% direct throughfall, 2–8% stemflow and 2–3% direct overland flow. Of the 82–88% precipitation reaching the ground, about 97% infiltrated into the soil. Infiltration measurements in very deep, porous, humic soils in the CERP-Tropical Forest Ecosystem Study at Bawangling, Hainan, showed that even a very heavy typhoonal downpour will be absorbed instantaneously, until the soil becomes saturated when overland flow occurs (unpublished research data of the Institute of Forest Hydrology, Hann-Münden, 1990, and final project report to the Ministry of Research and Technology, Bonn, 1991). Destruction of soil structure and loss of porosity at the soil surface reduce infiltrability and increase overland water flow and, consequently, soil erosion.

The broad global figures that Bruijnzeel (1987) reports for evapotranspiration from lowland tropical rainforests, which never experience soil moisture shortages (as determined by the water balance method), average 1460 mm ± 27% (SE) with 14 mm ± 2% interception and 1–2% stemflow. This is within the range of values reported by Heuveldop from the International Amazon Ecosystem Study at San Carlos. He estimated the annual evaporation in the zonal tropical rainforest (tierra firme forest) as 183 mm, the transpiration as 1759 mm, and the runoff as 1722 mm (Fig. 1.19). Effective protection of the soil against loss of porosity and infiltrability requires a protective, complex, vertically closed canopy with ample leafage and litter fall. Reduction of water runoff requires an aerodynamically rough canopy with high transpiration and high potential evaporation rates. Potential annual evapotranspiration rates under the climatic conditions of Kuching (Fig. 1.16) are class A pan evaporation 1970 mm, regression over sunshine hours 1740–2016 mm, the Penman equation 1800 mm or, amended by Monteith to account for canopy roughness, about 2000 mm. The calculation by the Holdridge formula, using forest stature and a supposed layering into seven (MDF) to four (simple KF) storeys and assuming unlimited water supply, produced similar values for the potential evapotranspiration in West Sarawak (see Fig 6.8):

TROPICAL RAINFOREST

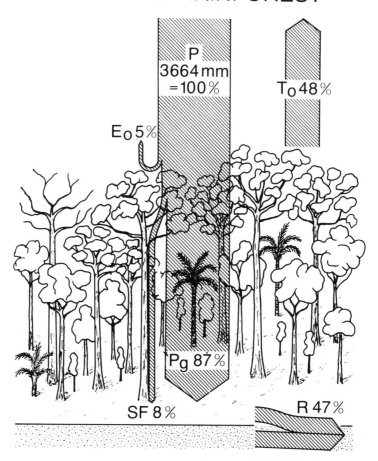

Fig. 1.19. The water balance in the zonal lowland rainforest on oxisol in the MAB International Amazon Ecosystem Study, San Carlos de Rio Negro, Venezuela. Rates of input, throughput, output and runoff of water are high. The high transpiration rate reflects high rates of net radiation, canopy roughness, the 'giant wick' nature of the emergent (A-layer) and codominant (B-layer) crowns, and high water-holding capacity of the soil. Interception and evapotranspiration are lower in the low-stature, simple sclerophyllous and xeromorphic Caatinga and Kerangas on shallow soils with low water-holding capacity, ET_0 may be as low as 50% at the right hand of the catena in Figs 1.11 and 1.18. P, precipitation; Pg, precipitation reaching the ground; E_0, evaporation from surface; T_0, transpiration from plant surface; SF, stemflow; R, runoff. (From Heuveldop, 1978.)

1. Mixed Dipterocarp forest (height > 45 m, complex): 1700–2000 mm
2. Kerangas
   height > 40 m, complex: 1520 mm
   height 35 m, moderately complex: 1000–1500 mm
   height < 28 m, simple, smooth: 800–900 mm

3. Kerapah (height < 25 m, simple, smooth): 740 mm
4. Mixed Ramin Peatswamp forests (PC1): 1800 mm
5. *Shorea albida* Peatswamp forests
   Alan (PC2): 1600 mm
   Alan bunga (PC3): 1400 mm
   Padang Alan (PC4): 1000 mm
6. Mixed Padang and open Padang (PC5 and 6): 800 mm

During the Kerangas study in Sarawak a simple tentative model was constructed to simulate soil-water depletion during dry periods in MDF and KF research plots on deep, medium and shallow soils. The state variables were plant-available water in the biomass and soil (difference between field capacity and wilting point in the actually rooted soil space, Table 1.5) and moisture vapour saturation deficit. The rate variables were transpiration rates and increasing soil water suction tension or decreasing soil water potential. Differences of stomatal reaction between fast-closing mesophyll MDF tree species and slow-closing drought-accustomed, strongly sclerophyll tree species of the KF were considered by different rates of transpiration reduction as water potential in the soil decreased. Rainfall amounts and distribution were taken from real meteorological observation data. The results showed that wilting points were reached in Kerangas forest after drought periods of 20–50 days. The MDF required periods in the order of 100 days to exhaust the plant-available water of the soil and the biomass. A test calculation of the course of water depletion in Kerangas forest RP 15 (KF) and RP 16 (MDF) in Bako National Park used real daily rainfall and other weather data during the 4-month period March–June 1965 (monthly rainfall P = 361, 239, 101 and 145 mm). A period of low rainfall began on 9 April and lasted to 20 June. This period created dry conditions hardly evident in the monthly rainfall figures (in brackets above). Kerangas RP 16 is a marginal MDF on medium-deep humus podzol, 70-cm rooting depth, very close to the plot on sandy ultisol described by Ashton and Hall (1992). RP 15 is a typical KF on a medium humus podzol but with shallower rooting (40 cm). The unexpected result was that the MDF in RP 16 exhausted the plant-available water within 46 days in spite of five intermittent rainfall events. The nearby KF in RP 15 approached exhaustion at one time after 70 days (18 intermittent rainfall events), but then heavy rain on 20 June and 22 June saturated the soil again. Obviously, the water requirements of the complex, aerodynamically rough and tall MDF could not be supplied and the type of canopy could not be supported by the moisture capacity of the soil. The explanation for the edaphically aberrant existence of MDF on a KF soil was that additional water, and probably nutrients, were supplied by surface and subsurface water flow from the plateau and escarpment further up-slope (Bruenig, 1971a).

An open, bunched architecture of the crown, such as in rainforest trees and temperate oaks and pines, improves ventilation and facilitates air exchange even during almost calm, bright weather by active free and forced convective air exchange. Rapid removal of the atmospheric boundary layer is facilitated by small, longish and more-or-less upright leaves. These features have a dual function. They reduce heat stress on bright days during drought and during noon peaks of radiative heat load and water saturation pressure deficits, but they also permit very high rates of transpiration if water is freely available. The high transpiration potential of drought-resistant sclerophyllous evergreens, which naturally occur on nutrient-poor, droughty sites, is well known. Examples in temperate and tropical forests are species of pine and eucalypt. These species transpire luxuriantly if water supply is ample, but survive without damage if there is shortage. High rates of transpiration help with the uptake of nutrients on the nutrient-poor sites on which most species of pines and eucalypts, as well as Bornean Kerangas and Amazonian Caatinga trees, naturally occur.

In conclusion, the leaves, tree crowns and the canopy of the tropical rainforest are so designed that very high transpiration and photo-synthesis rates can be achieved under favourable conditions, but at the same time short- and long-lasting periods of water deficiency can be endured during bright, hot weather. The root system is accordingly designed to obtain nutrients and water efficiently to support the high rates of net primary production and minimize nutrient losses during wet and moist periods and to exploit plant-available water at the greatest possible soil depth during prolonged dry periods to maintain production and prevent drought damage. Recent research results support our earlier suggestion from Sarawak (Bruenig, 1966, 1971a) that deep rooting in tropical rainforest is an important asset during drought. Drought stress on trees can induce their roots to increase exudation of carbon compounds. This process provides additional food for microorganisms and may be of considerable ecological significance, possibly affecting the mycorrhizal association 'as a fundamental regulator in plant water relations' (Killham, 1994) especially for water supply during prolonged periods of drought from deeper soil layers. According to Shuttleworth and Nobre (1992), citing recent research by Hodnett *et al.* (1992), 'Amazonian forests seem capable of accessing soil water to considerable depth, certainly to a depth of 4 m and quite possibly to depths of 6–8 m'. They conclude that this would be an important asset during drought.

The effects of plant-available mineral macronutrients in the mineral soil, in the SOM and in the decaying dead organic matter on the distribution and performance of rainforest tree species are difficult to assess. Many factors interact, the risk of determining pseudo-correlations is high, and the aut-ecology and physiology of the rainforest species and genotypes are little known. Nutrient contents in fine (mainly leaves, twigs, fine roots) and coarse (trunks, branches, roots) litter vary between species and individuals and the rates of litter production vary with time and mortality occurs patchily. Consequently nutrient availability would be heterogeneous over an area and variable over time. Simple relationships may be established for some parameters under otherwise relatively uniform conditions (e.g. Baillie *et al.*, 1987), but not if conditions are more heterogeneous (Ashton and Hall, 1992). Vitousek (1984) considered phosphorus to be the most important limiting nutrient element in most rainforest. Högberg and Alexander (1995) state that, while N-fixing nodule-forming tree species (NOD) should, in theory, be limited by phosphorus, non-nodulated tree species could be limited by either N or P. They quote Gartlan *et al.* (1986) and Newbery *et al.* (1988) that in the rainforest at Korup, N:P ratios were high in all three symbiotic groups (ECM, VAM and NOD + VAM), which supports the idea of P being the limiting element there. The authors have previously speculated that in Korup 'ectomycorrhizas (ECM) might offer advantages over vascular mycorrhizas (VAM) in the utilization of organic N, but all the new data, the higher foliar % P, the lower N:P ratio and the similar $\delta^{15}N$ values, argue against that hypothesis and for the notion that ectomycorrhizas are more important in P nutrition in that forest'. In the data on species composition in KF in Borneo and Caatinga forest in Amazonia, there is no indication that NOD–legumes had any particular advantages in nutrient supply. The mechanisms, dynamics and ecological significance of the nutrient deposits in the roots, stems and branches are as yet little understood, but need to be considered in nutrient cycling models (Noij *et al.*, 1993). Burslem *et al.* (1995), in a pilot experiment with seedlings of rainforest tree species, confirmed the existence of a wide variation of responses to fertilizer applications between species. Phosphorus, taken up in excess of immediate requirements, was transferred to the long-living stems rather than short-lived leaves. Rainforest trees habitually, like all leaf-shedding perennial plants, retrieve essential nutrients from the leaves before abscission. Both processes would support the hypothesis that trees are capable of building up an internal slow-release stock of nutrients. In contrast, moisture can be stored in the living biomass to meet the needs of only a few days. The storage in the soil may supply the demand of weeks to months (Table 1.5). When the store in the phytomass and soil is exhausted, critical strain develops (Bruenig, 1966, 1971a). Riswan (1991) analysed the contents of six mineral nutrients in living sclerophyll leaves of Kerangas trees. He found that P, K and Ca were higher than in MDF, and N, Mg and Na lower. There was considerable seasonal fluctuation and variation between trees and species. In Kerangas, Caatinga and Peatswamp forests, it seems the hydrological conditions and SOM are more distinct and probably more critical, but certainly less open to manipulation by the vegetation or by management, than the nutrient conditions in the mineral soil.

The deeply weathered parent materials of rainforest soils have few mineral nutrient reserves. Replenishment of nutrients has to come from the very variable particulate and gaseous contents of nutrients in the atmosphere that enter the forest ecosystem (Fig. 1.20). Episodic or seasonal

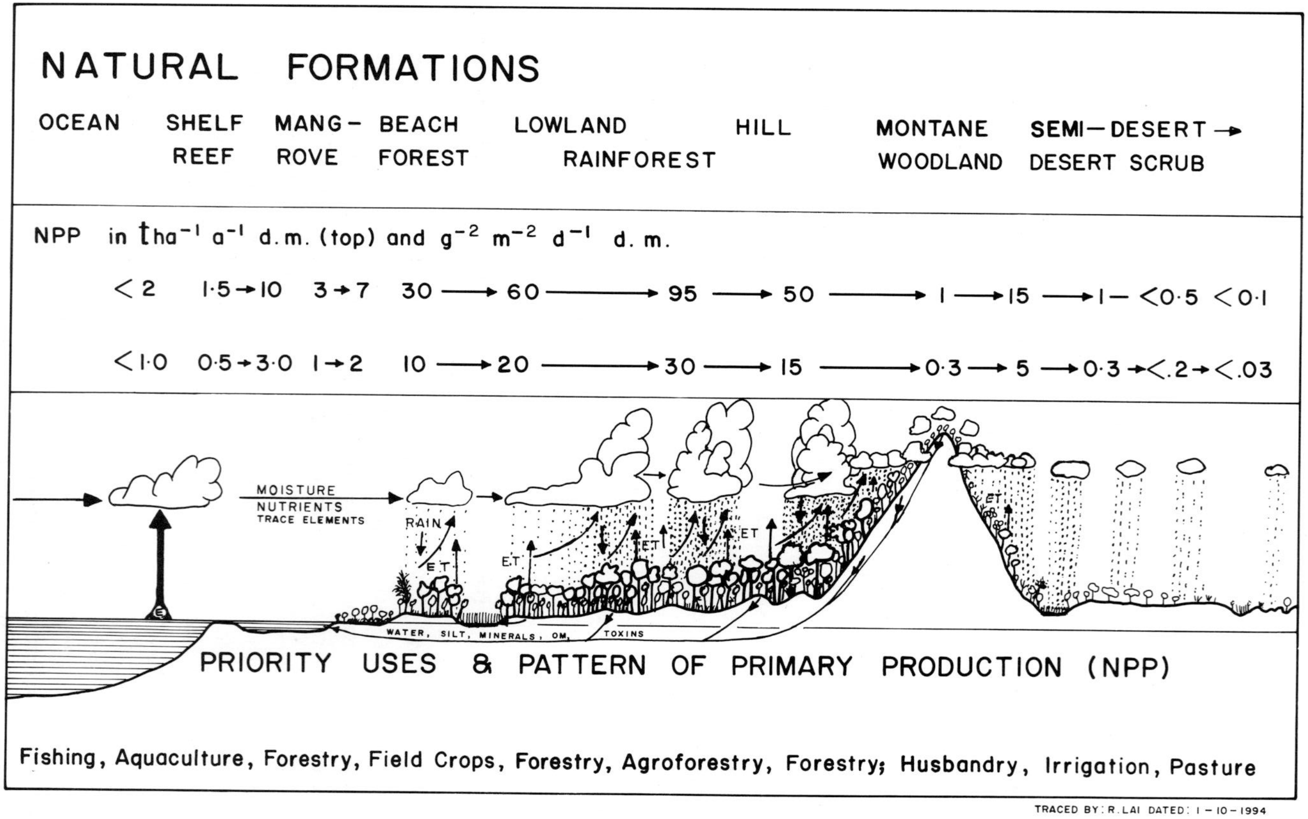

NATURAL FORMATIONS
OCEAN
SHELF REEF
MANG-ROVE
BEACH FOREST
LOWLAND RAINFOREST
HILL
MONTANE WOODLAND
SEMI-DESERT →
DESERT SCRUB
NPP in t ha⁻¹ a⁻¹ d.m. (top) and g⁻² m⁻² d⁻¹ d. m.
<2 1·5→10 3→7 30 → 60 → 95 → 50 → 1 → 15 → 1 – <0·5 <0·1
<1·0 0·5→3·0 1→2 10 → 20 → 30 → 15 → 0·3 → 5 → 0·3 → <.2 → <.03
MOISTURE
NUTRIENTS
TRACE ELEMENTS
RAIN
ET
WATER, SILT, MINERALS, OM,
TOXINS
PRIORITY USES & PATTERN OF PRIMARY PRODUCTION (NPP)
Fishing, Aquaculture, Forestry, Field Crops, Forestry, Agroforestry, Forestry; Husbandry, Irrigation, Pasture
TRACED BY: R. LAI DATED: 1-10-1994

Fig. 1.20. Generalized trend of net primary productivity (NPP) and climatic functions in the natural zonal lowland and montane vegetation. Nutrients and trace elements originate from the sea, volcanic eruptions, fires and desert storms and anthropogenic air-polluting sources.

events, such as large fires, desert and oceanic storms, eruption of volcanoes, change the atmospheric chemistry and thereby the nutrient and trace-element input into the rainforest. Data on nutrient inputs (Jordan *et al.*, 1980; Hilton, 1985) suggest that the rate of deposition of macronutrients and trace elements can be substantial and balance losses from natural leaching to which the primeval rainforest has adapted. Wet and dry deposition may restore losses from selective felling within adequately long felling cycles, except perhaps for phosphorus. However, there may be depletion in the course of several felling cycles unless high levels of phytomass stocking are maintained to store and cycle nutrients. Heavy rates of erosion and leaching as a result of selective logging, especially on steep slopes and with heavy machinery, and short felling cycles will very probably cause serious nutrient and microbiological depletion in the soil. Areas with naturally poor soils, high relief energy and intensive rainfall such as Borneo are particularly prone to depletion and lasting degradation. The relative poorness of soil nutrient availability and low nutrient status of the litter reported in MDF, KF and Alluvial forest in Gunung Mulu National Park in comparison with other regions (Anderson *et al.*, 1983) indicates a problem that urgently requires in-depth field and laboratory studies, while forestry practice, in the mean time, has to adopt a precautionary approach. The problem is seriously increased by conventional selective logging. The large amounts of dead biomass that are left behind (Section 2.8; Tables 2.3 and 2.4; Fig. 2.15) release large amounts of humic acids, polyphenols and tannins for at least 20 years (Section 1.3, decay rate), which leach the soil and pollute the water bodies, causing a multiple and heavy loss to the ecosystem.

1.8 Species Richness and Evenness of Mixture

The number of tropical plant and animal species is most likely around 3–5 million, but there is considerable uncertainty especially with respect to insects and other invertebrates (Whitmore and Sayer, 1992, especially pp. 8–9). Existing taxonomic and ecological knowledge of soil-dwelling animal, plant and microbial species in most forests of the world is fragmentary. The easily accessible plants are better known. For example, the regional programme Flora Malesiana can build on the results of more than two centuries of botanical collection and research. It possesses an unequalled network of participants including more than 150 scientists. Started in 1947 by Professor van Steenis, the programme was expected to cover up to 30,000 flowering plant species which could be documented and revised by the end of the century. The current estimate is 42,000 flowering plant species and completion is expected by 2020 or perhaps 2040 (Roos, 1993). If plants take so much time, the cataloguing of mobile animals and microbial organisms will take centuries more. The basic features of global recording and long-term monitoring of biodiversity are only just emerging (Vernhes and Younes, 1993). Sufficiency and representativeness at local and regional levels (Mawdsley, 1993) are particularly difficult to achieve in the species-rich, heterogeneous, constantly changing, complex and dynamic tropical rainforest. The problem of species monitoring for sustainable management is mentioned in Section 10.2.

The following section focuses on aspects of species richness and diversity that are directly relevant to sustainability of management and conservation. A primary assumption is that the most practical and rational strategy is preservation of quality and size of habitat. The habitat in a forest is shaped by the trees in the A and B layers of the canopy (Fig. 1.17 and frontispiece). They are the key elements in the viability and functioning of forest ecosystems that determine the microclimate, the qualities and quantities of stocks and flows of matter and energy, and the architectural or geometric structure of all layers of the growing stock. They are the objects of harvesting and the major tools of silvicultural manipulation towards adequate self-regulation, self-sustainability, economic viability and social values. The A- and B-layer tree species are the 'structural species on which the interstitial species depend' *sensu* Solbrig (1991a). Manipulation of the A- and B-layers is the most sensitive and effective means to regulate species richness and diversity. The state of the canopy determines the amount, quality and diversity of living space offered by the forest ecosystem above and below the ground surface. Therefore, the floristic and geometric structure of the tree canopy determine the ecological and economic diversity and diversity-related features of the rainforest ecosystem. This section therefore concentrates almost exclusively on the tree-species component, because trees are the primary structural elements with which foresters manipulate, produce and preserve, and trees are taxonomically and ecologically better known than other biological components of the rainforest.

The long-term evolution of the richness of plant, animal and microbe species in the equatorial rainforest is facilitated by the biological barriers between genetic populations, the multitude of breeding systems, genetic isolation mechanisms, heterogeneity of soil and climate, and the dynamics introduced by tectonic movements, sea-level changes and multiscale climatic variation and changes. In the present, diversity is maintained by habitat variation due to crown break, tree mortality, windthrow, lightning, pests and diseases forming gaps from a few square metres to hundreds of square kilometres (Figs 1.3 and 1.4; Browne, 1949). Species richness and biodiversity are naturally dynamic and not static. Accordingly, biodiversity must be managed and monitored by procedures that take account of this dynamic nature and the features of the ecosystem of which they form part (Chapters 6 and 10). Tree-species richness, in this book, will be expressed as number of species in 100 randomly selected trees above a certain stem diameter. This measure has the advantage that it is unaffected by the stature of the trees and the heterogeneity of site and forest. The evenness of mixture within the 100-tree batches will be expressed conveniently by the McIntosh index of diversity (Bruenig, 1973c; Bruenig and Huang, 1989). The values of diversity indices, including the McIntosh index, vary widely between sampling points within a forest community (Bruenig and Schneider, 1992; Prevost and Sabatier, 1993). Differences between natural forest communities (tree-species association groups, forest types) are statistically difficult to prove. However, the indices are useful for broad comparisons between distinct forest communities at medium and large scale of area.

Current species richness in mature, more-or-less untouched (Section 2.1) primeval rainforest is the result of the specific history of the site and is closely related to the present edaphic, climatic and atmospheric conditions. This indicates ecological saturation and sufficiency levels of tree-species richness in relation to the carrying capacities and stress factors of different sites, and could be a guide for target species mixtures in silvicultural management. Saturation and sufficiency levels are still uncertain in the much more intensively researched and managed temperate forests in species-rich North America and Far East, or naturally species-depleted Europe. Global change very likely will complicate the problem by changing stresses and carrying capacity differently on different sites. Tentatively, the actual species richness in mature, untouched primeval forests on a specific site could be taken as an indicator of the upper limit of sufficiency and carrying capacity. The *Dryobalanops beccarii* MDF (Ashton, 1995) in Sabal RP 146 had 241 tree species with diameters > 10 cm per hectare, while *A. borneensis* KF and *Casuarina nobilis* KF (Bruenig, 1974) together had only 166 (Table 1.6; Droste, 1995). This indicates the general trend of the ecological gradient correctly, but is too coarse for distinguishing more subtle differences for silvicultural decisions. The area-related data are too strongly affected by forest structure and by the minimum diameter of the chosen sample population. In the *A. borneensis* KF block 84 (50 × 50 m) of RP 146, Sabal, 86 species were enumerated with diameters > 10 cm and 260 with diameters 1–9.9 cm on 0.25 ha. These figures are high in comparison with MDF in Mulu and Lambir National Parks. Proctor's 1-ha MDF plot in Mulu National Park (DF in Fig. 1.8) had 223 tree species with diameters > 10 cm, and two randomly selected 1-ha MDF plots in Lambir National Park had 160 and 190 tree species with diameters > 10 cm and 314 and 344 tree species with diameters 1–9.9 cm per hectare (Research Branch, Forest Department Sarawak, Lambir dataset, figures supplied by William Then, 29 May 1995). To avoid the effects of forest stature and tree density, we had chosen to measure tree-species richness among batches of 100 trees with diameters > 1, 10 and 15 cm along ecological site gradients in Borneo, South China and Amazonia, using a randomly selected centre tree and its 99 nearest neighbours. The results (Figs 1.21 and 1.22, Table 1.7) show a close association of species number with soil type and latitude. The number of species declines with soil quality. The maximum is on deep ultisol/acrisol

Table 1.6. Species richness in RP 146, Sabal F.R. (Section 1.8) and growing stock above ground and growth (Section 1.12) in Lowland Mixed Dipterocarp Forest (MDF) and Kerangas forest (KF). (a) Sabal RP 146. Growing stocks > 10 cm diameter in 1963 (primeval, except for native collecting Gaharu by felling *Aetoxylon sympetalum* (Steen. and Domke) Airy Shaw) and 1990–1991, probably 12 years after selective logging MDF and creaming part of KF (forest type with *Agathis borneensis* Warb.) The tree trunk volume increment 1978–1990 in MDF is 10 m^3 ha^{-1} a^{-1}. (b) Mulu RP 142 B, MDF, and D, KF. Diameter increment in primeval MDF per diameter class and per crown position and social class, indicating the same trends as shown in Fig. 6.2 by *Shorea* spp. (Section 6.3). Crown scales are defined according to Dawkins (1959, 1963) in Synnott (1979). *N*, number of stems; *G*, basal area; V_{tw}, volume of thickwood (> 7 cm) of trees > 10 cm diameter; PM_{tw}, phytomass of thickwood volume; PM_{to}, total phytomass; Spp., species; st, standing; f, fallen to the ground; *d*, diameter 'breast height'; *i*, increment. The MDF soil in Sabal RP 146 is sandy–loamy humult ultisol, and in Mulu RP 142 B clayey udult ultisol. The KF soil in Sabal RP 146 is medium-deep to shallow humus podzol with some aqui-spodosols along streams and miniature woodland peat bogs in depressions, and in Mulu RP 142 D medium-deep humus podzol over Pleistocene gravel (Bruenig, 1966, 1974). $V_{tw} = 0.5$ $(d \cdot h)$.

(a) Sabal RP 146

	MDF – *Dryobalanops beccarii*		KF – *Agathis borneensis* (and) *Casuarina nobilis*	
	1963 (primeval)	1990/91 (post-logging)	1963 (primeval)	1990/91 (creamed)
Living PM				
N (ha^{-1})	642	693	850	891
G $(m^2\ ha^{-1})$	29.2	26.6	31.7	32.3
V_{tw} $(m^3\ ha^{-1})$	410.3	359.3	425.1	427.8
PM_{tw} $(t\ ha^{-1})$	246.2	215.6	255.1	256.7
PM_{to} $(t\ ha^{-1})$	451.3	395.2	467.6	470.6
Spp. (ha^{-1})	–	241	–	166
Spp. $(100\ N^{-1})$	69	70	53	55
Dead woody PM				
N_{st} (ha^{-1})	–	37	–	63
V_{st} $(m^3\ ha^{-1})$	–	15.0	–	34.3
PM_{tw} $(t\ ha^{-1})$	–	9.0	–	20.6
PM_{to} $(t\ ha^{-1})$	–	16.5	–	37.7
V_f $(m^3\ ha^{-1})$	–	167.5	–	67.9
PM_f $(t\ ha^{-1})$	–	100.5	–	40.7

(b) Mulu RP 142 B + D

	Stem diameter class in cm (d)							
d-increment in mm per d-class in cm	10–20	20–30	30–40	40–50	50–60	60–70	70–80	> 80
B. MDF								
id $(mm\ a^{-1})$	1.0	1.6	1.7	3.2	3.1	1.9	3.4	1.5

	Stem diameter class in cm (d)							
N (observed)	433	131	76	39	19	17	13	18
								18

D. KF

id (mm a^{-1})	1.1	1.6	2.0	2.5	3.2	3.5	4.7	3.7
N (observed)	300	93	43	16	13	8	4	16

Crown position class	5	4	3	2	1
Canopy layer	D	C	C/B	A/B	A
Social position	Suppressed	Suppressed	Oppressed	Dominant–codominant	Emergent

B. MDF
d-increment in mm

id (mm a^{-1})	1.0	1.3	2.2	2.5 (max. 6.9)	1.3 (max 4.6)
N (observed)	170	122	103	29	9

D. KF

id (mm a^{-1})	0.7	1.1	1.6	3.1	3.0
N (observed)	111	146	161	61	13

Source: Droste (1995, 1996).

in a tierra firme forest adjacent to the W. Egler Forest Reserve near Manaus (Klinge *et al.*, 1974) and on humult ultisol in transitional MDF in Sarawak (Sabal Forest Reserve, RP 146). Species richness consistently declines through deep, medium and shallow humus podzol and related soils in lowland to minima in montane KrF and KF in Sarawak, Brunei and South China, and in Bana in Amazonia. The rainforest analogues in China have, on soil of equal quality, a consistently lower species richness (Bruenig, 1977b, 1986a; Bruenig *et al.*, 1978, 1979, 1989). The forests near San Carlos are slightly less species-rich than their analogues on equivalent soils in Sarawak and Brunei. Typical Bornean MDF on equivalent soils and with moderate disturbance would probably match or surpass neo-tropical forest stands, such as MA (Fig. 1.22; Klinge *et al.*, 1974), or the plots in Costa Rica, Ecuador and Peru (Whitmore, 1990, Figs 2.27 and 2.28). The logged 7-ha MDF and KF in Sabal RP 146 did not lose species. The 241 tree species with diameters > 10 cm per hectare in 1990 (Table 1.6) is also higher than the 230 species which were recorded as the highest value in line 6 of the 52-ha primeval MDF plot in Lambir National Park (Chai, E.O.K. *et al.*, 1994). The creamed KF contains 166 species > 10 cm d ha^{-1} and 260 species 1.5–10 cm 0.25 ha^{-1} (Droste, 1996). The tierra firme plot MA had the flattest tree-species dominance–diversity line known at the time (Klinge *et al.*, 1974; Bruenig and Klinge, 1977).

An example for the within-area and between-area variability of species richness and floristic composition in MDF is given by Manokaran and Kochummen (1991). They compared the floristic structure of the tree

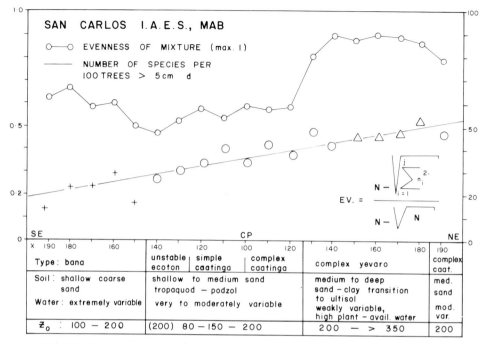

Fig. 1.21. Change of tree-species richness (number of species in a batch of 100 trees > 5 cm diameter, lower line) and evenness of the mixture of tree species > 5 cm diameter within forest types or species association group communities (α-diversity level, upper line) and differences between them (β-diversity level) in the International Amazon Ecosystem Study and MAB Pilot project, San Carlos de Rio Negro. The batches were selected systematically along a transect from the southeast corner to the centre and to the northeast corner. The values were calculated for centre trees spaced at 10 m intervals on the transect, and their 99 nearest neighbours. The evenness (EV) is expressed by the McIntosh (1967) index where in this case N = 100 and n is the number of individuals of each species in 100 N; Z_0 is the aerodynamic roughness estimator. (From Bruenig and Schneider, 1992.)

population > 10 cm in permanent 2-ha research plots in MDF in three locations in Peninsular Malaysia. The locations were Pasoh Forest Reserve (P 1850 mm, four plots), Menyala Forest Reserve (P 2286 mm, 30 m a.s.l., coastal granite-derived alluvium) and Bukit Lagong (P 2481 mm, 460–550 m a.s.l., granite-derived soil). They concluded that tree-species richness in the three forests varied within narrow limits (Pasoh: 264, 235, 261, 276, G (stand basal area, $mm^2\ ha^{-1}$) = 29.1 m^2; Lagong: 253, G = 41.1 m^2; Menyala: 232, G = 31.8 m^2). There was no clear distinction between hill and lowland forest. Ordination by the detrended corre-spondence analysis DECORANA, as would be expected, showed distinct floristic differences between the hill forest at Lagong and the two lowland forests at Menyala and Pasoh. Within the lowlands, Menyala differed somewhat from Pasoh, but the floristic variation between the four plots at Pasoh was as great as the difference between Menyala and Pasoh (Manokaran and Lafrankie, 1991, Fig. 5).

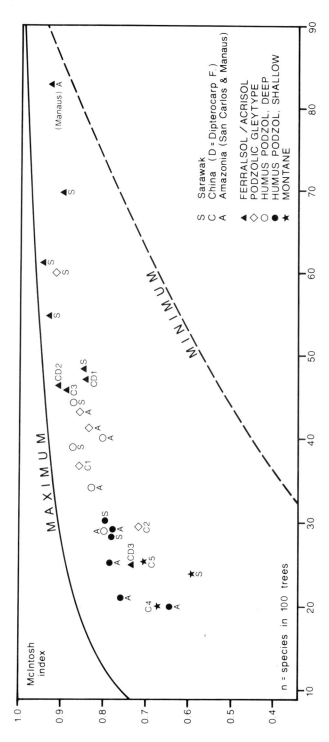

Fig. 1.22. Mean species richness and mean evenness of mixture or diversity among batches of randomly selected trees and their 99 nearest neighbours in predominantly evergreen lowland and montane forests on a variety of different soil types in rainforest types in Sarawak and Brunei (S), Hainan (C) and Guangdong (CH) and Amazonia (A). Species richness is closely related to soil type, diversity (Fig. 1.21) is more variable and less closely related to soil type–vegetation unit and the canopy roughness and disturbance regime. The series in the monsoonal climate of Hainan and Guangdong is shifted to the left in the coordinate matrix where soils are more unfavourable (hydrologically!) in the equatorial forests. (From Bruenig and Huang, 1989.)

Table 1.7. Mean number of tree species and evenness of mixture (diversity) in randomly (105 points) or systematically (diagonals, 37 points) selected sample batches of 100 individual trees above 1.5 and 13 cm diameter at breast height in 10 ha Amazonian evergreen caatinga in the International Amazon Ecosystem Study MAB pilot project near San Carlos de Rio Negro. The forest types are groupings of tree-species association groups classified by a nearest-neighbour monothetic divisive programme FANTASMB. The speci... richness (spp./100N) and the evenness of tree-species mixture (diversity index for 100 trees according to McIntosh, see Figs 1.21 and 1.22) are closely related to soil type, landfo... and intensity of gap formation. The roughness of the canopy surface, expressed in terms of aerodynamic roughness by the dimensionless estimator z_0, is also closely associated with the soil type, landform, frequency and kind of disturbance (all forms of mortality), and consequently with species richness. Yevaro, *Eperua purpurea*; cunuri, *Micranda spruce...* yaguacana, *E. leucantha*; tamacuari, *Caraipa densiflora*. Piapoco, media luna, concha amarillo are not reliably identified. The collection of *c.* 800 fertile reference specimens were lost in the herbarium in Caracas.

| | | | Number of species per 100 trees | | | | Mixture evenness | | Canopy roughness 10 ha area z_0 |
| | | | 10 ha area, over diameters d = 1, 5, 13 cm | | | Two diagonals | 10 ha area >13 cm | Two diagonals >13 cm | |
Forest type	Tree-species association group	SAG	>1 cm	>5 cm	>13 cm	>13 cm			
Complex transitional caatinga	Yevaro–hina	I	43	46	49	46	0.88	0.89	200–350
	Yevaro–cunuri	H	41	41	38	n.d.	0.82	n.d.	100–250
Complex caatinga	Yevaro–yaguacana	K	40	51	26	27	0.70	0.66	100–200
	Yevaro–cunuri	J	29	30	21	21	0.61	0.76	100–200
Simple caatinga	Cunuri–yucito	L	34	25	22	22	0.59	0.77	80–150
	Piapoco	M	n.d.	21	17	15	0.55	0.59	50–100
	Cunuri–yucito Cunuri–piapoco	N	29	31	25	23	0.69	0.67	80–150
Ecotone caatinga → alluvial	Yaguacana–tamacuari	O	n.d.	19	14	16	0.55	0.64	80–200
Ecotone caatinga → Bana	Yucito–media luna Cunuri	P	n.d.	24	18	19	0.63	0.67	80–200
Bana	Yucito–concha amarilla	Q	n.d.	21	15	19	0.73	0.4–0.6	100–200
Mean		I–H–Q	36	35	28	26	–	–	–
Number of points			13	46	52	37	105	37	–

Source: Bruenig and Schneider (1992); Bruenig *et al.* (1978, 1979).

Fig. 1.23. Taxonomic and structural stand diversity: *Dryobalanops beccarii* – Meranti – Keruing MDF on sandy–clayey loam humult ultisol on low hilly terrain, Selang F.R., Mattang, new Kubah N.P., Sarawak.

The data from Borneo, Amazonia and China indicate that tree-species richness is related to, and limited by, soil conditions along a gradient of declining rooting depth, aeration, moisture and nutrient status and humus quality. This trend is analogous to similar trends in temperate forests. The data from West Malaysia confirms the existence of some correlation between species richness and site, irrespective of floristic composition. The very similar tree-species richness in natural forest on equivalent site types in Borneo (Sarawak and Brunei) and Amazonia (San Carlos), and the constancy of evenness of mixture in mature natural primeval rainforests, suggest that these features may have some relevance for self-sustainability of the forest ecosystem. The information on trends of species richness in the primeval forests could tentatively be used as an indicator for designing site-specific sustainable silvicultural systems for natural forest management (Chapter 6) and for restoration (Chapter 7) or recultivation (Chapter 8), until we know more about the critical lower limits of species richness for self-sustainability.

1.9 Floristic Changes and Distribution Patterns

Silvicultural systems cannot be well designed and their results cannot be judged if the significance of natural floristic patterns and changes are not understood. The analysis of tree distribution patterns at small spatial scale in Sarawak (Newbery *et al.*, 1986) and Liberia (Poker, 1992) has indicated that tree-species associations are probably ephemeral and shifting. The statistical probabilities for a number of neighbouring tree species to form a persistent association in the form of a calculable 'eco-unit' (Oldeman, 1989, 1990) are very slim. Divisive association analyses in the Sabal (Weiscke, 1982) and in San Carlos (Bruenig *et al.*, 1978) research areas produced, at a relatively high level of division, a close association between species ranges and landform-soil units (LSU). The same held true for tree-species association groups and the regularly zoned LSU in San Carlos (Bruenig *et al.*, 1979) but not in the edaphically and topographically more irregularly heterogeneous Sabal area (Weiscke, 1982). In Menyala Forest Reserve, Peninsular Malaysia, the tree-species composition over 34 years in a 2-ha ecological research plot is unpredictable. In 1947 and 1981 (in parentheses) the plot had 244 (244) species among 538 (484) trees with diameters > 10 cm per hectare; 48 of the 244 species with diameters > 10 cm recorded in 1947 were not found in 1981, while 48 new species were added; six species invaded after 1947 and disappeared again before 1981. The lowest recorded level of species richness was in 1971 with 229 species, 6% below the level in 1947 and 1981. Of all species, 95% are primary forest species, 5% are secondary forest species; 24 species with 100 (82) trees are potential emergent-layer species, more than half of these being Dipterocarps (Manokaran and Kochummen, 1987).

Newbery *et al.* (1986) analysed patterns in the 20-ha RP 146 in Sabal Forest Reserve, Sarawak. The soil and site mosaic is small scale and heterogeneous, and accordingly so is the pattern of the Lowland Mixed Dipterocarp–Kerangas forest tree species mosaic. The 64 most abundant species representing 13,154 trees were selected from the 16,063 enumerated trees with diameters > 10 cm (diameter at breast height). All 64 species had individuals in the upper canopy; 30 of the 64 species showed pattern in the form of clumps. The most frequent scales of clump size were between 35 and 55 m across or 0.01–0.3 ha. These 30 patterned species were the less abundant species in the plot, had a greater proportion of smaller (10–20 cm diameter) trees and a lower ratio of upper to lower canopy trees than species without pattern. The scale of pattern matched the common size of gaps (Bruenig, 1973b; Bruenig *et al.*, 1978, 1979; Fig. 1.3). It is hypothesized that patterned species are light-demanding and grow from seeds in or near gaps, whereas non-patterned species are shade tolerant and persist and grow relatively well in the shade below the closed canopy. A preliminary comparison of the tree-species lists in 1963 and 1990 produced no evidence of species loss, but some additions occurred through ingrowth, e.g. the extremely rare *Falcatifolium angustum* de Laub which previously was considered to be extremely endemic and restricted to two sites near Bintulu.

The existence of persistent associations between species within tree groups or mosaic-units of the same age was tested with data from 20 1-ha long-term silvicultural research plots in Liberia, West Africa, and six paired 0.5-ha plots in the Philippines (Poker, 1992). The hypothesis was that the trees form a mosaic of phasic eco-units (see Ashton and Bruenig, 1975; Whitmore, 1975a) ranging in size from 0.01 ha (single tree crown size) to about 0.3 ha (five to seven tree group size); such units would be persistent and could be defined quantitatively. In Liberia, all trees were classified into four height classes (< 25, 25–50, 50–75, 75–100%) relative to the maximum attainable by the species on that site. The mosaic was computed for each 1-ha plot by classifying the species as emergent (A-layer), intermediate (B-layer) or sub-canopy (C-layer) species. The potential maximum height of each species was determined from the literature (Vorhoeve, 1965) and the recorded maximum height in or around the 1-ha research plots. Finally, each tree was classified according to its height in one of the four height classes and digitized. The height class distribution map is reproduced for one plot in Fig. 1.25. The distributions in all plots show the existence of some phasic units or equal height class clumps in a matrix of random scatter, despite the very small size (1 ha) of the individual plots. Periodic remeasurements indicated that the mosaic units in the lower height percentage classes rapidly change and shift with time. Some members retain their growth rate, others accelerate growth and yet others reduce it. Eventually new ephemeral associative mosaic groups are formed. A similar study of mosaic structure was made in smaller sample plots (50 × 100 m) in MDF in the Philippines. Two enumerations of diameter and height, 6 years apart, showed a slight change of pattern indicating possibly that mosaic structure may shift noticeably within such a short period of time. This constant oscillation and change between states of order (structurally complex mature phases) and disorder (structurally simpler, more chaotic decay and regeneration phases) in a small-scale, shifting mosaic pattern on a very heterogeneous soil and often also heterogeneous micro-relief (landform – soil units) may facilitate the co-existence of many tree species (structural elements of biodiversity). Kohyama (1993) linked such dynamics to one-sided competition related to cumulative basal area of the light-depriving larger trees around a subject tree, irrespective of species and assuming constant rates

Fig. 1.24. Well-illuminated interior of a simple Kerangas forest on medium to shallow podzol, coastal terrace in Dalam F.R., near Miri, Sarawak. A/B layer species regenerate well in contrast to MDF (Fig. 1.23), but invading MDF species cannot survive and die as seedlings or saplings.

of gap formation and of mortality with stand age. This model accords too little with reality because it does not account for the irregularity of the distribution of events over space and time that affect the patterns of growth and mortality. The strong random element of distribution and dynamics at the spatial scale of gaps (0.1–0.3 ha) forces silvicultural research and monitoring to use plot sizes of at least 10 ha.

At medium scale of spatial distribution pattern (a few hectares to many square kilometres) there is generally a close association between the distribution of tree species and the pattern of clearly separated LSU

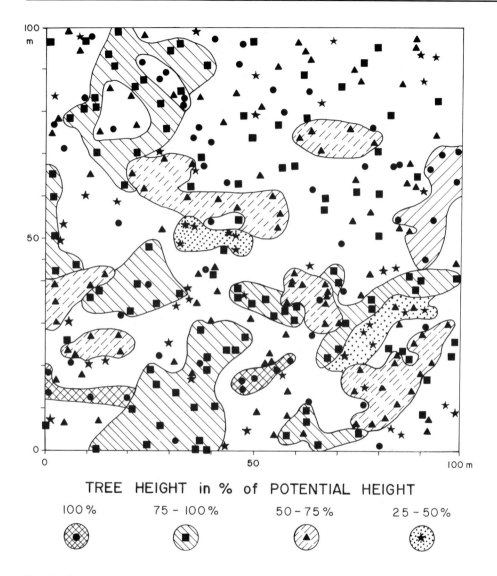

TREE HEIGHT in % of POTENTIAL HEIGHT

| 100 % | 75 – 100 % | 50 – 75 % | 25 – 50 % |

Fig. 1.25. Example of the small-scale mosaic structure of the trees in a tropical semi-evergreen rainforest in Grebo National Forest, Liberia. The mosaic units comprise trees which have reached 25–50, 50–75, 75–100 or 100% of their specific maximum height which they could reach on that particular site. The mosaic units are apparently not composed of trees of equal age and relative vigour and change and shift rapidly with time (Poker, 1992, 1995) (compare Fig. 6.5); they also do not represent guild associations as described by Taylor *et al.* (1995).

(Anderson, 1961a, 1983; Ashton, 1964; Bruenig, 1966; Bruenig *et al.*, 1978; Newbery and Proctor, 1984; Huang and Bruenig, 1987; Droste, 1996). This simple association becomes obscured if the LSU form a mosaic of small-scale patches, such as in RP 146 (Weiscke, 1982). The overlay of variation due to stand dynamics and site heterogeneity obscures either pattern (Weiscke, 1982; Newbery *et al.*, 1986). Pattern at large scale may become obscured if species are replaced by vicariants (Bruenig, 1966, 1974; Ashton, 1988b). A tree species may characterize a forest type on an LSU in one area and be absent in the same LSU in another region. The peculiarly disjunct and irregular LSU coverage by *S. albida*, *Dryobalanops fusca* V.Sl. and *A. borneensis* in Sarawak is an example (Bruenig, 1974, pp. 150–154). Ashton gives examples of vicariousness at a wider scale of spatial separation. The arboretum in Semengoh Forest Reserve and the ecosystem research area in Pasoh (1000 km apart) have some of the most abundant Dipterocarp species in common, but

> *S. acuminata* Dyer and *S. parvifolia* Dyer are vicariants. Others such as *S. dasyphylla* Foxw. and *S. parvifolia* change their rank order. We believe this is due to edaphic differences between the two sites, because the most abundant species are remarkably constant on the same soil in different localities in northwest Borneo.
>
> (Ashton, 1988b, p. 364)

The abundant and very site-tolerant *S. albida* in KF, KrF and Peatswamp forest is not very constant, nor are other abundant, site-tolerant or intolerant Kerangas tree species, such as *A. borneensis* and *G. bancanus* (Miq.) Kurz. (Newbery, 1991). Species distributions are determined by such a multitude of physiological and ecological factors, history and chance that generalizations and predictions on probable site requirements, species occurrence and compatibility for purposes of conservation and silvicultural management are hazardous.

The problem is even more difficult for rare species. In 1959 the conifer genus *Falcatifolium* de Laub. species *F. angustum* de Laub. was first collected on a Kerangas ridge-crest in Niah Jelalong Protected Forest (P.F.) north of the Merurong Plateau. The new species was thought to be extremely rare, site-intolerant and strictly endemic. In the same year, it was recorded 80 km further west from Kerangas terraces in the Binio basin near Bintulu, some years later 420 km further west on Gunung Santubong in western Sarawak and in 1990 in Sabal RP 146 on medium deep humus podzol near a stream. Rare and very rare tree species may be either restricted to very rare sites or have a wide site tolerance but are scattered at very low frequencies (distances between mature trees 200–330 m (10 ha) for rare and > 330 m for very rare species). Inventory of these species for the purposes of conservation is extremely difficult and expensive. A low-intensity sampling with 0.1–1 ha sample plots has little chance of yielding useful statistics on rare and very rare species. Sampling of occurrence and association of common, characteristic tree species by low-intensity survey poses the same problems. Newbery (1991) re-evaluated 38 small (0.1 ha) Kerangas plots, which were scattered throughout Sarawak and Brunei in a pilot study in the 1950s. The dataset was 21,727 trees with diameters > 2.5 cm which contained 637 taxa.

Various analyses produced no evidence of consistent differences in species associations and physiognomic–structural features between plots and regions in spite of their visually obvious existence in the field. Tree-species inventories at regional scale must combine a few large core plots (i.e. 20–50 ha in MDF, 10–20 ha in KF and Peatswamp forest) with many small plots in the order of 0.1–1 ha. The latter covers the regional variation of sites and random species associations; the large plots cover a large proportion of the regional tree flora, including some of the rare species, and can be used to monitor dynamics. The recording of very rare species will require additional site-focused, subjective reconnaissance.

1.10 *Forest Canopy and Animal Life*

Browne (1949) reported the most drastic recorded change of rainforest canopy over large tracts of land by natural causes. A devastating tropical cyclone had felled probably more than 1000 km^2 of mostly Meranti–Keruing type of MDF in Kelantan in 1883. In 1949, this 'disaster tantamount to clear-felling has resulted either in a complete victory or in utter failure (in about equal proportions) of a new dipterocarp stand'. Some previously abundant species, such as *Neobalanocarpus heimii* (King) Ashton, had become rare. Browne did not comment on animal life, but self-sustainability obviously had not been lost and restoration of the canopy was well underway in the 'new dipterocarp stand' 70 years after the disaster. The recovery of the canopy by the MDF high forest species meant that also the animals, including pollinators and seed distributors, could be assumed to be in the process of recovery.

Little is yet known of the pollination systems of individual tree species but it appears so far that exclusive specialization, such as in figs, is rare. Pollination by animals would benefit from habitat diversity and animal species richness if there are several pollinators for one species and the various pollinators have different habitat requirements. A recent study of the pollinators of a widespread climbing rattan palm species, *Calamus subinermis* Bl., in Sabah, listed insect visitors from at least 21 genera during 48 hours of the pollen-release period. Microlepidoptera seemed to be the main pollinator (Lee *et al.*, 1993). An effect of the foraging range of pollinators on genetic evolution of species richness in out-breeding species was suggested by Appanah (1987). He argued that the short-ranging, rapidly propagating species of thrips, which are the favoured pollinators of Dipterocarp species, may have been a factor in the evolution of the species richness in the family.

Apart from pollinators and seed distributors, forest self-sustainability also depends on the herbivorous consumers and detritus decomposers in the canopy, on the soil surface and in the rooting sphere of the soil. Frass and decay of leaves affect the rates of photosynthesis and respiration. Net primary production may increase or decrease depending on whether herbivores eat young and productive or old and consumptive leaves. Herbivory pathways were studied for five species of trees in the subtropical rainforest of New South Wales, Australia: 21% of the leafage of the crowns of the five species was annually consumed by herbivores. This amount is very high and, if the five species are considered representative, would indicate the great importance of canopy herbivory in this forest (Lowman, 1992). Values for tropical rainforest in the literature are generally much lower and range between 2 and 10% but may reach 100%, for example the mysterious *ulat bulu* epidemic in the uniform, single-species *S. albida* forest, PC3, in Sarawak Peatswamp forests (Anderson, 1961b). Large spatial and temporal variation is probable, but not yet well documented. Proctor *et al.* (1983b) estimated eaten leaf area of fallen leaves in four forest types in Mulu. They stratified the frass in '20%-area removed' classes and counted leaf frequencies in each class of fallen leaves. The results showed rather uniform values in the four forest types (Proctor *et al.*, 1983b, table 8). The frequency distribution indicates an overall reduction of leaf area by frass of 7–8%. There is a weak ranking Alluvial > KF > MDF > Limestone forest. Herbivory in the generally animal-poor KF would have been expected to be substantially less than in MDF because of the microclimatic and productivity limitations on microbial and animal life. However, the concentrated food source in the simple, species-poor KF probably favours contagious pest expansion.

The more complex, vertically more integrated structure of an aerodynamically rough canopy is a more diverse habitat and can harbour more species of animals and plants than a smooth, simple canopy. A greater number of tree and other plant species in a rough canopy would also support more species and a larger number of animals, pollinators, decomposers, pests and predators. The traditional concept of permanent forest (*Dauerwald*) uses this linkage to maintain high levels of biodiversity and self-sustainability. The main criteria of self-sustainability in forestry practice are the balance between litter production, decomposition and humus formation, and the effectiveness of biological regulation and pest control. Integrating key indicators are the humus type, reflecting the state of soil biology and ecology, the non-tree and tree ground flora and the diversity of mammals, birds and insects (Moeller, 1922, 1929). Also possibly important, but largely unknown in their role as indicators, are epiphytes and dead organic matter in the canopy (including humus) (Paoletti *et al.*, 1991) which provide habitat for animals. There are no quantitative studies on the complex relationships between the floristic and spatial structure of the canopy and animal life, nor is there much information on the interaction between α- and β-diversity in the rainforest. To understand this relationship is important for designing sustainable harvesting and silvicultural management systems, including refuge habitats, in forests, and integrated land use in landscapes. The results of the trapping in the four forest types in Gunung Mulu National Park (Proctor *et al.*, 1983b) and results of comparing the fauna in successional stages of the restoration sequence of ecosystems on barren coastal lands in the CERP ecosystem study (Bruenig *et al.*, 1986b) suggest that interaction at β-diversity level exists. There is a well-known relationship between forest structure and birdlife in temperate and tropical natural and managed forests (Bruenig, 1971c; Kikkawa, 1982). There is also evidence of ecological association between avian variables, such as bird abundance, bird species richness and diversity, and various measures of vegetation structure including plant species composition and life-form, vertical and horizontal stratification, canopy roughness, and patchiness (Kikkawa, 1982; Boettcher, 1987; Thieme, 1988; Ellenberg, 1993; MacKinnon

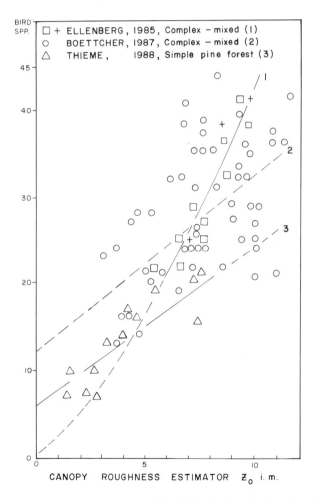

Fig. 1.26. The number of birds counted in censuses in broadleaf, mixed broadleaf–conifer and conifer forest stands in relation to the aerodynamic roughness z_0 of the canopy, estimated from canopy stature (tree height). All three regression lines are statistically highly significant. (From Boettcher, 1987 (47 diverse stands, size range 6.3–25 ha); Ellenberg, 1985 (15 diverse stands sampled by 25 1-ha plots); Thieme, 1988 (17 uniform stands of Scots pine, size range 13–18 h.)

and Freeman, 1993; Fig. 1.26). 'The main group of phytophagous birds and animals live in the top of the canopy, the other zones contain fewer primary consumers.... Studies in greater detail have led to the discovery of considerable subtlety in niche differentiation' (Whitmore, 1975a, p. 38). Canopy roughness and complexity could be indicators of this subtlety. Self-sustainability and sustainable forest management and conservation, to a large measure, depend on the preservation of this subtlety. Carefully adapted harvesting and appropriate silviculture can maintain or even increase the floristic and geometric structural diversity needed for the niche differentiation that may be essential for adequately functioning biological self-regulation and regeneration.

1.11 Small-scale Dynamics, Regeneration

The floristic, faunal and spatial structure of the tropical rainforest are extremely heterogeneous, variable and constantly changing. Gap formation is equally heterogeneously distributed over time and space, depending on site, canopy structure and fluctuations of the causal factors (Bruenig, 1966, 1973b; Bruenig *et al.*, 1979; Hall, 1994). The composition, distribution and successional development of tree-species associations are determined by many interacting biotic and abiotic factors of site, internal gap-mosaic dynamics and external disturbances (Section 1.6). Flowering, fruiting, seed viability and dispersal are affected largely by climatic and faunal factors. The coincidence of favourable or unfavourable conditions of weather, pollination, pests and diseases during flowering and fruiting influence the amount and quality of seeds. Weather, soil conditions, transmissibility of light for crowns and canopy, and predators determine the subsequent chances of germination, establishment and growth of the seedlings. Microhabitat conditions and their haphazard changes determine the chances of survival and growth of the saplings. The rate of mortality from seeds to overmature tree proceeds theoretically by a negative exponential curve of population decline (Sheil *et al.*, 1995). In reality, mortality occurs in unpredictable leaps and bounds related to accidental combinations of mostly non-linear relationships of causal factors which affect different species differently. The negative exponential selection–forest tree frequency curve and stand-table projection are, therefore, idealistic generalizations and little more than a theoretical background concept in a silviculture management system (SMS). Reality requires more sophisticated models (Bruenig *et al.*, 1991). Summary reviews of the ecological aspects of fruiting, seed dispersal, gap formation, establishment and growth of trees in the rainforest are given by Whitmore (1975a, 1990) and Longman and Jenik (1987). The periodic fluctuations of abundance of seedling regeneration in closed tropical rainforests are rather similar to conditions in temperate broadleaf forests, but the spatial heterogeneity of conditions within a forest stand is much greater in the species-richer rainforest. Flowering, fruiting, success of establishment and subsequent growth are even more unpredictable than in temperate forests. The high rate of light extinction throughout the year in the evergreen mesophyll canopy (Fig. 1.18) make seedling establishment, survival and growth more hazardous than in seasonally deciduous forests. Consequently, regeneration requires either natural gap formation or fairly drastic opening of the A- and B-layers of the canopy to provide more light, and also more nutrient for rapid growth and more water to survive erratic drought. Conditions are fundamentally different in sclerophyll/microphyll forests where the illumination at ground level is several times stronger than in the 'zonal' rainforest (Figs 1.18 and 1.24).

In zonal but particularly in edaphic forest formations (Table 1.2), in ecotones and in successional forest, the absence or presence of seedlings of the canopy species, or the presence of invading species, are not absolutely reliable indicators of the dynamic trends in the community. Well-documented cases are the initial misinterpretations of observed absences of regeneration of *S. albida* in Alan bunga PC3 and the presence of *G. bancanus* in the understorey of Alan PC2 (Section 6.7) in Peatswamp forests, and of the presence of invading MDF species in KF (Anderson, 1961a, 1964, 1983; Wood, 1965; Bruenig, 1966). It is safer and more in accord with the natural dynamics to work with trees in the 'great period' of growth (*c.* 40–80 cm trunk diameter) in SMS, and leave seedling regeneration to nature (Sections 3.5 and 6.4). The following scheme gives a simplified overview of common situations in primeval or slightly disturbed tropical rainforest in South-East Asia, especially in Borneo (compare profiles in Figs 1.11, 6.6 and 6.7).

1. Complex, mesic 'zonal' rainforest on 'zonal' soil, regeneration patchy and fluctuating, but on the average and over larger areas, adequate (e.g. majority of lowland MDF types, such as the mosaic of 'Red Meranti–Keruing' and 'Balau' types in Pasoh described by Manokaran *et al.*, 1991). The stocking of A-layer species in the intermediate size classes (C and B layers) depends on the local disturbance regime (gap formation). The subcanopy layers B and C and the ground vegetation D play an important role as matrix through which the light-demanding A-layer species struggle to the top when gaps are formed naturally or by harvesting or silviculture. This reduces the tendency to overcrowding especially after mast years. This filtering effect of the understorey matrix is an essential feature of natural regeneration mechanisms in primeval and in managed forests in the species-rich and diverse zonal forest formations such as the Malesian MDF.
2. The same forest formation, but consistently (except immediately after seed years) deficient in regeneration, intermediate size classes of A-layer species poorer than in (**1**), sometimes completely lacking (e.g. Hill Mixed Dipterocarp *Shorea curtisii* forest in West Malaysia, also most African rainforest types in a state of successional development).

Fig. 1.27. Impact area of the crown of a large wind-felled *Eperua purpurea* in high Caatinga forest (Fig. 1.18, Yevaro-Yaquacana forest ecotone). MAB International Amazon Ecosystem Study, San Carlos de Río Negro, Venezuela.

3. Moderately complex, less mesic and more sclerophyll forests with more-or-less pronounced dominance of one or several A/B-layer species, seedling regeneration mostly sparse, but present as scattered singles or in patches, intermediate size classes patchy but adequate (e.g. complex tall Kerangas with *A. borneensis, S. albida, D. beccarii, D. fusca*; complex, tall Caatinga with *Eperua purpurea* Benth. (Figs 1.27 and 1.28), Yevaro; Mixed Ramin-bearing Peatswamp forest, PC1), regeneration, survival and growth of seedlings and saplings regulated by the dense C/D-layer matrix, and by occasional drought conditions.

4. As in (**3**) but more sclerophyll/xeromorph, intermediate (younger) size classes of the dominant species distinctly deficient or completely absent (e.g. tall and dense Kerangas–Kerapah ecotone with gregarious *S. albida* or

Fig. 1.28. Canopy gap caused by the fall of the tree in Fig. 1.27. The crowns of the edge trees close the gap within 10 to 15 years from the sides (A/B-layer trees) while the regeneration in the C/D layer rushes upward.

D. rappa Becc.), differentiation of regeneration in the well-illuminated B, C and D-layers is strongly affected by drought-related mortality.

5. Simple, xeric, sclerophyll forest with smooth canopy and good illumination to the ground (Figs 1.18 and 1.24), ample regeneration of the species of the A/B-layer (e.g. Kerangas forest *C. nobilis*, *Dacrydium pectinatum* de Laub., Caatinga with *Micranda sprucei*), regeneration primarily controlled by hydrological conditions.

6. Single-species dominant forest, regeneration of seedling and sapling size rare or absent, only sporadically and ephemerally dense after mass fruiting, inadequate intermediate size classes, densely closed, uniform main canopy of uncertain age structure and not even-aged (e.g. Peat-swamp forests with *S. albida*, FT36, PC2 and FT37, PC3, occasionally vicarious species, not occurring in Amazonia or Africa). The regeneration mechanisms in these forests are still a complete mystery. Tracts defoliated by *ulat bulu*, or that are clear-felled, do not seem to regenerate.

1.12 Forest Biomass

The primeval rainforest has been successful in coping with the variable and changing natural environment over millions of years on a wide variety of soils and sites. The amount, structure and physiognomy and dynamics of the primeval forest biomass should be expected to accord with the requirements of self-sustainability. Therefore, the primeval rainforest can provide baseline information for developing rational strategies of forest management and conservation, but data are scarce (see Whitmore, 1975a, 1990, 1991). Data on amounts and structure of forest biomass in small plots can only be meaningfully generalized if the position of the sample in relation to the pattern of spatial and temporal variance of the surrounding forest is known (Bruenig, 1973d; Kurz, 1983). Few cases fulfil this condition. Most biomass data are from small plots on some insufficiently defined site that were selected subjectively from an area of diverse, heterogeneous tropical rainforest with some purpose in view. In Sarawak, the 57 single plots in the Kerangas study 1955–1963 and one plot in Alan bunga PC3 were subjectively selected in mature and intact patches to assess the tree biomass above ground in a simple management-orientated manner (Bruenig, 1966, 1974). The International Amazon Ecosystem Study at San Carlos was the first to assess biomass in tropical rainforest in a manner that integrated biomass assessment with large-area structural analysis in a range of sites along an ecological gradient with 10 different forest types (species association groups (SAG) H to Q in Bruenig *et al.*, 1979). H. Klinge harvested 1543 trees and their fresh- and dry-weight biomass was determined. Various biomass regressions were developed and applied to the population of 13,920 trees in the 10-ha research area (Kurz, 1983), which were enumerated according to the harmonized MAB-IUFRO methodologies (Bruenig and Synnott in Bruenig, 1977b; Synnott, 1979). The regressions of the dry-weight of the tree trunk and for branch weight are reproduced in Fig. 1.29. Trunk and branch weight add to total biomass but separate regressions were necessary because the best fits were obtained with different independent variables (d^2h for trunk weight; d^2A for branch weight, where A is crown diameter × crown length). Stratification by forest type did not improve the fit. The regressions were then used to compute the standing above-ground dry-weights and leaf areas of the 13,920 trees in the 10-ha research plot. Biovolume (BV = $\sum 0.5\ g \cdot h$) was calculated, stratified by canopy layer. Biovolume is quick and easy to calculate from enumeration data and is a useful indicator of biomass, if the differences in timber specific gravities between forest types are considered. The results for the various forest types (association groups) are shown in Table 1.8. The amounts of biomass

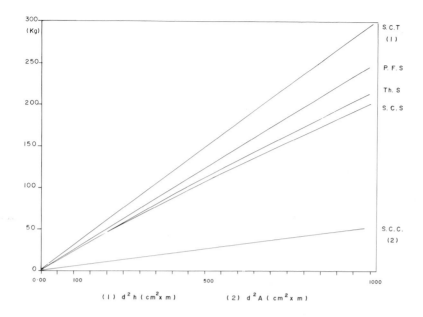

Fig. 1.29. Linear regression of tree biomass over structural tree parameters *d* (diameter of stem), *h* (total height of tree) and *A* (sectional crown area). (From Kurz, 1983.) SCT: San Carlos total wood biomass above ground; PFS: Pasoh forest, MDF, stemwood (Kira, 1978) (1); Th.S.: Thailand semi-evergreen, stemwood (Ogawa *et al.*, 1965) (1); SCS: San Carlos caatinga and bana, stemwood (1); SCC: San Carlos caatinga and bana, crownwood (2).

in the diameter classes 1–5 cm and 5–13 cm (C- and D-canopy layers) are inversely correlated ($R^2 = 0.94$) with the biomass densities in the 15–20 and > 30 cm diameter classes. This makes it possible to simplify biomass assessment considerably by subsampling procedures. Leaf area index and leaf weight are correlated with the independent variables d^2h ($R^2 = 0.756$) and d^2A ($R^2 = 0.746$). The tree height h and the crown sectional area A in turn are positively correlated with the aerodynamic roughness. These relationships can be used for pre-stratification and assessment on aerial photographs of biomass stocking and of any changes over time.

The biomass assessment approach of Klinge *et al.* (1974), Klinge and Herrera (1978), Kurz (1983), and the work of Ogawa *et al.* (1965) and Kato *et al.* (1978) have produced regression equations which make destructive complete biomass harvesting in tropical rainforest unnecessary. Total biomass and the proportions of the various components, including roots if the soil type is known, can be estimated from parameters that are relatively easy to enumerate. The total-tree biovolume > 1 cm diameter was accordingly calculated from the enumeration data of nine of the 57 kerangas plots and measured in one clear-felled plot in *S. albida* Peat-swamp forest, PC3. All plots were fully stocked patches in mature-phase

Table 1.8. Basal area (G), above-ground oven-dry tree biomass of stem, crown and twig wood (SCT), leaf area index (LAI) and leaf area per biomass unit in the ten tree-species association groups (SAG, landform–soil type–plant sociological (trees and ground vegetation) units) in the International Amazon Ecosystem MAB-Pilot project near San Carlos de Rio Negro, Amazonia, Venezuela.

Forest types and species association group (SAG)						
Forest type	SAG	G (m^2 ha^{-1})	SCT (t ha^{-1})	LAI (m^2 m^{-2})	LAI/SCT (m^2 kg^{-1})	Deadwood > 10 cm (t ha^{-1})
Complex transitional caatinga, yevaro SAG	I	41.2	330	3.7	0.11	35.0 (10.6%)
Yevaro–cunuri SAG	H	38.5	270	3.6	0.13	
Complex cunuri caatinga (2 SAG)	J	37.5	236	3.7	0.16	26.0 (10.5%)
	K	39.1	259	3.5	0.13	
Simple cunuri–yucito caatinga (3 SAG)	L	35.9	208	3.7	0.18	
	M	37.5	233	3.6	0.16	28.2 (10.1%)
	N	40.9	278	3.8	0.14	
Wet caatinga → alluvial	O	36.9	232	3.3	0.14	33.7 (14.5%)
Ecotone caatinga → Bana	P	33.8	136	3.1	0.22	31.7 (23.3%)
Bana (open woodland)	Q	16.8	57.7	1.4	0.24	6.6 (11.4%)

Source: Kurz (1983) table 541–2; dead-wood distribution data previously published only as map in Bruenig *et al.* (1978, 1979).

stands representing the maximum biomass carrying capacity of their respective site. Biovolume was converted to dry-weight by specific gravity values obtained from weighing wood samples (*S. albida, D. fusca, A. borneensis*) and from the literature. The total tree dry-matter above ground of trees > 1 cm diameter varied widely between forest types. The lowest value was in a Kerapah forest with 215 m^3 wood volume > 5 cm and 246 t total plant dry-weight above ground per hectare (basal area G 27.3 m^2 ha^{-1} and top height 30 m). A biovolume of 1279 m^3 ha^{-1} was enumerated in MDF on deep rich basaltic soil (SP10, h 62 m, rough canopy, G 55.5 m^2 ha^{-1}). A biovolume of 1269 m^3 ha^{-1} was determined by complete harvesting in *S. albida* Peatswamp forest PC2/3 (h 42 m, moderately rough to moderately smooth canopy, G 38.2 m^2 ha^{-1}) and 1381 m^3 ha^{-1} estimated from tree enumeration in *S. albida* Kerangas/Peatswamp ecotone on groundwater humus podzol (h 41 m, closed dense canopy, G 67.7 m^2 ha^{-1}). The mean biovolume of the 53 Kerangas and Kerapah survey plots, all in mature, closed forest patches, was 590 m^3 ha^{-1} (mean top h 38.5 m, mean G 36.5 m^2 ha^{-1}). The actual biovolume and biomass stocking over larger tracts in these forests are lower as a result of young gaps (1–5 year old gaps average 0.5–2% of area) and low biomass density in the subsequent building phases. The variation of basal area of the various statistical tree populations in the 20-ha RP 146, Sabal Forest Reserve (data of 1963, mean G 32.5 m^2 ha^{-1} > 10 cm, mean top h 38.5 m) suggests that the mature-plot values had to be reduced by approximately 25–30% to arrive at an estimate of the most probable mean value for large tracts of natural primeval forest (Bruenig, 1966, 1973d, 1974).

The biovolume and biomass of dead wood (coarse litter including standing dead trees and stumps) have been assessed in San Carlos de Rio Negro in 1975 (Bruenig *et al.*, 1979) and in the Sabal RP 146 in 1988–1990. The proportion of dead wood is consistently, and with little variation between forest types and regions, about 9–15% of the total above-ground live and dead tree biomass, or 10–17% of the live tree biomass. The data for San Carlos are shown in Table 1.8. The dead tree biomass > 10 cm diameter was enumerated, mapped and assessed in the newly enumerated 16-ha of the Sabal RP 146. The standing dead trees on the 7-ha MDF, selectively logged in 1978, in 1990–1991 amounted to 15.0 m^3 ha^{-1} biovolume and 16.5 t ha^{-1} total dry-weight woody phytomass. Fallen tree trunks and branches were 167 m^3 ha^{-1} and 63 t ha^{-1} (Table 1.6). The total dead-wood biomass > 10 cm diameter was 117.0 t ha^{-1}, amounting to 22.9% of the total living and dead tree biomass above ground of trees > 10 cm diameter of 512.2 t ha^{-1}, or 29.6% in comparison to the biomass of living trees of 395.2 t ha^{-1} (Bruenig and Droste, in press; Tables 2.2 and 6.2). In the KF, creamed mainly for *A. borneensis* in the *A. borneensis*-type, the standing (37.7 t ha^{-1}) and fallen (40.7 t ha^{-1}) dead woody phytomass totals 78.4 t ha^{-1}, or 14.3% of the total phytomass and 16.6% of the living phytomass (Table 1.6). Klinge *et al.* (1974) reported a dead-wood dry-weight biomass of 25.8 t ha^{-1} from a 0.2-ha plot in tierra firme forest adjacent to the W. Egler Forest Reserve near Manaus (forest stand MA in Fig. 1.22). This amounted to only 4.4–5.7% of the 450–500 t ha^{-1} total dry tree biomass. They also reported a dry-weight of 250 t ha^{-1} (25 kg m^{-2}) of SOM which gave a total of 750 t ha^{-1} dry-weight of living and dead phytomass. Yoneda and Tamin (1990) assessed the states and flows of the standing dead-wood matter of trees > 10 cm and fallen wood > 10 cm in two plots in upper–lowland Dipterocarp forest, both 1 ha, in West Sumatra. The enumeration was repeated six times during five years. The living standing tree above-ground biomass over bark > 10 cm in one plot was 519 m^3 and 408 t ha^{-1}. The dead-wood biomass > 10 cm diameter in the same plot was 116 m^3 or 39 t ha^{-1}, equalling 22% of the volume and 9.5% of the weight of living standing biomass > 10 cm diameter, or 18% and 9% respectively of the total living and dead standing tree biomass of 635 m^3 and 447 t ha^{-1}. The conversion factors for volume to phytomass used by Yoneda and Tamin obviously differ from those used by us, but the values are within the range of data from RP 146, Sabal, Sarawak (Table 1.6).

The structure and amount of the phytomass of the trees in the rainforest has manifold functions. The phytomass provides a multiple-purpose protective semi-transparent and interceptive filter, is the

Fig. 1.30. Ecotone between lowland *Shorea albida*-bearing Kerangas on podzol (Fig. 2.9) and Kerapah on peat. High mortality, dead-wood stocks and species diversity. Pueh F.R., Sarawak.

exchange interface between atmosphere and the soil, functions as a medium- to long-term store of nutrients and supplies organic matter to the soil. Modification and manipulation of structure and amount of tree phytomass by harvest and silviculture will affect these functions. It is the science-based art of the forester to find the range within which the phytomass can be safely manipulated without unacceptable risks (Section 6.2). The phytomass is a constantly flowing slow-release (wood) and instant (leaves, exudates) source of organic matter which mulches the soil at ecosystem scale.

Fig. 1.31. Ecotone, edaphically similar to that in Fig. 1.30, between *Agathis*-bearing Kerangas and *S.albida*-bearing Kerapah on the Merurong plateau, 750 m a.s.l. Mortality and dead wood stock, and tree-species diversity are higher in the ecotone than in the adjoining forest types.

1.13 Productivity and Growth

Productivity of the ecosystem in this context is the rate of annual net primary productivity (NPP), which is the apparent gross photosynthesis (GPP) minus respiration from leaves, branches, stem and roots of a tree or a tree community. The NPP is allocated to various parts of the tree, metamorphized and stored or used as a source of energy for maintenance and growth. Phytomass is constantly converted to litter and exudate and released in and above the soil. The biological net primary productivity and production (BNP, annual rate or sum over a period) supply the material for primary economic productivity and production (PEP) of commercial products. How much of this is actually merchantable and extractable will be determined by the current economic and political environment and the safe range of phytomass manipulation (Section 1.12). For timber, the rate of effective PEP of the potential varies from 80% (full utilization in reduced impact logging (RIL)), to 20–40% (selective logging), to less than 5% in native customary collecting of speciality timber for sale or own use. Growth in this context is the change over time of tree height, stem diameter, stem form and of roots and consequently of biovolume and biomass, and PEP.

Conventional timber-orientated inventories of forest growing stock and studies of growth and yield give little useful information to assess the potential or actual BNP or PEP. Firstly, information on site (soil, hydrology, biology; atmospheric, physical and chemical conditions, climate), vegetation and tree growing stock is usually inadequate. There are rarely data to relate site conditions to tree performance. Increment and yield are usually expressed in units of some arbitrarily defined merchantable volume content and based on the means of tree populations which are not representative and not relevant to the problems in forestry practice. Yield tables are based on unrealistic assumptions of uniformity, stability and equilibrium of the forest and its natural and economic environment. Often simplistic linear regressions of growth are used that do not accord with the range of complexity and variability of reality (see also Kimmins, 1988). The merchantable volumes change if standards of grading and utilization change in accordance with law enforcement and market conditions (Vanclay, 1991c; AIFM, 1993). An early step towards improvement of PEP assessment was the variable timber volume table for *S. albida* by Haller (1969). The PEP estimates of merchantable timber volume were flexible according to different grading standards and timber defects. Essential improvements for estimating the growth of trees and stands were the various types of model simulation for tropical rainforest tree and stand growth, and merchantable yield (Bossel, 1989; Vanclay, 1989, 1991a,b,c, 1994; Korsgaard, 1992; Ong and Kleine, 1995) but these tools are rarely applied in practice. The conventional yield tables, based on simple or multiple linear regression, and stand-table projection models still prevail, but cannot cope satisfactorily with the heterogeneity of sites and forests, the variability of growth dynamics of trees in complex mixed forest stands and the effects of variable utilization standards. Therefore, they are of limited value for productivity assessment even of single species, and even less so of mixed forest stands. Trees in the aseasonal evergreen tropical rainforest exhibit marked rhythms of stem growth (diameter or basal area growth) that vary widely between trees of the same species within a community, between different radii within one stem, and between species in a manner which is difficult to explain.

Superimposed are the effects due to variation of flowering and fruiting, leaf fall and flushing, climatic events, damage, mortality and gap formation (Bruenig, 1971c; Wrobel, 1977). To overcome the information gap, a workshop on age and growth rate determination for tropical trees recommended solving the problem by means of an international network of study plots with harmonized recording of ecological and growth data (Bormann and Berlyn, 1981) but the problem still exists.

The root compartment within the soil probably consumes a large proportion of GPP and NPP. Measured fine root production rates in Queensland rainforest were on fertile basalt soil 3.05 t ha^{-1} a^{-1}, on granite soil 1.72 t ha^{-1} a^{-1} and on alluvial soil 1.60 t ha^{-1} a^{-1} (Maycock, 1993). It is possible that the total amount of root litter and exudates may produce equal or even more organic matter for SOM formation than the leaf fall, if data from temperate forests can be used as an indication (Fahey and Hughes, 1994). The tropical database is especially small and improvements are expensive and difficult (Kurz and Kimmins, 1987). Kimmins (1988) recommends, for PEP assessments, a combination of the traditional linear-regression analysis of increment and yield ('historical bioassay') with analyses based on growth, increment and yield processes ('ecophysiological approach') in an ecosystem simulation procedure. His 'hybrid' forest production simulation model FORCYTE-11 provides such a modelling framework. The model includes factors that are of particular relevance to growth and yield assessment in tropical rainforests such as nutrient and moisture inputs and fluxes, litter fall, decomposition, soil compaction, erosion and harvesting. The same principal process-based approach is adopted by Bossel (chapter 5 in Bossel and Bruenig, 1992). The unexplained and unpredictable variation of BNP and PEP of stands and trees above ground and the lack of simple linear relationships between growth and growth factors are the enigma of forestry growth and yield research and management planning. It is largely solved in temperate forests, but progress is slow in tropical rainforest research and practice.

After the Second World War global and regional assessments of forest resource production potentials became fashionable. Climatic parameters, such as rainfall, rainfall distribution, moisture, temperature and net solar radiation, were used as independent variables to estimate potential gross and net phytomass production (Paterson, 1956; Weck, 1960, 1970). Since the 1960s, data processing has improved and more field data have become available for modelling NPP (Munn, 1981, 1984; Lieth, 1984). The review of gross and net primary production in UNESCO (1978, chapter 10 by F. Golley) quoted theoretically derived rates of gross primary productivity (GPP) in rainforests between 56 and 89 t ha^{-1} a^{-1} on a wide range of soils (Bruenig, 1966) and between 82 and 95 t ha^{-1} a^{-1} on very good sites (Weck, 1960; Fig. 1.20). Measured values were 53 t ha^{-1} a^{-1} (Mueller and Nielson, 1965) from a plot in seasonal evergreen, probably very old secondary forest in Côte d'Ivoire, with high respiration, 119 t ha^{-1} a^{-1} (Odum and Pigeon, 1970) from lower montane forest at El Verde, Puerto Rico and

144 t ha^{-1} a^{-1} (Kira *et al.*, 1967) from Dipterocarp-bearing seasonal lowland moist evergreen forest at Khao Yong, Thailand. The amounts and variation of light and dark respiration in leaves and of respiration in the other parts of the plant are still factors of great uncertainty. In MDF in Pasoh, Malaysia, Whitmore (1990) added an estimated 50.5 t ha^{-1} a^{-1} respiration to an NPP of 29.7 t ha^{-1} a^{-1} which gave 80.2 t GPP and came very close to the classic estimate of Weck (1960). Whitmore's NPP estimate for Pasoh consisted of measured, estimated or guessed values for tree biomass increment (7.1 t), fine and coarse above-ground litter and root mortality (10.3, 3.7, 4.0 t), and grazing (insects measured 0.3 t, mammals and birds guessed 4.4 t).

Golley in UNESCO (1978, pp. 241–242, table 3) gave five NPP values from tropical rainforest: Yangambi, Zaire 32 t ha^{-1} a^{-1}; Khao Yong, Thailand 29 t ha^{-1} a^{-1}; Pasoh, Malaysia 22 t ha^{-1} a^{-1}; El Verde, Puerto Rico 16 t ha^{-1} a^{-1}; Côte d'Ivoire 13 t ha^{-1} a^{-1}. Whitmore's value for Pasoh, 29.7 t ha^{-1} a^{-1} NPP (Whitmore, 1990, fig. 9.7 and table 9.8) was higher than the 22 t given by Golley (1978, citing Bullock's publication in Gist, 1973). Three different methods used in Pasoh to estimate GPP gave rather similar values (summation 80.2 t, canopy photosynthesis 87 t and CO$_2$ flux 89 t GPP; again corroborating Weck). The CO$_2$ flux method gave an NPP value of 49 t ha^{-1}, much higher than any of the other values. Measured coarse litter (branches and trunks or stems > 1 cm) fall per hectare between May 1991 and February 1973 (1 year 9 months) was 10.8 t with a variation between subplots of 3.7, 12.4, 10.1, 3.5, 24.3. This added to the existing litter stock of 52.6 t. Golley (UNESCO, 1978, table 4) listed the leaf litter and total litter fall in 48 tropical forests; 12 of these are rainforests with a mean leaf litter fall of 5.1 t ha^{-1} a^{-1}. Whitmore (1990) recorded 10.3 t ha^{-1} a^{-1} leaf litter and 3.7 t ha^{-1} a^{-1} coarse litter fall in a 0.8-ha plot in Pasoh. Coarse litter fall is naturally erratic and more variable in space and time than fine litter fall. The long-term value would probably be somewhat higher than 3.7 t. Judging from the measured stocks of dead wood and the estimated decay times (Section 1.5) it may even be as high as twice the amount actually recorded by Whitmore (1990).

For the purposes of estimating the order of size of sustainable yield, we may tentatively assume that the potential NPP above ground ranges between 20 and 40 t ha^{-1} a^{-1} on zonal soils and average lowland sites (Table 1.9; Fig. 1.20). From this about 20% has to be deducted for leaf litter and 30% for woody litter. The remaining 50% or 10–20 t tree biomass consists of crownwood (20–25%), bark (10–15%) and trunkwood (60–65%). The merchantable proportion of the trunkwood varies between 40 and 80% depending on site (e.g. hollowness), quality utilization (waste) and market conditions. The upper limit of PEP would then lie between a low of 6 t and a high (maximum) of 13 t on average lowland sites in natural rainforest. Natural damage and inefficiencies in management cause losses of about 20–30% in production and 15–25% in harvesting. We then arrive at a realistic estimate of potentially harvestable yield of 3.4–4.1 t ha^{-1} a^{-1} at

Table 1.9. Tree biomass (dry matter) above ground and annual net primary productivity (NPP) in tonnes in various tropical forest formation classes including the Malesian Mixed Dipterocarp forests: (a) values apply to the zonal forest formation; (b) values include the whole range of edaphic conditions (Bruenig, 1966, 1977b; Kurz, 1983; Lieth, 1984). Harvestable refers to current market conditions and the proper utilization in an orderly harvesting system (reduced impact logging, RIL) with extraction by tractors or long-distance cable.

Forest formation class, resp. forest formation	Tree biomass stocking (t ha^{-1})		Total tree biomass on 1.6×10^9 ha (t $\times 10^9$) (1990)	Annual NPP (t ha^{-1})	
	Mean	Range		Average	Range
Predominantly evergreen wet (rainforest)	450	(a) 300–800 (b) 50–1500	180	25–30	(a) 20–40 (b) 3–40
Mixed Dipterocarp forests: harvestable volume in SMS	550 50	400–1500 20–100	– –	30–35 5–10	20–40 2–15
Predominantly rain-green, seasonally moist	300	200–500	160	15–20	10–25
Rain-green, dry, closed	150	50–250	70	7	3–12
Predominantly rain-green, semi-arid, open	30	10–60	~ 10–20	1	0.5–2

a low production level and 6.8–8.3 t ha^{-1} a^{-1} at a high production level. How much of this can be harvested depends on external forces. Markets determine what species and timber grades are merchantable; political, social and private preferences determine intensities and efficiency of harvesting. Ecological considerations may demand the retention of part of PEP, especially of lesser used species (LUS) and non-commercial species, and of moribund giant trees for litter and seed production and as habitat. This, and the establishment of nature reserves and totally protected areas (TPA), may reduce the yield potential by another 10–30%. The comparatively higher global net radiation balance and the species dominance pattern in Malesia make it likely that the potential PEP is generally higher than in Africa, possibly even in Borneo, in spite of the nutrient scarcity assumed for Borneo by Anderson *et al.* (1983) (Section 1.7). Potential PEP is possibly lowest in Amazonia.

These estimates of potential PEP apply to the late building–early mature phase forest on deep-rooted, porous clay–loam or sandy–loamy clay with good crumb structure in the lowlands up to 700–1000 m a.s.l. Forest ecosystems consist of a mosaic of developmental phases from young gaps to overmature patches. One-year-old gaps occupy about 0.2–0.5% of the forest surface, while 1–10-year-old gaps (regeneration below 5 cm diameter and 7 m height) account normally for 1–5% of the area (Bruenig, 1973b; Bruenig *et al.*, 1979; Heuveldop and Neumann, 1980). Building and mature phases cover 60–75% but are difficult to distinguish, and overmature patches may cover 20–30% of the area, depending on the past history of disturbances. In the Kelantan storm forest the pre-storm dominance of *N. heimii* indicated overmaturity over much of the area. The species almost completely disappeared after the storm of 1883. Most of the forest developed into the building phase, but in parts remained arrested in the early gap phase throughout the 66 years to 1949 (Browne, 1949). Overmature phases, failure of regeneration and delayed regrowth in the gaps caused by large-scale natural disaster or by selective logging may, in large management units, reduce NPP and PEP by another 15–25% on average. A plausible long-term sustainable-yield estimate for lowland mixed natural forest, such as MDF, would be 2.6–3.6 at a low production level and 5.2–7.2 t ha^{-1} a^{-1} at a high production level, or roughly between 3 and 5 t ha^{-1} a^{-1}, respectively 6.5 and 8.0 m^3 ha^{-1} a^{-1}, provided all species are commercial which reach the A and B canopy layers (Table 1.10). Of these, perhaps only one-third to a half are more-or-less constantly merchantable in Malesia, but currently much less than that in Africa and America under prevailing market conditions.

The productive forest in the permanent forest estate (PFE) will include a large proportion of steep terrain or rocky sites, especially in South-East Asia–Pacific, and disadvantaged soils. Yield assessment for forests on these less productive sites suffers from almost complete lack of data. Long-term ecosystem studies that include assessment of GPP and NPP do not exist. An exception are the growth data from RP 146, Sabal Forest Reserve, Sarawak, which are supported by comprehensive datasets on site and forest conditions (Table 1.6, Sections 1.8 and 1.12). A preliminary estimate of the mean annual stemwood increment over bark in *D. beccarii* MDF from 1978 (earliest year of selective logging) to 1990 is 10 m^3 ha^{-1}. The potential increment in KF is estimated to range from 5 m^3 ha^{-1} (*Casuarina nobilis* KF) to near 10 m^3 ha^{-1} (*A. borneensis*-type) after proper selection felling (Section 6.3) (Droste, 1995). The large Caatinga research plot at San Carlos de Rio Negro had only once, in 1979, been remeasured. The annual diameter increments 1975–1979 in the Caatinga forest were very small, 0.9–1 mm in trees > 30 cm diameter, less than 1 mm in trees 13–30 cm diameter, and just above 0.2 mm in trees 1–13 cm diameter. This is only a fraction of average values in zonal forest and points to very low levels of PEP in the Caatinga under a moderate disturbance regime with 0.8% mortality, 0.5% first-year gap area, and 2.6% older gaps (Bruenig *et al.*, 1979; Heuveldop and Neumann, 1980). There are few tree increment plots in, and growth assessments of, Kerangas and Peatswamp forests and their conventional, simple diameter growth data are not very suitable for assessing NPP and PEP (Section 6.6). Indirect deduction of GPP and NPP by comparing biomass parameters with MDF is not generally possible. Very tall canopies, high leaf area index, very high densities of basal area and biomass stockings are not always useful, reliable indicators of the relative or absolute GPP and NPP. Extremely oligotrophic soils can carry very high biomass stocking provided water supply is

Table 1.10. A. Mean annual periodic increment of trunk-base diameter, basal area and merchantable volume, 37-year-old *Dryobalanops aromatica* Gaertn.f. from natural regeneration, unthinned and with heavy thinning, FRIM research plots 85 and 86 (Abdul Rahman, K. *et al.*, 1992). B. Annual volume increments of selected élite trees, other dipterocarp and non-dipterocarp trees in logged and thinned (TSI) and logged and untreated stands of natural Mixed Dipterocarp forest in the Philippines. The annual volume increments are the means and in parentheses the minima and maxima in the range between plots, calculated from 5-year periodic volume increments (Manila, 1989). *I*, increment of stand per hectare; *G*, basal area per hectare; *V*, volume per hectare. See also Fig 6.3, p. 169.

Increment parameter	Unthinned		Thinned	
	85	86	85	86
A. FRIM RP 85 and 86				
Id (cm)	0.4	0.5	0.6	0.6
IG (m^2 ha^{-1})	0.31	0.28	0.41	0.31
IV (m^3 ha^{-1})	5.4	5.1	7.2	5.5

	Logged, no TSI	Logged and TSI
B. MDF Philippines		
IV (m^3 ha^{-1})		
Élite dipterocarp trees	1.0 (0.4–1.9)	2.4 (1.1–4.1)
Other dipterocarp trees	2.5 (1.4–3.6)	4.1 (2.2–5.2)
Non-dipterocarp trees	3.2 (1.0–4.4)	1.9 (0.3–3.4)
IV total (range)	6.7 (2.3–9.3)	8.4 (3.6–12.7)

adequate (e.g. *A. borneensis* Kerangas on giant podzol (haplic arenosol) or on medium deep podzol over clay-loam, and Peatswamp forest PC2 and 3). NPP and PEP are higher in the less tall Mixed Peatswamp forest, PC1, than in the much taller PC2 and PC3 where the phytomass density culminates (Fig. 1.11).

In contrast to Peatswamp forest, in the Caatinga and Kerangas forests stature and soil types within the forest formation are associated in a manner that appears to relate to productivity. If this is so, relative productivity may be assessed from canopy height. In Caatinga and Kerangas forests episodic drought strain and frequent alternation between waterlogging and dryness are probably the primary growth-limiting factors, possibly more so than soil nutrients. Bruenig (1974) estimated the BNP for stem-wood on the best sites in MDF as 16 t ha^{-1} a^{-1} (total NPP 40–50 t ha^{-1} a^{-1} minus coarse and fine above-ground litter fall of 16 t ha^{-1} a^{-1} and root litter 16 t ha^{-1} a^{-1}) on deep soils well supplied with water, and an evapotranspiration (ET_0) of 1800–2000 mm. Assuming tree height and ET_0 to be reasonable indicators of NPP (Bruenig, 1971c, 1974), the tall forests on deep, mainly sandy-silty podzols could produce 20–25 t NPP and 5–8 t stemwood BNP (ET_0 1200 mm). The forests on medium deep soils could produce 3–5 t stemwood BNP (ET_0 900 mm) and the low forests on shallow sandy podzols, bleached clays and lithosols 1–3 t (ET_0 700–800 mm). From the BNP, 30% must be deducted for unavoidable and unpredictable losses, 20% for harvesting loss and 20–25% for defects, or a total of 70–75%. This would leave about 1.5–2 t for PEP in the tall Kerangas types with *A. borneenis, S. albida* and *D. fusca* or *D. rappa* and just enough PEP (2–3 $m^3 ha^{-1} a^{-1}$) to qualify as potentially commercially productive. Most of the medium and all of the poor Kerangas and Kerapah forests fall out of the category of commercial timber production forest. The generally 20–30% lower canopy height of the corresponding types of Amazonian Caatinga indicates that the NPP is generally lower, perhaps by more than 20–30%. Probably only the Yevaro forest type (Bruenig *et al.*, 1978; Fig. 1.1) could be considered commercially productive.

Rainforest Use: Wisdom, Folly, Ambivalence   2

2.1 Forest-dwellers and the Rainforest

Life on earth has been, since its origin, essentially expansive and acquisitive. Life has reached out towards the limits of the carrying capacity of habitats, and continues to do so. Local and global scarcity of resources is the inevitable consequence. Scarcity provokes competition, evolution and conquest of new habitats. Natural 'genetic biotechnology' increases the efficiency to acquire and assimilate foreign substance to one's own. Human biological and social evolution are no exception (Markl, 1986, p. 19). The early abode of the precursors of the human species most probably was the aseasonal to weakly seasonal tropical (rain) forest. An indication is the association of the fossils of early Pliocene hominids at Aramis, Ethiopia, with faunal and floral fossils of a closed forest, which obviously was the habitat in which the hominids of 4 million years ago lived and died (Woldegabriel *et al.*, 1994). The rainforest canopy provides the only type of habitat on earth in which the physically weak generalist could develop into a mentally dangerous, inquisitive and aggressive brain specialist, ready to take off to conquer the perilous world (Bruenig, 1989d). However, the species-rich, diverse and abundant vegetable growth and animal life in the dynamically changing rainforest create no paradise. Food and other products useful to human beings are thinly and unevenly spread in space and time; abundance is localized or happens episodically. Scarcity occurs, but something edible is always available. The unpredictability of abundance and of scarcity, and the confidence that something to eat can always be found in order to survive, are not incentives to reduce the risks of scarcity, for example by caringly tending and purposefully enriching existing resources. The early roving forest-dwelling human beings were, and had to be, exploitative and acquisitive, but did not have to be effectively creative or caring. Their small number and their weak biotechnological capabilities prevented them from destroying their habitat and resource. Such a lifestyle is biologically sustainable, but not culturally. The general trends of the lifestyle changes among the forest-dwelling peoples of the rainforest show the innate urge to abandon arboreal and forest-bound lifestyles in favour of cultural advancement if alternatives are within reach and safe to take.

In Sarawak, at most 400–500 Penan are still genuine forest-dwellers (Langub, 1991). They live nomadically in large tracts of mostly primeval, in parts very old secondary, Mixed Dipterocarp forest (MDF) in northeast Sarawak. The quality of life and the chances of survival of the Penan are determined by the great species richness and the heterogeneity of forest types and site conditions (Figs 1.8, 1.9, 1.10, 1.13, 2.1 and 2.2) with widely dispersed and thinly spread useful plants, such as sago palms, rattan palms, fruit trees, poisonous and (assumedly) medicinal plants and vegetables. Small-scale natural disturbances that break the canopy cover maintain the diversity and productivity of the food resource in easy reach of the ground-living human beings. Episodic and sporadic extreme events, such as drought and subsequent fires, may cause large-scale fluctuations in the vegetable and animal food sources that are unpredictable, unexplainable and uncontrollable for the Penan. The richer types of forest, especially alluvials and wet, narrow valleys, offer more affluent and steadier resources, but also higher risks of pests and diseases to humans.

The state of health of the forest-dwellers is poor and life is short. None of the risks, hazards and tribulations can be assessed and counteracted effectively without scientific knowledge and an understanding of cause-and-effect relationships, which the forest-dwellers do not have. The unexplainable sporadic and episodic, almost always unpredictable, events of disaster or of windfall, nourish superstitions that provide the scenario for the richness and diversity of the omenology and religion of the Penan, which Brosius (1991) mentioned. Superstitions and experience without knowledge are no help in coping with a present and future that are essentially inexplicable, unpredictable and uncertain. Even their only certainty, that there will be forest in future and that this forest will produce something useful, is now being challenged from outside. The uncertainties in the life of the forest-dwellers have increased as a result of modern demography, migrations and economic development. For example, the staple food of the Penan (Low, 1848; Kedit, 1982), the sago-producing palms of the genus *Eugeissona* (Whitmore, 1975b, p. 59), flourish on exposed sites and in naturally disturbed and carefully logged forest, but succumb to deforestation. In the region occur large areas of old and very old secondary forest on land that had been farmed by migrant shifting cultivators in the distant past. This old secondary forest is a rich source of fruit and game, but is still under real or pretended native customary rights (NCR) of the descendants of the original farmers. The forest-dwelling Penan face an unavoidable dilemma. The free range and

Fig. 2.1. The high β-diversity of the Penan landscape produces a life-support system with manifold food sources which vary independently with time in the various ecosystem types.

carrying capacity of their habitat declines, life in the forest becomes more difficult and uncertain, while contacts with the outside world bring new ideas, attitudes and aspirations that change the Penan view of the world and of themselves.

The ecosystem regulates Penan group size and population density by such mechanisms as availability of food and abundance of pathogens (Fig. 2.3). To increase food supply and improve health (Chen, 1990; also chapter 17 in UNESCO, 1978) in the natural rainforest would require sophisticated biotechology (Appendix 1) and technically demanding management

Fig. 2.2. Diversity at β-level between smooth-canopied Kerangas on sandy podzols or clayey ultrabasics and rough-canopied MDF. Usun Apau N.P., Sarawak.

which are beyond the capabilities of the roving forest-dwellers. A combination of improved health care, longer life and education with traditional forest life is a contradiction. It is neither technically possible nor psychologically acceptable or sociologically feasible. The only sustainable options are either to continue the secluded nomadic lifestyle unchanged, or to abandon the forest-dwelling lifestyle altogether and emigrate out of the system (Fig. 2.3). Forest-dwelling Penan, as any other forest-dwelling people in other parts of the world, are not the caring and sustainably managing stewards of the forest ecosystem, but rather utterly dependent and poor subsistence tenants. In the MDF, survival at this level of the biological struggle for existence leaves precious little energy to invest in the advancement of biotechnology or for cultural evolution. The exodus into open land is the only feasible and attractive option now in Borneo, as 4 million years ago in Aramis.

In conclusion, the tropical rainforest, particularly the MDF and alluvial forest, is a tremendously active and efficient biochemical factory with extreme diversity of products. These products are typically widely dispersed and thinly spread. The annual yield from the resource is variable. The availability of foods for forest-dwellers changes unpredictably between abundance and scarcity according to seasonal, periodic and random patterns. Forest-dwellers, such as the Penan, cope in a painful and costly manner by continued trial and error. The biological existence of human beings in widely dispersed small nomadic groups in the Bornean rainforest is possible but at excessive costs of individual suffering and with severe constraints on economic and social development. The Penan share the biological urge and physical and spiritual need of the human species to explore and acquire. The basic human motivations and hopes for fulfilment, long life and a better life quality are also theirs. As a result they, as any other human beings, want change and improvement. They too want to enjoy what they see, or they believe, other people enjoy. However, they also share the counteracting human aversion of risk and uncertainty, and the reluctance to abandon an accustomed material or spiritual status quo ante for the unknown. Typically, forest-dwellers believe they are left out and deprived, and they suffer from it. However, they do not have a very clear conception of what they could and should want and what they could do to get it. Their own experience is confined to within their ecosystem, their view of the outside world is narrow, fragmentary and biased, and has recently become even more distorted (e.g. Hong, 1987; Manser, 1992). Dr Charles Hose, 'one of the earliest white men to have any real intercourse with them' (1927, p. 72) described the Penan as:

> shy but friendly, and ready at any moment to fight for (his) life … defend himself adequately … – a fact which … procures him a certain immunity from wanton attacks … wandering, but attached to an undefined area … any tribe … welcomes the Penan … with a friendly sense of proprietorship … and jealous of interference from other tribes. Such are these strange people – isolated, not even self-supporting; unresourceful, yet able to make a livelihood at others' expense, simple, yet wise, untrained, and with a fear of the unfamiliar, yet knowing instinctively what is right; a type of civilization, and a civilization of no mean kind.
>
> (Hose, 1926, pp. 38–44)

The issue now is for the society at large to open the door for them to enter the next phase of advancement and change without loss of pride and identity.

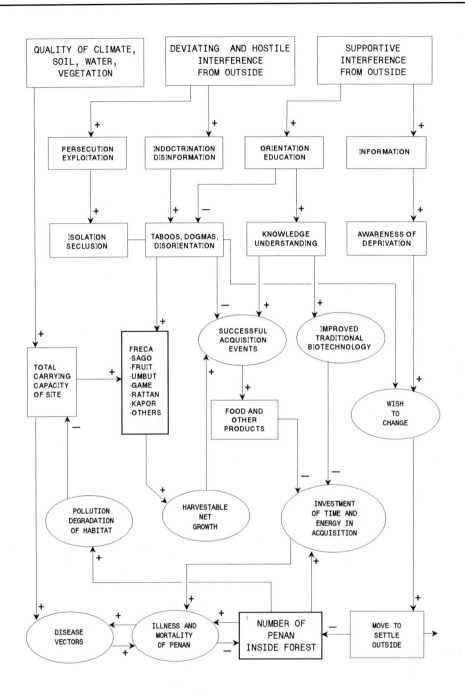

Fig. 2.3. (*opposite*) The natural and human ecological factors which interact within the life supporting ecosystem of a possibly primeval, or more likely secondary, hunter and gatherer population in the tropical rainforest. The example illustrates the ecosystem of the nomadic Penan in the Mixed Dipterocarp forests of Sarawak. The feedback loops regulate the population density and the activities of the nomadic Penan. Use leads unavoidably to food source depletion. Scarcity requires more investment into acquisition. This and the effects of natural disasters impair health and fertility. Information and education reduces time spent on gathering and hunting, and also the motivation for forest life. Information creates the wish to abandon the traditional lifestyle. The Penan ecosystem is essentially open and would otherwise not be viable. FRECA, free capacity available for acquisition.

2.2 Shifting Cultivators and the Rainforest

Technological advancement in the open savannah habitat during the Pleistocene gave human beings the capability to cut trees efficiently and clear the forest for purposes of hunting, husbandry and agriculture, and also the expertise to domesticate forest trees and grow them in enriched forest gardens. Slash-and-burn practices most likely developed in the seasonally dry deciduous tropical forests where felling of the big trees is not necessary for drying slash to get a good burn. Eventually, with better tools, the rainforests became accessible to slash-and-burn agriculture in the form of a rotational crop–fallow system with periodic forward migration when the productivity had declined. The change of the landscape in the course of pioneer colonization, expanding shifting cultivation and gradual change to more intensive land use in South-East Asia is illustrated in Fig. 2.4. In tropical America, conversion of rainforest is mainly caused by immigrant poor and landless small farmers, land speculators and big agro-industrialists and ranchers. In tropical Africa, traditional shifting cultivators clear small isolated pockets in the rainforest and soon move on. More contagious clearing is caused by population increase, politically enforced migration and state-supported plantation schemes. Worldwide the small shifting agriculture (shag)-farmers are the main cause of deforestation in the tropics (Heske, 1931a,b; Bruenig, 1989e). In the decade 1981–1990, forestry caused 2–10% of the deforestation, agriculture 86–94%, shag 41–49%, permanent agriculture 21%, pasture 24% (Brazil only), and others 4% (mining 1%, dams 1%) (Amelung and Diehl, 1992, table 31). The pioneers at the agricultural deforestation front are the actors, but they are not the culprits but rather the victims. The real culprits have to be sought in the world of big money and power politics. The same can be said for the loggers who, while naturally profit-orientated and greed-propelled, take advantage of irresponsible policies of governments in producer and consumer countries.

The rain-fed slash-and-burn cultivation or shag is exotic and not native to the rainforest. With repeated fallow cycles productivity declines as a result of nutrient loss, erosion, growth of weeds, pests and diseases. The Iban pioneer communities in Sarawak rarely practise soil protection

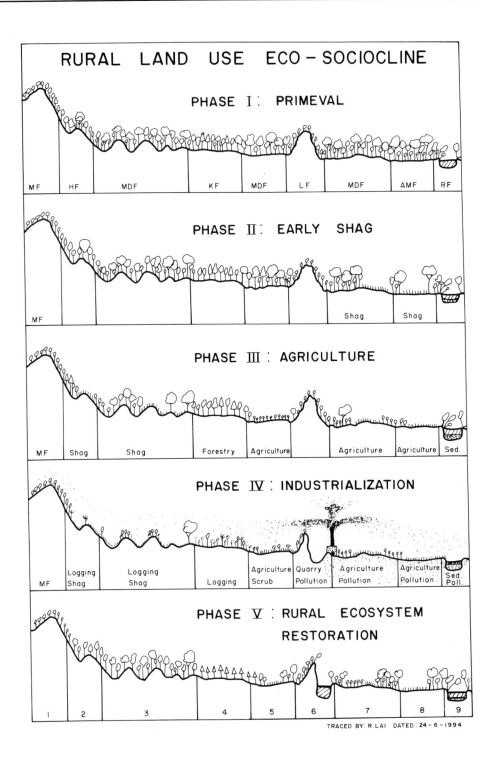

RURAL LAND USE ECO-SOCIOCLINE

PHASE I: PRIMEVAL

MF | HF | MDF | KF | MDF | LF | MDF | AMF | RF

PHASE II: EARLY SHAG

MF | | | | | | Shag | Shag |

PHASE III: AGRICULTURE

MF | Shag | Shag | Forestry | Agriculture | | Agriculture | Agriculture | Sed.

PHASE IV: INDUSTRIALIZATION

MF | Logging Shag | Logging Shag | Logging | Agriculture Scrub | Quarry Pollution | Agriculture Pollution | Agriculture Pollution | Sed. Poll.

PHASE V: RURAL ECOSYSTEM RESTORATION

1 | 2 | 3 | 4 | 5 | 6 | 7 | 8 | 9

TRACED BY: R.LAI   DATED: 24-6-1994

Fig. 2.4. (*opposite*) The change of the rural landscape in the course of human cultural evolution and population increase. I. Hunter-gatherers leave the primeval landscape practically unchanged. II. Traditional slash-and-burn shifting cultivators deforest few patches on the most fertile sites, forest creaming does little harm and territories shift with migration so that landscapes can recuperate. III. Population increase and human inventiveness force agricultural revolution, intensive and expanding land use replaces shifting agriculture (shag), which is pushed to unsuitable, fragile sites; overlogging deforestation, pollution of soil, water and air from fertilizers, pesticides and weedicides occur locally, heavy erosion causes soil degradation and river-silting. IV. Industrialization and urbanization cause accelerating liquidation of renewable and non-renewable resources in forestry and agriculture, industry invades rural landscape, heavy agricultural and industrial pollution and soil degradation, very heavy erosion, forestry reclaims devastated sites, but tends to maximize the production function to meet urgent industrial and societal needs for timber and wood; river, soil and air pollution and biodiversity become vital problems at national scale. V. Balanced matching of land use and site, restoration of degraded ecosystems, creation of balanced and healthy forests and landscapes by integration of absolute protection (1, 2 part of 7, 8 and 9) with site-specific forms of naturalistic silviculture (3 and 4), mixed farming and agroforestry (5, 7), intensive agriculture (8), recreation and amenity (6, also 1, 2 and 4), principle of containment of pollution at source.

and vegetation restoration (Halenda, 1985). In Indonesia alone, shag had produced 17 million ha grassland by 1965 (Dilmy, 1965) and over 40 million ha by 1990. Grassland (*Imperata cylindrica*) generally occupies the more favourable and productive soils and sites, which are farmed and degraded first. Virgin forest land is constantly needed to compensate for degraded land, which unavoidably means shifting into new territories. More optimistic descriptions are quoted from the seasonal tropics (Hong, 1987) where the natural site conditions are fundamentally different. Parnwell (1993), in his intensive survey of 13 Iban longhouse communities near Bintulu, Sarawak, found that two-thirds of the padi (rice) producing households considered that 'levels of rice self-sufficiency had fallen over the last fifteen years, with the over-exploitation of land resources (38% of the answers) being given as an important explanation'. Also 'crop diseases (80%) and pest infestations (12%) were cited as principal causes'. The traditional response would have been to move into fresh territories. This alternative is now no longer available and the option is irrigated-field cultivation on flat land, supplemented by husbandry and agroforestry, but this requires capital, skills, initiative and, above all, suitable soils and sites.

The Iban of Borneo, formerly also called Sea Dayak (early sources are Low, 1848; John, 1862; Bock, 1881; Hose, 1926, 1927), are the prototype migrant pioneers at the agricultural slash-and-burn front. 'They regard the forest as something outside the essential habitat, as strange, mysterious, threatening, potentially dangerous.' For them the rainforest is 'certainly an environment to be entered with caution and circumspection' (Dunn, 1982) and a resource which must be domesticated.

> The fundamental difference between *appropriation of spontaneous resources* as practised by hunter-gatherers and the *production of*

domesticated resources by swidden cultivation lies in a major change in the relationship between man and the other components of the ecosystem to which he belongs.

(Sautter *et al.* in UNESCO, 1978, pp. 414–451, cit. p. 443)

It also entails a major change of the whole spectrum of spiritual and material culture including social values, perception of resources, attitude to customary claims, usufruct and land rights. The Iban practise shag with great vigour and aggressiveness. Forests, for them, are a challenge and forested land a free common resource, unowned and available for clearing. Territory occupied and claimed by other ethnic groups is available for the taking (Sandin, 1956). In contrast the Land Dayaks (Geddes, 1954) regard all land as owned, and have a concept of stability of territory associated with a given village. A review of the subject in the South-East Asian context is given by Kunstadter (UNESCO, 1978, pp. 319–350). All shifting cultivators in Sarawak are in rapid cultural transition from migratory shag to settled hill-rice cultivation, mixed farming and, finally, intensive agriculture. At the same time the rapid economic development in the state of Sarawak already draws the young generation into the next phase of cultural change, away from the longhouse and agriculture into a semi-urbanized existence.

Parnwell (1993) found that the people of the 13 longhouse communities in transition from shag to settled cultivation and from largely autonomous to dependent economy have many diverse and contradictory, sometimes very fanciful, bizarre and unrealistic views and opinions of the state of the natural and social system and the environment they live in and have to cope with. The people feel backward and uneducated, and suffer from a sense of inequality. The dependence on natural forest resources is declining but wildlife, plant and land resources continue to be overused. New technical capabilities for exploitation and outside interference, especially from loggers, quicken the decline. The rural natives themselves generally share in the over-exploitation, which causes shortages of non-timber forest produce (NTFP), naturally durable timbers (Taylor *et al.*, 1994) and productive land. Their traditional esteem for the value of natural resources, forests and environment is diminishing as the desire for better living heightens. Accordingly, taboos, customary communal claims to territory, individual rights of forest usufruct and rights to cultivate land are losing relevance with the declining dependence on natural resources. Interest in the forest, however, quickly revives with opportunities for land speculation and profits from logging. According to the villagers the effects of timber licencing reduce land availability and aggravate the general trend towards degradation of natural forest and land. The spread of exploitative timber logging, the establishment of forest-based industries and of plantation agriculture fundamentally transform the native culture. The young move away from the longhouse and become urbanites. The issue of logging rights by government to outsiders restricts the availability of land to the locals. In the progress of logging, village territory will inevitably be encroached upon. Often this is inadequately announced to the longhouses and is viewed as equivalent to the former intrusions by hostile tribes. These could and would be fought in battle. The loggers are legalized intruders and cannot be fought. This offends, hurts and cuts right across all traditions and ideals of the *adat* and contradicts the sense of communal territorial rights, individual owner-ship and general propriety. Among tribal people, it leaves a deep-seated trauma of tribal and personal defilement. Their spiritual world is desecrated and their forest raped. The forest within the claimed territory was undisputed acquired legacy and the source of material and spiritual subsistence for the longhouse community. The profound social, spiritual and economic degradation in which the villages are left after the logging bonanza is over is partly rooted in this trauma. An example is vividly described in Marsh and Gait (1988) (see Section 2.5).

The advancement of agricultural technology reduces land require-ments, potentially to less than one-tenth. There is no more pioneering conquest into virgin lands and no more tribal wars and warrior ideology which go with it. The pressure on the rainforest is reduced, but the whole traditional spiritual orientation is lost. Communities that traditionally usurp and fight for large territories suddenly find themselves without this motivating challenge. Material and spiritual interdependence weaken and so does social cohesion, as do essential elements of the traditional *adat* ideology (legends, lore, code of conduct, rules, regulations, law). The social mechanisms of conflict avoidance, which traditionally helped to cope with social and material emergencies, disappear. In contrast, the new lifestyle enforces individualism, materialism and opportunism that are already inherent in the pioneering personality (Sutlive, 1992) and which were essential for success and survival at the pioneer front, but were constrained, bridled by the forces of social coherence and *adat* in the longhouse community. While pressures on the rainforest and the communities are reduced at the agricultural front, new material pressures and social threats arise at the social front of industrial development and urbanization.

The sustainability of tropical forestry is conditional on the successful introduction of sustainable forms of settled agriculture. Traditional and new agroforestry systems have been hailed as a panacea capable of solving the conflict between agriculture and forestry. The long-term trend of change is not towards a closer integration between farming and forestry. The agroforestry type of integration is transitional from tradi-tional to new lifestyles and therefore ephemeral. In this respect it plays a similar role to 'extractivism' (Section 4.6, Table 4.1) at an early stage of cultural transformation. The change continues towards completely new social goals and aspirations. Employment is preferred outside forests, farming and forestry. This brings social and cultural side-effects that often have not, in the past, been fully realized by the people and governments.

The case of the Iban in transition shows that nothing less than fundamental cultural transformation is involved.

Employment in logging camps and sawmills, as well as in agricultural
and forestry plantations, creates new material dependencies and social
problems. This can only be mitigated by the people themselves if they
wish to preserve the cultural heritage as a treasured social value while
new qualities of life develop. The hope that the traditional collecting of
NTFP in the forest could be an attractive alternative to logging is
wishfully utopian. It takes more than that to save forests and stabilize the
rural economy. My personal experiences among all ethnic groups in
Sarawak and Brunei and the findings of Parnwell agree well with the
hypotheses of Godoy and Bawa (Anon., 1994b) that rich households use
less NTFP, poor people derive a larger share of income from NTFP, and
that forest opportunity costs for NTFP are about DM80 or RM130 ha^{-1} a^{-1}.
These figures appear rather incompatible with potential annual net-
returns of US$1257 ha^{-1} and US$2830 ha^{-1} respectively from collecting
NTFP calculated by Grimes *et al.* (1994) based on two forest plots in
Ecuador. NTFP gathering in the natural forest is an activity that in
Sarawak and Brunei has proved to depend on poverty and non-
availability of alternatives. The same applies to 'extractivism' in Amazonia (Clüsener-Godt and Sachs, 1994; Section 4.6). It is an important socially
integrating factor in the traditional community and may play an important role during transition, but disappears as an economic factor with
higher material standards and improved quality of life. The economic
value and social importance of the rainforest for the people undergo a
profound change during the transition from nomadic forest lifestyle to
migratory shifting agriculture. The change is no less profound from shag
to settled agriculture in an urbanizing and industrializing environment
into which the young generation eventually migrates.

2.3 Native Customary Rights and Forestry

Utilization of the forests and use of forest lands has been, since the
beginning of civilization, traditionally regulated by customary rights,
which were developed by the people in response to the natural and
cultural conditions and the material needs of the people. Individual
community and tribal claims had to be defended against contenders, if
necessary by force. Hostile tribal invaders, colonial usurpers or native
political power élites supplanted their own concepts of NCR. The scope
and contents of NCR of no tribe have ever been static. Tun Jugah anak
Barieng, the last paramount leader of the Sarawak Iban and eminent
expert in *adat* of such culturally different tribal people as the Iban, Kayan,
Kenyah and others has tried to elucidate and codify adat in Sarawak. Adat
denotes native norms and lore of what is desirable, undesirable, right and
wrong, including NCR and penalties for transgression. Tun Jugah tried

but eventually had to give up (Sutlive, 1992). The subject is intractable, fluid, diverse, heterogeneous and disputable within and between tribes. Colonial governments rarely ventured further than generally requiring observance of NCR in all projects of development involving native lands. Traditionally, if disputes between communities or tribes over territorial claims could not be settled by negotiation, there was fighting and the weaker tribes were subjugated or pushed out. Forest-dwellers, as the weakest group, were ignored altogether in disputes over land and usufruct. The miserable social and legal status (Cranbrook, in Gill, 1993) of the Penan in this respect became known to the outside world towards the end of the nineteenth century (Bock, 1881; Hose, 1926, 1927). In response the Sarawak Government instituted the *tamu*, an officially supervised trading day to bolster status, equality and prestige of the Penan, at least in trading (Brosius, 1988, 1991).

In the absence of clearly defined, universally accepted and codified native land rights and legally supportable claims, governments filled the void and in many countries assumed full authority over the forest lands and the land-based resources. In Sarawak, as elsewhere in Borneo, any unoccupied land was unowned until the government issued the Land Regulation 1863, supplanted by the Land Orders 1920 and 1931, and by the Land Code Ordinance, 1958. NCR were specifically recognized, titled land ownership introduced (but sluggishly implemented) and unoccupied land under primeval forest declared stateland. Generally in the tropics, with exceptions such as Papua New Guinea, state bureaucracies assumed centralized fiscal control of forest and land resources and their utilization, and consequently new conflicts of interests arose. Local tribal populations saw infringed what they considered to be their rightful property according to their traditional customary rights. The relationship between the emerging forestry and NCR has been and still is more precarious than any other sector of the economy, be it infrastructure, mining or agriculture. With the establishment of a permanent forest estate (PFE), large tracts of land and forest are excised from the territory that has been freely accessible to communal–tribal activities. In Sarawak claims and customary rights are negotiated before establishment of the PFE and, as a rule, admitted. In totally protected areas (TPA), such as a national park or wildlife sanctuary, any claims and NCR are also negotiated, but admission is the exception and compensation is the rule. In Mulu National Park, for example, only the local group of Penans retained a right to make use of the forest in their traditional way, but not the other claimants. However, whatever the result of the negotiations is, the local community will consider the establishment of a PFE as a disturbing inroad into NCR and their habitat that takes much land and gives relatively little in return. The most painfully felt interferences are logging that passes through NCR land and the establishment of PFE and TPA. At the same time, NCR land is everywhere gaining potential value for land speculation and timber logging, both for the NCR holder and the nation. This adds new dimensions to the problem. Manser (1992, p. 236) rightly emphasizes that a depletion of natural resources 'would not be a matter of the "Urvölker" (aborigines) only, but affect the whole nation'. This is exactly the crux of the NCR problem: claims and rights are embedded in the totality of the societal environment and the national socio-economic and political fabric. Any alteration of NCR practices has to be viewed in the context of national interests and preferences.

There are, then, two major problem areas that universally require action by strategic approaches. One concerns the notorious problem of finding a compromise between NCR claims and logging; the other concerns the long-term problem of accommodation of NCR with the establishment of PFE and TPA. The first requires inclusion of NCR in the comprehensive pre-logging fact-finding surveys as a database for fair and pragmatic negotiations with the affected people before the logging and road-building plans are finalized and operations begin. The second demands a permanent solution and requires much more factual knowledge on NCR of the various ethnic groups and localities than most governments have. National surveys and action-orientated monographs on NCR are needed but there are few knowledgeable people who could do them. Without this basic information, adequate planning and negotiating of the constitution of PFE and TPA is not possible. This information is needed equally for the preparation of management plans. Mutually accessible information may bridge the gap of communication and understanding between the parties. Cleary and Eaton (1992, pp. 175–189) see this gap as a major cause of material conflict and human disregard, and of the widespread destruction of resources and environment through greed and ignorance.

An alternative that is often advocated is to grant additional or even exclusive usufruct rights to those who happen to hold or claim NCR in a certain area. However, such a policy may be constitutionally and legally inadmissible; it also is impracticable. The gathering of non-timber produce in the forest is no longer very attractive except in areas of poverty. Granting exclusive rights is tantamount to the issue of permanent timber licences to the claimants, which may not even be the local communities. The experience with the establishment of exclusive rights for indigenous Indios in reservations in the Brazilian rainforests indicates that this is a very risky strategy if the targets are preservation of culture, lifestyle and sustainable forestry. The Environmental Protection and Indigenous Land Reclamation Association (APARAI), a non-governmental organization in the state of Rondonia, Brazil, states in the project application PD 15/1992(F) to ITTO:

> Indiscriminate timber logging by Indios and local communities has already been well documented. Approximately 905,000 m^3 of high-quality timber is reported to have been logged during 1987–90 in only eight areas of the State of Rondonia, where there are twenty

indigenous communities. Another clear example is the Mequens Indigenous Area in Southern Rondonia, where 90% of hardwood species have already been logged. These results are alarming for they reveal that 50% of the communities are currently selling hardwood timber species, while the remaining 50% are not doing it simply because they no longer have any. This problem is not only limited to the indigenous areas; it also extends to the colonists, who often ignore timber resources, burning them and deforesting areas indiscriminately. Therefore, there is a need to develop appropriate forest settlement policies to regulate colonist settlements in the forest, while avoiding massive urban invasions and the resulting consequences of this migration. Studies have shown that since 1985 the income of colonists has gradually been decreasing, partly because of the timber shortage.

The lesson from this and similar experiences in Papua New Guinea and southeast Asia is that the greed-motivated drive for acquisition under such circumstances is universally overwhelming. It can only be contained and diverted to sustainable development of community-based forestry and forest industries by intensive, qualified and long-term extension services. Such developments leave the realm of traditional NCR and move into the next stage of cultural development. There are many examples of successful transition into this stage in all parts of the tropics. The Pilot Forestry Project of Quintana Roo, Mexico, is one of the community-based comprehensive forestry and forest industry development projects that has achieved a high level of natural, technical, economic and social sustainability (Janka: 'Experiences in forest management in the estate of Quintana Roo, Mexico', case study 3 in Bruenig and Poker, 1989). While traditional in concept, such projects, to be sustainable, must widen the scope from simple usufruct to comprehensive development. The young generation of Sarawak Iban, Kayan, Kenyah, Kelabit, Bidayuh and other inland ethnic groups of settled farmers, as well as coastal Malays and Melanaus, are drifting out of the purely agricultural village environment and village economy. The NCR in their original function become meaningless to them. Already, young people find it difficult to describe the NCRs of their parents and to identify the areas in the landscape to which their NCR apply.

2.4 *The Enigma of Conventional Logging*

A high social value of forest production and service functions is the most effective safeguard against forest degradation and deforestation. Commercial timber harvesting from single-handed creaming (Appendix 1) to mechanized large-scale logging is motivated by the greed to acquire and is essentially profit-orientated. The creamer and logger have no stake in sustainability unless the society puts a value on sustainable conservation and management of the forests and enforces conditions from which creamer and logger also benefit. Otherwise, forest utilization degrades to overuse, abuse and misuse. Sustainability in multipurpose forestry is technically and economically no great problem. It is fully achieved in many traditional forest countries. Heske (1931a,b) lamented in the 1930s that sustainable management had not yet been achieved in the boreal and tropical forests. Today, we are not much better off. Logging still destroys forests as resources and ecosystems, desecrates national heritages, ruins beauty, and damages the environment. Throughout the world, the physical, mental and spiritual well-being of forest-dwellers and rural and urban populations is threatened by forest overuse and misuse by loggers. The causes for failing to conserve and preserve the natural and traditional values of the tropical rainforest and respect old and new material, spiritual and aesthetic values are many. Management systems for sustainable management have been developed over the past 100 years. They are technically adequate and legally sufficiently supported, but are not implemented in practice as a result of lack of public awareness, conflict of interests, greed-propelled profiteering and irresponsible policies at home and abroad. Foresters are often dogmatic and conventional, non-traditional, and lack the capability and motivation to grasp the concepts of natural and social ecosystems and consequently are poorly equipped to argue their case convincingly and effectively. Even more critical is the common lack of professional expertise among timber concessionaires, employees, workers and government servants. Fiscal and private interests in money and land tempt politicians and administrators to ignore the long-term needs of forestry and land development, to condone illegal logging and illegal land occupation by locals and migrant people and by their own clique, and to ignore environmental protection and conservation of species, nature and ecosystems. Finally, the lack of law enforcement and rampant collusion between political and economic power 'élites' at home and abroad, and between loggers, local people and traders overrule the principles of sustainability.

The major causes obviously lie in the social and political environment (Figs 11.1 and 11.7). The notorious flouting of rules and disregard of the law by loggers at local operational level and by managers and concession holders can be overcome only if the political and societal will and strength are there and put into action. The technology and principal guidelines on how to apply biotechnology and mechanical technology in forestry are available. The pantropical failure to enforce implementation and the persistence of the conventional practice of selective logging creates conflicts with the principles and with the ecological, social and economic criteria of sustainable management and conservation. Unique beauty and valuable habitats and their inhabitants are destroyed. Water courses are denatured by siltation from eroding roads and trails and polluted by leachates from soils, rotting debris and abandoned timber. Waterways become impassable by heavy siltation. Site and forest productivity are destroyed by criss-crossing skidtrails, and by unskilled tree felling and log extraction. Overlogging removes more than 50%, and up to 80–90%, of the canopy, completely altering the structure and function of the ecosystem and its features as a productive, renewable resource, as a habitat for flora and fauna and as an aesthetic and ecological component of the landscape. Cutting and smashing of the intermediate trees, which are the fastest growing part of the growing stock, and destroying a large proportion of the regeneration reduces tree increment far below the site potential. Overlogging enhances the naturally high heterogeneity of growing stock by clear-felling mature patches and leaving overmature patches and less attractive species untouched thus affecting adversely future primary economic productivity (PEP) and management. Wasteful harvesting and so-called volume adjustment and illegal removal cause the forest areas to be logged two to three times faster than necessary. In balance, the national economy suffers a monetary loss that is two to three times larger than the value added to the gross national product (GNP). The natural resource account for forestry consequently declines. These practices contradict the very basic principles of sustainable forestry. Selective logging in the tropical forests is, as with clear-felling in Siberia, Alaska and Canada, deeply entrenched in the vested interests of international timber buyers (Section 2.6) and local profiteers (Section 2.7). This causes social and socio-economic costs for the nation and for future generations that far exceed the profits of the contemporary beneficiaries.

2.5 *Conventional Selective Logging and the Community*

Currently, costs and benefits from conventional selective logging (definition in Appendix 1) in tropical rainforest are neither transparent nor equally shared. The nation bears the cost of the inadequacies, socially and fiscally. Damage to the environment, mainly by flash-floods, pollution and silting of water bodies, and damage to the productive resource, mainly by wasted timber, destroyed trees and eroded soil, are quantifiable costs that are exclusively borne by the nation. More difficult is the assessment of the present value of losses of future increment and yield caused by overuse, misuse and mismanagement, and of the consequent reduction of employment and economic activity in the future. Reduction of the high growing stock volume in the mature and overmature phases of the primeval (primary, virgin) forest (Fig. 2.12) is an ecologically acceptable and economically essential instrument of naturalistic selection silviculture. It is not a social cost factor. The reduction should be balanced by increased increment in the appropriately harvested and treated forests and by the value of investment in the forest-based industry. In selective logging, this is not the case. The balance of the natural resource account of selective logging in the PFE is not positive, but decidedly negative. The net proceeds from the liquidation of the merchantable growing stock in the stateland forests must be fully invested in development of productive enterprises and human resources to meet the conditions of sustainability. In reality, this is hardly ever the case. Illegal logging and transfer of proceeds from logging out of the producing country by means of volume adjustments and transfer pricing arrangements are dark chapters of contemporary forestry in many, not only tropical, countries.

Export orientation of resource use generally leads to domestic instability and often to high costs of environmental deterioration and resource degradation and exhaustion. GNP as a simple measure of progress obscures actual cost and benefit relationships and disguises the social costs of resource, environmental and social degradation. Export orientation and GNP juggling also encourage rapid over-exploitation of the country's natural resources (Goodland and Ledec, 1987). Conditions differ between countries and regions. Low-yield logging in the more strongly successional African rainforests is, in every respect, a fundamentally different matter from high-yield logging in Malesian MDF or Peatswamp forest. Selective logging in Africa has little effect on the ecological conditions of the forests which retains its resource value, assuming that a wider range of tree species will become marketable in future. In the Malesian MDF selective logging of the whole merchantable growing stock unavoidably depletes the natural resource and reduces the phytomass below safe standards of sustainability (Sections 6.3 and 6.4). High-yield conventional selective logging and the loggers deprive local communities of clear water, game, fish and customary-use timber and NTFP. These effects may be ephemeral, even in very heavily overlogged and ravaged forest areas, but still are resented by the people. For them these effects are immediately economically important and emotionally annoying enough to pose a political and social problem that cannot be disregarded. The effects are visibly negative and directly felt by the local people. No research is needed to prove that. Other effects are less obvious, vary between locations and are differently perceived by the villagers. Evidence is contradictory, and individual and communal opinions and reactions are conflicting on such matters as the effect of logging on the accessibility of land, local climate, future availability of forest product, employment and livelihood generally (Parnwell, 1993; Taylor *et al.*, 1994). The depletion of the timber stand is obvious, but the decline of the natural primary and secondary (especially game) productivity and yield potentials are not immediately recognized, either by the government or by the people. Both immediate resource depletion and the long-term decline of potential are of no concern to the short-tenured concessionaire. The consequences at national level will only be felt long after the revelry or the timber bonanza is over and the people who revelled in it are gone. The declining resource will cease to provide a sustainable source of secure jobs, revenue and income. Repair of the damage to restore the potential will be costly, but needs to be done if the forest is to retain its essential role as a renewable multiple-use natural resource for sustainable development.

Logging creates not only health hazards by the high accident rates, which are four to six times higher than in Germany for instance, but also creates new contacts with disease vectors, especially if there is no qualified health care in the camps. Diseases are transmitted from the forest canopy into camps and villages, then to urban centres and abroad. Malaria and yellow fever are well-known examples of diseases which have resurfaced (Cooper and Tinsley, 1977; Fig. 2.5). Jacobson's disease is another growing threat. The virus was originally carried by a mosquito which avoids human beings. Development projects brought infected cattle to the newly cleared ranches in the Amazonian rainforest. The local

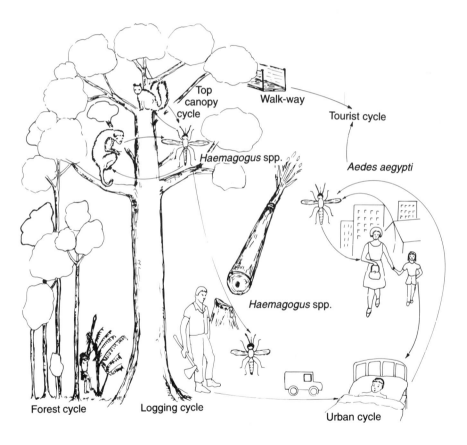

Fig. 2.5. Health hazards of exploiting tropical rainforest. Wild animals are reservoirs of yellow fever in the upper rainforest canopy in South/Central America and in Africa. Cutting down the large trees lowers the canopy and brings humans in contact with the vectors of yellow fever, and with other identified and unidentified 'jungle fevers'. (Adapted and amended from Cooper and Tinsley, 1977.) The tourist cycle opens a new outlet for the diseases directly from the rainforest canopy to the urban centres and abroad.

forest-living and cattle-feeding mosquito, *Aedes albopictus*, sucked infected blood and, habitually feeding on human beings, transmitted the virus to the local population. AIDS is the most publicized example of transmission of a rainforest arboreal virus to people outside the rainforest and of the efficiency of tourism as a disease vector.

An example of the profound, manifold and, as a whole, disruptive effects of careless selective high-yield logging on rural communities is given by the comparative study of two villages in central Sabah (Marsh and Gait, 1988). Both villages lie at the fringe of a large concession area held by the public company Yayasan Sabah (Sabah Foundation). The physical and biological effects of selective logging were a reduction in nut-producing Dipterocarps (engkabang or illipe nuts) as a result of (illicit) felling of the trees, which are totally protected by law. Rattan suffered badly from waterlogging of its prime riparian habitats caused by poor standards of road-making and drainage. Game species reacted differently: generalist grazers increased (deer, elephants, wild boar), dietary specialists decreased less massively. The streams and rivers carried heavy silt loads, the water turned muddy and the stream beds were silted, possibly affecting aquatic life. Socio-economic effects were equally varied. At the beginning, the community benefited from improved mobility, communications, access to amenities, health care and schooling, and from chances for trading during the period of logging. Game was depleted by the increase in illegal spotlight hunting by loggers and outsiders along logging roads, including poaching of rhino for sale to Hong Kong, Taiwan and mainland China. Local affluence and wealth increased through wage-earning in logging and through trading with the camp, initiating a change from subsistence to cash economy while the timber bonanza is on. Economic expectations were raised, especially among the younger people, which led to a habit of over-exploitation of available opportunities and resources, or to frustration if opportunities fell short of expectations. The locals became aware that urban businessmen, officials and the government are the prime beneficiaries of the timber boom, while the villagers only enjoy a petty share for which they have to work hard and risk their health. Eventually they lose out. Consequently, the villagers also wanted timber licences of their own which they could sell to contractors, who then have to cut as fast and as rigorously as possible to get their money back, repay bank loans and make a profit. The long-term net result was that the community became less self-reliant, social coherence weakened, and out-migration became more attractive. The community became more dependent on officials and politicians, whose actions are often inappropriate and self-serving. In conclusion, the authors identified three states of change of village life in relation to conventional selective logging:

1. *Before logging*: relatively stable, mildly cyclical, traditional subsistence economy.

2. *During logging*: local economic boom during and immediately after logging; the villagers' share is temporary and relatively small.

3. *After logging*: dislocated subsistence economy, depleted natural resource base, reduced options for forest-based village trade and industry, trend to emigrate and settle permanently outside, dependence of the more vigorous and usually younger villagers on the outside economy for jobs and quality of life. Lasting degradation of the community life, loss of social coherence and persistent economic decline. None of the initial benefits survived in the two villages to initiate sustained economic and social development.

Deculturalization is a worldwide process which has fully engulfed the tropics (Levi-Strauss, 1978). Selective logging is spearheading this process in the rainforest biome. Sabah, the Philippines, Thailand, China, Papua New Guinea and a few West African states (e.g. Côte d'Ivoire, Nigeria) are examples of countries where the socially and socio-economically disastrous effects of forest resource depletion are apparent. The bonanza has left the economy in these countries with a legacy of high environmental and social costs which have to be borne largely by the rural areas, which enjoyed only a small fraction of the benefits during the bonanza. The profits are gone and the high costs and difficult problems of restoration have to be borne by natives and funding agencies who did not benefit. The young and coming generation has to pay for the greed of its predecessors. The conclusion is that the current system of selective logging is socially harmful and requires fundamental changes of harvesting, management and infrastructure, and the upgrading of moral attitudes and professional performance of concession owners, managers, technicians, labourers and government. Such changes are slow and tedious, and have to progress against many obstacles arising from vested interests and prevailing preferences for the comfortable status quo ante. Sustainable change also needs consensus among all strata of society, ethnic factions and interest groups involved. Many examples show that, as long as individual, tribal and ethnic selfishness, reckless exploitation of political and social privileges, greed, ignorance, power gamble, corruption and foreign pressure groups dominate the social, economic and political scene, the degradation of environmental, land and human resources will be difficult to stop. Examples among others are given by Barnett (1989) for Papua New Guinea and by the non-governmental organization APARAI for Rondonia in Amazonia (see Sections 2.3 and 11.1). In Papua New Guinea, the government eventually succeeded in introducing new guidelines to control timber exports and to curb rapacious plundering of the forests 'only after a bitter and bruising battle with timber companies, dominated by foreign (largely Malaysian) timber industry groups which say they are defending the rights of investors and (native) landowners.' (*New Straits Times*, 24 May 1994, p. 30).

2.6 Timber Production, Trade and Demands

Exploitation of tropical rainforests for fuel-wood, fibres, bark, wax, resin, oil, medicines, dyes and other essential or luxury items for domestic use and export has a history as long as human civilization. Tropical timbers found their way from equatorial Africa and southern Asia to Egypt and Europe many thousands of years ago (Gonggryp, 1942). Since time immemorial, Chinese traders bartered delicacies, pottery, beads and medicines for Belian (*Eusideroxylon* spp., the Bornean ironwood), incense, ebony, rhinoceros horn, hornbill beaks, bezoar stones, birds' nests, spices and camphor. From the sixteenth to the nineteenth centuries, the rainforests supplied Europe, first with small amounts of highly valued cabinet woods, later with a wide variety of timbers for naval stores, buildings, furniture, car fittings and heavy marine construction, such as harbour piling for which Bongossi (*Lophira* spp.) and Greenheart (*Chlorocardium rodiei* (Schomb.) R., R. W., syn. *Ocotea rodiaei*) were used. Initially, Mahogany for high-class furniture in Europe was the dark and beautifully ornamental wood from forks and big branches of *Swietenia mahagoni* (L.) Jacq. The large trees were felled, but only the ornamental pieces in the crown region were utilized. The trunk was left behind to rot (Eggers, 1890). Local use of Mahogany was limited to naval stores and tobacco boxes. The rising mass consumption in Europe from the middle of the nineteenth century introduced the logwood of traditional *S. mahagoni*, added *S. macrophylla* King and extended uses to interior fittings and mass-produced furniture. Later, wood from other genera of the *Meliaceae* (*Khaya* spp., *Entandophragma* spp.) from Africa was also traded as Mahogany. Since the Second World War the demands shifted from a narrow market of highly specialized luxury timbers to a less discriminating market of commodity timber (Fig. 2.6).

In 1990, the total non-coniferous roundwood production from tropical forests for all uses was assumed to be 1.4×10^9 m^3; 275 million m^3 were reportedly used industrially, 86 million m^3 were exported in round or processed form (IIED, 1993). About half of this comes from tropical rainforests. The actual removals and exports are, in reality, probably much larger. Accurate statistics are not available. However, even if the actual annual cut is twice the reported figure, it will not exceed 1.0% of the above-ground timber biomass of the tropical rainforest ecosystems. Contemporary commercial timber removal and direct consumption are below the potential of the rainforests to produce commercial timber. Assuming that another 200–300 million ha rainforest will be deforested, the surviving 500 million ha will contain about 300–350 million ha productive forest and 150–200 million ha should remain non-productive forest and TPA. The sustained yield potential with a 60–80-year felling cycle under the selection silviculture management system (SMS) will be about 1×10^9 m^3, plus thinnings. If current conventional selective logging continues, at least in Malesia, the supply potential would be reduced to

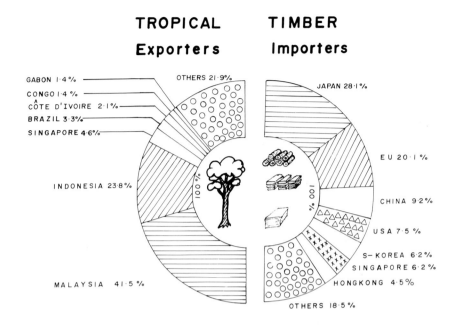

Fig. 2.6. Shares in the export and import of tropical timber and wood-based products. The EU is the main high-price-paying importer of processed timber and wood-based products. Japan, Hong Kong, Taiwan, China and Korea are low-price markets for mainly roundwood of high grade, but China and Korea also of medium to low quality logs of mixed species. The share of the EU has been declining since 1988 as a result of the anti-tropical forestry and timber campaign, which caused consumption of tropical timber to decline while the total timber consumption in Germany and EU increased. (Source: ITW, Berlin.)

half or one-third of the sustainable potential under SMS (Section 6.3). Global consumption and regional demands for timber in the tropics are likely to continue to increase (Grainger, 1985, 1986, 1989; World Bank, 1991a,b). To meet these future demands, and to obtain the greatest possible social benefits in the producing tropical countries, it is crucial to upgrade conventional selective logging to sustainable harvesting immediately and to stop further depletion of the forest resource (Section 2.7). The shift in the international tropical timber trade from underpriced logs towards processed timber and high-value timber products will continue. Imports into the European Union (EU) since the Second World War have been consistently shifting from logs to processed products, including high-quality furniture and especially window frames, doors and flooring (International Trade Centre UNCTAD/GATT, 1990; ITTO, 1993c; ITW, 1994b–d). The shift would have been faster if product quality and recovery rates in the processing industry in the tropics had been better and if the anti-tropical timber campaign had not intervened (Chapter 10). As a result of this campaign the volume of Meranti sawnwood sold in Germany dropped from about 400,000 m^3 in 1988 to about 110,000 m^3 in

1994. Window frames and other high-grade tropical timber products will be in great demand for reconstruction in the former communist countries in Europe. This high-price market niche needs attention and care in order to keep it secure and open by good market-orientated sales policies (Chapter 10). The future of tropical timbers in the Japanese market is more difficult to forecast than in the case of the EU. Many other sources compete against tropical timber in the sawn lumber sector. Competition is tough from North American hardwood lumber and softwood veneer. Resources in the USA for both types of timber are very large and production sustainable at very high levels. High-quality plywood, veneer, building materials and furniture are most likely the best and most secure options for tropical timber in the Japanese market (John V. Ward Associates, 1990) and elsewhere, as soon as the anti-tropical timber bias of the buyers can be overcome.

The low-price market sectors of construction timber and timber-based boards could be adequately served with lower-grade timber from improved complete-trunk utilization, thinnings for timber stand improvement (TSI) and milling residue, now wasted. Opening additional resources by expansion of the list of commercial species in species-rich rainforests to include what used to be called secondary species, then lesser-known species and now lesser-used species (LUS), has been advocated especially by the Food and Agriculture Organization (FAO) since the 1950s, but the wisdom has been questioned. Utilization of LUS may 'release primary high-grade species more and more for higher grade requirements' (Rule, 1947) which would particularly apply to cases such as Japan and other East Asian buyers who buy high-quality timber at low prices for low-quality uses. In Africa and Amazonia cutting intensities are low. Addition of LUS would help to open the canopy more strongly, and stimulate regeneration and growth. In South-East Asia, expansion of the list of commercial species should be discouraged because the present selective logging system already opens the canopy too much, at least in patches (Sections 2.4, 2.7 and 6.4). Felled trees, not the forests, should be utilized more thoroughly. Recovery rates are badly in need of substantial improvement. The timber must be better graded at source to be sold in the most lucrative markets. At present, high-grade tropical timber is sold in low-price markets as commodity timber in competition with cheaply produced temperate and subtropical plantation and natural-forest hardwoods and softwoods. In all markets, at home and internationally, tropical timber has to compete with substitutes, such as cement, aluminium, steel and synthetics, which are sold at low prices that do not reflect the social cost of consuming non-renewable energy and raw material (Table 2.1) and of damaging the environment during production, use and final disposal of these materials.

The recorded export of industrial roundwood and sawnwood from tropical forest countries accounts for only 11% of the total recorded production of roundwood and sawnwood from the tropical forests. The

Table 2.1. Energy used in kilowatt (kW) equivalents, and metric tonnes CO_2 emitted in the production of A, 1 tonne of construction material, B, 50 m^3 window frame and C, general construction timber in buildings. In the case of tropical timber, untreated meranti is used as an example that needs no preservative application but other durable tropical timbers, such as teak or afzelia, would show the same relations. The carbon fixed in the wood is not subtracted, otherwise the balance would be still more favourable for wood and could even be negative.

Material	Energy used (kW)	CO_2 emitted (t)
A. 1 t construction material		
Tropical timber	300	0.2
CVP	7500	1+
Bricks	450–700	0.1
Cement	2000–3000	0.3
Concrete mix	300–750	0.1
Iron	3000	2.5
Glass	6000	?
Aluminium	15,000	25
Synthetics	8000–20,000	variable
B. 50 m^2 window frame		
Tropical timber	500	0.3
CVP	12,500	2+
Aluminium	40,000	65
C. Construction timber in buildings	100–150	0.1

Adapted from sources: ITW, Berlin, 24 February and 29 March 1994; Frühwald and Wegener (1994).

percentage is higher for wood-based panel (59%), wood pulp (23%) and paper and paperboard (13%). The proportion of export is large for some of the rainforest countries, such as Malaysia (75%), Indonesia (60%), Congo (62%), Côte d'Ivoire (57%), Gabon (78%), Ghana (49%) and Liberia (64%) (Varangis *et al.*, 1993). In these countries, the international timber trade is strong enough to influence forestry development, favourably or harmfully. Trade policies of exporting and importing countries should consider the needs of sustainability and socio-economic development in the tropical forest country. These needs are best served by a policy of change from log export to high-quality products for sale in high-price markets both abroad and at home. In practice, trade policies in the producer countries are often indifferent and in some major importing countries simply designed for procurement of raw material at low prices. Low value production of the PFE and the forest-based industry means low social benefits to the nation and the local population. Inevitably, this will strengthen those who, for their private gain, wish to see the tropical rainforest regarded as a wasted asset. This opens the way to convert forest land to other, immediately more profitable uses, but also provides the continued opportunity to supply the illegal tropical timber trade cheaply with unrecorded or volume-adjusted timber (Callister, 1992), partly from conversion land for which there is no incentive to place it under sustained-yield management.

2.7 Rainforest Abuse or Use: Exploitation or Integrated Harvesting?

Until the Second World War, commercial timber exploitation in the rainforest was predominantly by creaming the forest for speciality timbers of high value (Section 2.6). Single trees or, at most, small patches of trees of one or two species were felled and extracted manually or with the aid of draught animals along rough tracks. Occasionally, mechanized extraction, usually by trolleys on light railway tracks or by ex-army lorries fitted with winches, was introduced in flat to low hilly terrain. Damage was patchy and insignificant, regeneration was no problem. Locally, overuse (Fig. 2.7) and such peculiar traditional native practices as *baring* (rolling logs cross-wise) in Peatswamp forests (Fig. 2.8), caused more extensive damage to the forest. Highly selective creaming of large trees by natives (Fig. 2.9) in the upland forest and extracting by *kuda-kuda* generally had little effect on the forest structure (Fig. 2.10) but was labour intensive (Fig. 2.11). As timber utilization expanded, felling intensity was checked and clear-felling prevented by a minimum girth limit on felling. Clear-felling was first introduced into the rainforest by the Japanese army during the Second World War in Malaya. Surprisingly, it resulted in satisfactory regeneration of Dipterocarps and paved the way for the Malaysian Uniform System (MUS) (Sections 3.5 and 6.4). After the war, mechanization of logging MDF spread rapidly in the South-East Asian region. In the Philippines, non-adapted high-lead and other cable systems were introduced from the USA. Mechanized logging in Peninsular Malaysia started with demobilized army lorries on which winches were mounted. Sabah adapted technology from the Philippines and also introduced light railways in flat terrain. Sarawak experimented with various manual and mechanized techniques and even with elephants after the Second World War, but settled for railway in the Peatswamp forests and crawler tractor in MDF. Indonesia, Kalimantan first, and Papua New Guinea adopted railway or tractor extraction.

When the global timber-rush developed in response to the buoyantly expanding post-war world economy, the tempting markets for tropical timber instigated malpractices by the logging industry and the trade. This in turn provoked national and international reactions from non-governmental organizations (NGOs) and the media. Tropical timber and forestry in the rainforest were ostracized in the minds of an environmentally conscious public. The devastatingly harmful high-yield logging practices in the northern softwood forests since the end

Fig. 2.7. Strongly modified ecotone between lowland MDF (*Dryobalanops becarii – Shorea* spp. – *Dipterocarpus borneensis*) and KF (*Dipterocarpus borneensis – Shorea ovata – Whiteodendron moultonianum*, type 6.1 in Bruenig, 1965a) on medium podzol (escarpment) to sandy podsolic humult ultisol, which has been impoverished by centuries of creaming of timber (boats, houses) and poles (fish traps) by the local fishing communities. A relic Meranti (*Sh. beccariana*) is surrounded by an almost pure stand of *Eugeissona insignis*. The villagers are now stout supporters of the park.

of the nineteenth century (Heske, 1931a,b) and in the timber-rich Malesian rainforest since the Second World War succeeded in perverting the image of timber from a raw material that is environmentally and socially friendly and can be sustainably grown and processed (Table 2.1) into a threat to the environment and humanity. Growing international anxiety spawned campaigns against the use of tropical timber

Fig. 2.8. Unsustainable traditional timber extraction: the 'Baring' method of rolling logs sideways along a very wide log track. Rejang delta, Mixed Peatswamp forest, Sarawak, 1960.

and rainforest management. Both were successfully discredited exactly among those people who should have advocated the use of tropical timber in order to promote sustainable forestry, save the rainforest, protect the atmosphere and support the local people in the tropics. The undeniable cases of reckless plundering of forests and callous infringement of NCR and basic human rights were used to bolster the campaigns. The uniquely socially and environmentally 'friendly' raw

Fig. 2.9. *Shorea albida*, Alan, emergent tree in Kerangas forest on medium podzol transition to humult ultisol, being felled in the native fashion during the 1954/54 inventory. Pueh F.R., now N.P., Sarawak.

material timber acquired the evil image of a threat to the environment and to the life of indigenous people (Section 10.1). All this happened while the available technology and expertise in timber harvesting and silviculture could have kept rainforest management on the road towards sustainability, if properly applied. Feasible techniques of integrating regeneration and harvesting in MDF (Appanah and Weinland, 1992; Section 6.4) and in Mixed Peatswamp forests (Bruenig, 1957; Lee, 1979, 1991; Hadisuparto, 1993; Section 6.7) are available. Harvesting would then

Fig. 2.10. Sustainable traditional timber extraction: 'Kuda-kuda' in Peatswamp, Kerangas and almost flat to undulating MDF. Example: *Agathis borneensis* – Kerangas near Bintulu, Sarawak, 1959.

conform with the criteria and principles of good traditional forestry of sustainability if four conditions are fulfilled. Firstly, harvesting has to accord with traditional norms and standards (ITTO, 1990b, 1992a,b, 1993a; FAO, 1994; ITW, 1994a; Heuveldop, 1994) to maintain the residual forest stand and soil in an appropriately functional state. Secondly, the trees in the great period of growth in diameter classes 40–80 cm must be retained as future crop trees and not be harvested. Thirdly, the

Fig. 2.11. The 'Kuda-kuda' method of dragging the logs on sledges in undulating terrain is forest-friendly but socially unsustainable. Semengoh F.R., Sarawak, 1955.

logged areas must be closed immediately after harvesting and effectively protected against re-entry by the loggers and encroachment by farmers and land speculators. Fourthly, sustainable forest management for timber must be socially acceptable to the local, regional and national communities. In most rainforest countries, it is the constitutional and political responsibility of the forest authorities to ensure that forest use complies with the standards of good traditional forestry (Chapter 10). If this is observed, tropical timber can recover its environmentally and socially positive image at home and abroad.

Poor standards in forestry and in the forest-based industry are also the result of poor motivation. Gartlan (1993) commented on the negative effects of short-sighted concession policies in Cameroon on the motivation of timber companies towards sustainability. Ill-conceived policies to issue concessions for short-term tenure of areas that are too small lead to under-capitalization. The results are substandard technology, insufficient planning, poor construction of badly aligned and poorly drained roads, excessive destruction of soil and forest in harvesting and wastage of timber in the forest and in the factories. As much as 65–75% of the felled tree commonly goes to waste. Under such conditions, even an annual allowable cut or yield of $1-3 \text{ m}^3 \text{ ha}^{-1}$ in the PFE can hardly be called truly sustainable. Conventional selective logging simply prescribes a minimum diameter above which all trees may be felled. In most Malesian forests with high stocking of merchantable timber, the result is excessive biomass reduction and canopy opening in patches (MDF) or over large tracts (Peatswamp forest). Correspondingly heavy is the damage to the residual trees, the soil and the habitat for plants and animals. Productivity, habitat value, environmental functions and values decline and regeneration lacks sustained seed sources. The damage is worsened by the universal habit of cutting well-shaped undersized trees and of premature or illegal re-entry to log previously non-marketable trees of large size and undersized residuals when prices are high. Recovery of such forest will take much more time than the usually prescribed short felling cycles of 25–40 years, and will more likely require 100–120 years. Extreme cases of degradation, such as the Demaratok forest in Sabah, will take even longer. Adherence to the well-known principles of selection felling, orderly extraction and proper utilization would have avoided this dilemma.

'Planning is the most essential function to be performed in logging business' (Conway, 1986). This change from exploitation under a selective logging system to orderly harvesting under a selection silviculture management system (SMS) requires fully professional planning and efficient implementation of harvesting (Vanclay, 1994), silviculture, and social functions. A key factor is the establishment of a permanent well-trained workforce at all manual and managerial work levels. Also indispensable is a comprehensive sensitivity analysis of the possible impacts of forestry on the local communities, and a long-term socio-economic development plan to avoid the bonanza phenomenon discussed in Section 2.5. The currently gross deficiencies in technical planning, skilled manpower and law enforcement contribute to the prevailing destructive nature of conventional selective logging operations in the tropical rainforests. Lack of understanding and motivation and indifference to the social dimension among the leaders add to the problem. Ingrained habits, distrust of change, preference for the accustomed, sometimes profitable status quo ante among all social sectors, and over-optimism and opportunistic wishful thinking among political leaders are further obstacles to improvement. The question is not one of choice between two apparently equally admissible options of forest use, traditional selection felling within SMS or conventional selective logging, but between sustainable traditional but modern forestry or intergenerational and social breach of trust.

2.8 Low-impact Harvesting Systems

2.8.1 Conventional practice of logging versus professional integrated harvesting

A conscientious choice among the many tree harvesting systems, state-of-the-art professional operation and compliance with traditional codes of forestry conduct (e.g. FAO, 1994) could avoid all the harm for which the prevailing logging practices are rightly blamed, and would create larger net stumpage returns. Timber harvesting is the one operation that has the most profound and lasting silvicultural effect on forest structure and ecosystem functioning. The felling of trees in the A- and B-layers of the canopy is the most effective way to manipulate the distribution of light and moisture in a stand of trees. Opening of the A- and B-layers shifts the net primary productivity (NPP) and PEP laterally to the residuals in the A-, B- and C-layers and downward to the regeneration in the D-layer. The intensity and kind of harvesting, therefore, must be fully compatible with the objectives of silvicultural stand management. It is a principle of selection forest management and related systems that harvesting is done in such a manner that the forest is self-regenerating (Fig. 2.12). In these systems manipulation of the forest is restricted ideally to harvesting and timber stand improvement in the upper C- and the B-layers. Harvesting ideally under optimal conditions equals silviculture. To approach this ideal, the residual stand must include the fast growers in the 40–80 cm diameter range and damage to the residual trees and the soil must be kept to the absolutely unavoidable minimum. Conventional selective logging does not do this (Fig. 2.13). Earth-bound extraction of logs requires access by road and trails for human beings and machinery to the place where the tree is felled. In the practice of conventional selective logging, roads and skidtrails and criss-crossing crawler tractors (Yeo, 1987) loosen, move and compact soil on 20–60% of the area. Hybrid ground–air long-distance cable systems require opening of cable tracks and helicopter systems require even larger canopy openings and some access by road, but avoid the damage to soil and forest caused by crawler tractors.

Harvesting can be carried out basically in two diametrically opposed ways (see Appendix 1), either by selection felling of single marked trees, or by the total removal of tree crop above a certain diameter limit either as selective logging or as a combination of selective logging and poison-girdling. Blanket removal of all merchantable trees of commercial timber species with a high minimum girth or diameter limit (selective logging) may approach selection felling, but with a low girth limit it may amount to clear-felling. Re-entry in short intervals amounts, in final consequence, to clear-felling. Examples in Sabah (e.g. Demaratok) show that successive premature re-entries inevitably

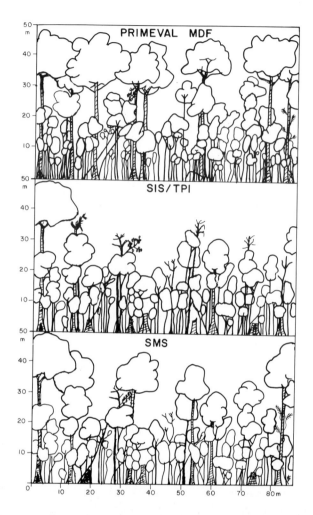

Fig. 2.12. Selection silviculture. Top: untouched timber-rich typical pristine (primeval, primary, virgin) lowland Mixed Dipterocarp forest, mature phase. Centre: the same forest after planned and controlled first selection felling under a Selection Silviculture Management System (SMS) with a uniform girth or diameter limit. This forest is not secondary forest, but natural forest modified by harvesting and manipulated by silvicultural treatment (Appendix 1). SIS, Philippine Selection System; TPI, Indonesian Selection System. The cut in the example is very heavy as a result of a low diameter limit for felling. Bottom: the same forest after harvesting only marked trees of commercial species for directional felling according to SMS prescriptions. Stippled: regeneration of dipterocarp and other commercial species. The tall, moribund giant tree at the left has been retained as a genetically superior seed source and habitat for epiphytes and animals.

remove smaller trees and eventually degrade the forest to a tangle of lianas and climbing bamboo with scattered tall relic trees, small high-forest trees and patchy pioneers (Fig. 2.14), rather similar to hurricane and cyclone forests in Queensland and the Caribbean. Selective logging

Fig. 2.13. Conventional selective overlogging and several times re-logging with very low diameter limit; also Malaysian Uniform System with poison-girdling. In the same Mixed Dipterocarp forest as in Fig. 2.12 the former causes depletion of the forest resource more severely than the latter. A, preferred species; B, desirable; C, acceptable; U, undesirable; stippled, regeneration.

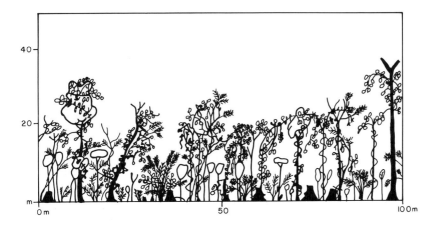

Fig. 2.14. The final climber-tangle stage of resource decline through overlogging after five to seven cycles with short (3–5 years) intervals. Since 1955 a highly productive, species- and timber-rich Mixed Dipterocarp forest has been converted during 35 years of overlogging to a scrub of pioneers, derelict trees, climbing bamboo and lianas, but still containing some regeneration of Mixed Dipterocarp species. Natural recovery of growing stock will take at least one or two centuries, much longer for species richness and biodiversity. Example: Demaratok, Sabah, restoration experimental area of the Malaysian–German Sustainable Forest Management Project, Malaysia.

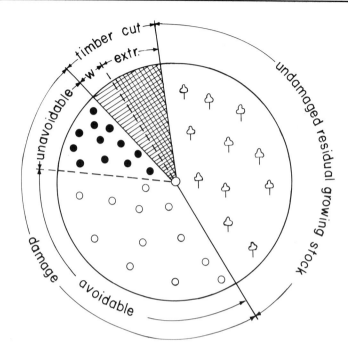

Fig. 2.15. Exploitation by selective logging in an average Lowland Mixed Dipterocarp forest. The disc (100%) represents the biomass above ground of commercial and potentially commercial trees species, including the growing stock between harvest diameter limits and 10 cm diameter at breast height. The sectors illustrate the excessive waste (w) of potentially merchantable timber and the excessive damage to the residual growing stock of actually and potentially commercial species. The causes are lack of motivation and expertise, reckless logging by unskilled personnel, poor or non-existent management, employment of apparently cheap and certainly unskilled, in Malaysia often illegal, foreign staff and labour, and lack of enforcement of sustainability. This kind of exploitation is rampant throughout the tropics. W, wasted timber (breakage, abandoned logs, bucking mistakes); extr., extracted logs. (From Bruenig, 1991, *Holz.Z.Bl.* 32, 518–19.)

with a low diameter or girth limit followed by a uniform poison-girdling of all non-desirable trees above a low diameter of 5–10 cm practically amounts to clear-felling. Naturally, damage to the residual tree growing stock increases with the amount of basal area cut. At the common cutting rate in MDF of 10–20% biovolume or biomass (Fig. 2.15), proper selection felling with extraction by tractor damages 20–30% basal area of the residual stand. The damage rapidly increases with intensity of cut, at 50% removal reaching about 70–80%. Beyond this, the falling crowns and skidding smash practically the whole residual stand. Conventional selective logging as currently practised causes 60–70% damage with only 10–20% basal area removal. However, while the socio-economic effects of waste of timber, reduction of productivity and disturbance, exposure and erosion of the soil are extensive and serious, the forest ecosystem as such will still be viable and recover and restore its functionality and diversity, but this may require very long periods, in the order of centuries. In the storm-thrown forests of Kelantan, Malaysia, the growing stock biomass had recovered in patches 66 years after the collapse, but about half of the area had still stagnated in an early successional state of low secondary woodland with poorly shaped trees. The tree species in the recovered patches seemed to be the same as before the catastrophe, but the proportions differed markedly from the original MDF (Browne, 1949). Generally, observations of clear-felled or storm-thrown, but not burned, rainforest on a variety of sites and soils (Browne, 1949; Heuveldop and Neumann, 1980; Riswan *et al.*, 1985; Lamb, 1990; Section 4.5) have shown that a single complete felling or windthrow of rainforests on ultisols/acrisols and humid podzols without disturbing the soil surface does not lastingly destroy the forest as a viable ecosystem. Regeneration of more or less the same species as the original forest can develop rapidly from the seed bank, coppices and invaders but, as the Kelantan storm forests in parts show, may take a very long time. Species dominance diversity, however, changes usually in favour of the more light-demanding fast-growing species. The drastic reduction of the complexity and value of habitat and growing stock make clear-felling an ecologically and economically undesirable option for sustainable management. Production shifts from competitive high-value big timber to small-sized commodity grades that are not competitive on the world market and fetch low prices at home. At the same time it destroys essential features of habitat and of the organizational complexity and self-sustainability of the tropical rainforest.

According to Kemp (1992)

> Economic and market forces have imposed primitive systems of management, centred on the harvesting of the principal timber species, with the regeneration of the next crop and the survival of the 'old-growth' or late succession species left to chance. However the renewed concern for the broad range of biodiversity in the forests, and for the long-term sustainability of the forest ecosystems, presents new opportunities to develop and apply systems of management based on ecological principles, to reconcile both production and conservation objectives.

Cash-flow orientated economics of management and demand-orientated market forces certainly are the backstage driving forces. The primitive systems of management, such as exploitative and careless selective logging with unskilled personnel and excessively heavy crawler tractors, are driven by more intractable and sinister factors in the rear. The decisive forces lie in the social and political sectors which favour fast cash-flow producing primitive logging to supply easy markets for unqualified raw timber. Within the harvesting sector, replacement of selective logging by selection felling is not only technically possible but also economically feasible. State-of-the art, traditional, well-planned and skillfully operated harvesting, more recently termed reduced impact logging, is the most promising and immediately effective strategy within forestry towards sustainability. It is cost-neutral, costs of planning and supervising being balanced by savings in operations and higher out-turns. The total direct internal and external costs of comprehensive selection silviculture management, reconciling sustainable timber production, harvesting, multiple-use forestry and conservation of habitats, species and ecosystems, are lower than the properly accounted total costs of current conventional selective logging.

2.8.2 Low-impact harvesting: earth-bound systems

Orderly harvesting in accord with the traditional principles of sustainable forestry, so-called reduced impact logging (RIL), has a long tradition in tropical and temperate forests. The essential condition is to integrate harvesting with management planning, execution, monitoring and control. The essential technical features of RIL are: pre-felling survey and mapping of topography, site and growing stock, technical planning of access and extraction, including roading and drainage specifications, pre-felling climber cutting, directional felling towards planned skidtrails and multiple impact zones but away from streams (Fig. 6.5), low stumps, efficient utilization of the felled trunks, minimized width of road and skidtrails, proper winching with arch, fairlead or pan, no criss-crossing by tractors, slash management to reduce fire hazards and water pollution, adequate safety and working conditions and general compliance with plans, rules and standards (Fig. 2.16). These are age-old traditional principles of orderly forestry. They have, more recently, been advocated also in connection with the currently fashionable carbon-offset campaign (Putz and Pinard, 1993). As doubtful and controversial as the value of forestry measures for the atmospheric carbon balance may be (Sections 2.9 and 2.11), the adoption of the principles of orderly harvesting makes social, economical, environmental and ecological good sense. The figures of Putz and Pinard (1993) on the effects of RIL on phytomass and costs are therefore highly relevant (Table 2.2). They estimated in their example a carbon store of 348 t ha^{-1} before logging. This agrees reasonably well with other estimates (Section 1.13; UNESCO, 1978). The felling and extracting of 80 m^3 ha^{-1} logs (approximately 22 t C and 61.1 t dry-weight phytomass) in conventional non-RIL logging damaged 50% of the residual trees and 40% of the ground surface, estimates based on the data of Nicholson (1958, 1979). Putz and Pinard assume death of 50% of the damaged residual trees and an annual increment in phytomass of 3.3 t ha^{-1} after conventional logging. In RIL all large lianas are cut before directional felling of marked trees. Consequently the damage is so much reduced

Fig. 2.16. Frequent forest inspections in the field jointly by senior (taking the picture) and junior foresters, the licencee, his operations executive and the local logging manager are indispensable for introducing and maintaining sustainable forestry. Rejang delta, Mixed Peatswamp forest, Sarawak, 1960.

that the phytomass increment is 4.8 t ha^{-1} or 45% higher. Regeneration on tractor trails amounts to 1.7 t in conventional logging and 0.8 t biomass production in RIL. The tendency is clear, RIL leaves in the order of 77 t more phytomass in the ecosystem and produces 5.6 t annual plant growth, which is 12% more than the 5.0 t in the conventionally logged area. In terms of PEP of timber, the relative advantage of RIL is considerably greater, at least 20–30%, because fewer crop trees in the

Table 2.2. Estimated carbon stocks in Mixed Dipterocarp forest before logging and after logging by conventional selective logging and by selection felling and orderly extraction (RIL). C, carbon; PM, phytomass. All values in t ha^{-1}.

| Component | Pre-logging | | Post-logging | | | |
| | | | Conventional | | RIL | |
	C	PM	C	PM	C	PM
Above-ground biomass (live)	200	450	83	210	128	280
Root biomass (live)	50	110	23	50	32	70
Soil organic matter	70	160	63	140	67	150
Litter (including roots)	28	60	52	120	40	90
Total carbon/biomass	348	780	231	513	267	590
Difference of PM						
to pre-logging				−267		−190
to conventional logging						+77

Source: Putz and Pinard (1993).

40–60 cm diameter class are destroyed or damaged so that more larger timber of high quality is produced in the next felling cycle. The relative advantage will not be greater on steep slopes, where tractor drivers keep to fewer tracks and refrain from criss-crossing. The 50% damage to the residual stand estimated by Putz and Pinard (1993) agrees well with 43% enumerated on 7 ha MDF by Phillips (1993) in Sabah. On average, 168 m^3 residual growing stock was damaged or killed to harvest 54 m^3 ha^{-1} logs. An enumeration of damage among 536 trees in the residual stand in lower-hill MDF in Sarawak showed 27% broken, fallen or dead trees. Re-entering the logged forest to harvest left-behind merchantable trees when the market becomes more buoyant causes even heavier damage (Korsgaard, 1985). The amount of basal area harvested (6.8 m^2 ha^{-1} in this case) equalled the basal area destroyed (6.9 m^2 ha^{-1}). If RIL reduces the damage by 50%, which is feasible, the economic advantages are obvious. RIL also includes full recovery of merchantable timber from the felled trees. Current selective logging practice causes 20–50% of merchantable timber to go to waste by poor standards of felling, bucking and grading, carelessness, poor management and lack of supervision in the felling areas (Fig. 2.15). A post-logging survey of 7 ha of selectively logged MDF and Kerangas forest (KF) in Sabal Forest Reserve, RP 146, measured 15.2 m^3 ha^{-1} merchantable log timber of good quality and sizes left behind. The amount extracted was calculated as 31 m^3 ha^{-1}. Accordingly, 33% of the merchantable felled trunks were wasted. This waste could have been avoided by RIL and either more timber could have been extracted, or fewer trees could have been felled for the same yield.

In conventional selective logging practice, not only the skidtrails but also the roads are notoriously badly aligned, drainage is poor and very ineffective, and the road trace clearing is excessively wide, between 50 and 100 m. Consequently, rates of erosion and runoff of muddy water are extreme and far in excess of unavoidable increases. In addition, destruction of growing stock and regeneration decrease productivity directly. The success of improvements depends primarily on creating skill and motivation among loggers and managers. 'Frequently … the tractor driver moves aimlessly around with his tractor …' (Mattson Marn and Jonkers, 1981). 'Tractor drivers bulldoze their way from tree to tree without planning and without considering the best extraction routes, causing particularly bad criss-crossing in easier country' (Yeo, 1987). In steeper, more difficult terrain, skidtrails tend to be fewer and winching distances longer, but skidding tracks are cut deeper into the soil and erosion is very heavy. A practical crawler-tractor logging experiment comparing improved harvesting on 57.2 ha and conventional logging practice on 34.0 ha in Sarawak showed that directional felling with 74% success rate (26% of the trees went off course, but this could be improved if labour training and liana cutting are introduced), marking of trees to be retained, and proper design and construction of skidtrails can reduce damage by one-third (Liew and Ong, 1986).

Replacement of the conventional practices of current selective logging by orderly and proper harvesting (RIL) reduces the logging costs per cubic metre extracted timber by 20–30% (see Section 2.8.4) and the damage to the residual stand and the soil by at least 25–30% of the area. This requires proper advance planning and skilful application of the appropriate technology. Crawler tractors should only be used on flat to moderately sloping (15° or 27%) ground, and should only use preplanned and demarcated skidtrails and winch with proper technology from preplanned points. That this is absolutely necessary and technically and economically feasible has been convincingly demonstrated by the Queensland Department of Forestry (1983).

2.8.3 *Hybrid and fully airborne systems*

On slopes of 15–25°, semi-mobile long-distance cable systems which carry logs through the air can be operated with advantage. Skyline logging has been tried in Sarawak in Alan bunga Peatswamp forest, PC3, in the 1950s, but results were discouraging. Stability of the spar trees was a constant problem. A technically very simple, home-made version of cable yarding is still in use by one licencee in Brunei Darussalam to log Alan bunga. The system works well technically and financially, but the problem is regeneration. It creates, with little damage, a strip pattern of logged and unlogged forest. On two visits, in 1958 and in 1993, I could not detect Alan regeneration in the logged strips. The former understorey developed into an impregnable, 6–10 m high thicket after logging, resembling conditions in the *ulat bulu* areas. In MDF, skyline and high-lead yarding have resulted in very heavy damage to the forest and soil in Malaysia and the Philippines. The Malaysian–German and Philippine–German Sustainable Integrated Rainforest Management projects have tried and adapted a skyline system that had been developed for alpine forestry in Europe. This semi-mobile long-distance heavy-duty cable system transports logs of 60–120 cm diameter off the ground along 3–5 m wide corridors (3–4% of the harvested area). The bigger logs have to be cut shorter than the present market is used to, but this is only a problem of adaptation to downstream processing. The advantages are minimum damage to soil and forest; the disadvantages are presently higher costs than conventional and RIL tractor logging (approximately DM33 or RM60 m^{-3}) and the need to employ well-trained, skilled labour and managerial staff. With skilled staff and established routine operations, the extra costs will disappear. The system is particularly promising for timber stand improvement (TSI) thinning and for the final harvest felling in the second cycle. Technical advances and training of crews will make its application in primeval forests feasible, especially on slopes. The system is proposed to replace tractor logging in management class 2 (b), slope 16–25° (Kleine and Heuveldop, 1993). In management class 3, slope gradient > 25°, earth-bound logging is unsuitable, hybrid cable systems are difficult to operate, and the areas should become TPA. Under certain conditions fully airborne extracting devices may be a feasible option.

Airships have been tried experimentally in West Africa but are not easy to operate especially in dissected mountainous terrain, turbulent atmosphere and bad weather. More promising are helicopters, which have been routinely used for logging in difficult terrain in combination with other harvesting systems in the European Alps, North America and Siberia since the Second World War. Major open questions in rainforest concern the effects of tree felling and large canopy opening on steep slopes with highly erodible soil, costs, timber wastage, environmental effects and the implications for conservation. The size of log is limited by the carrying capacity of the helicopter. There is bound to be heavy waste by weight-determined bucking. Helicopter logging can be useful on slopes between 25 and 35°, but encroachment onto even steeper slopes must be prohibited as it causes excessive erosion (Chai, 1993; Butt *et al.*, 1985). Felling on very steep slopes can cause ground surface disturbance, which may initiate gully formation. Spotting from the air and the need for obstacle-free lifting make it necessary to widen the felling gap by removing trees in the B- and C-layer. In a trial by a concessionaire in Sabal Forest Reserve, Sarawak, areas of more than 1 ha each were almost clear-felled to lift the logs of two trees (field visit by the author on 19 May 1995). The access roads to the log landings will most likely also pass through steep country with all the disadvantages connected with road building in terrain with long and steep slopes. Direct costs may run very high if the expertise of the crew and the quality of organization are inadequate and also if prices for petrol increase. Felling on very steep slopes causes shake and breakage of wood. The high costs of extraction by helicopter exclude defective and damaged timber from lifting, which increases timber waste. The very high fuel consumption and carbon oxide emission would weigh against general application. Logging by helicopter may carry logging into areas that otherwise would be maintained as TPA for ecosystem and biodiversity protection. Generally, control of operations is difficult and the system is only admissible if the principles of sustainability are firmly established.

Apart from these concerns, timber extraction by helicopter appears to be a feasible option for logging in difficult terrain that has the following distinctive features which are relevant to sustainability:

- Felling a single, large tree (50–70 m top height in the A-layer) with a large crown (20–30 m diameter and 20–30 m crown length) produces only a 0.05–0.1 ha gap in the canopy and on the ground (Figs 6.4 and 6.5), which is barely enough to stimulate tree growth in the B- and C-layers (Woell, 1989) and regeneration in the D-layer. Widening of the gap (see above) improves gap effects, but unskilled untrained personnel can cause even more damage to the canopy and residual stand than in earth-bound logging (see Sabal example of 19 May 1995).
- Flexibility with respect to yield levels per hectare (2–3 trees ha^{-1} (15–35 m^3) felled and lifted are still economically feasible if operations are efficient), to locations of the felled trees, and to wetness of ground conditions.
- Only few trees meet the high-quality standards and are felled, but often less than half the timber of the felled trees is lifted, incurring high socio-economic costs; a field study in Sarawak showed that about 10% of felled commercial trees and around one-third of cross-cut merchantable logs tend to be left behind on the ground, while noticeably defective but still partly merchantable large A-layer trees are left standing, equalling the number of trees felled (Chua, 1993), which confirms the essential need for close and effective control of management and operations.
- Soil disturbance is much less than with extraction by crawler tractor and the road network density may be cut by more than half.
- Reduced damage to residual trees by vertical lifting of logs without the lateral turning that is unavoidable in tractor ground skidding; field observations in Sarawak indicate that extraction-related damage to residual trees is drastically reduced, by up to 90%.
- The time between felling and delivery at road side is very short, so logs fetch a higher price for freshness and superior quality.

A month-long performance study in Sarawak indicated that

efficient timber extraction by helicopter is 35–100% more expensive than conventional logging by tractor (Chua, 1993), and possibly 50–125% more costly than efficient RIL. The higher private cost must be weighed against the private and public benefits which arise from reduced road construction and maintenance, fewer machine operators (but who are expensive and mostly expatriate staff), less damage to forest growing stock, soil, water bodies and long-term productivity. The forest habitat and fauna are less disturbed. With an average daily out-turn of 600–700 m^3 from 20–30 ha, logging passes on quickly and the time of disturbance is very much shorter than in the case of logging by tractor. Silvicultural measures may be needed to counteract the decline in quality (phenotypic and genetic), partly by stimulating growth of selected superior élite trees by stand improvement procedures. A-layer Dipterocarp species in the B/C- and D-layers may require more drastic canopy opening than heli-logging provides. One way to achieve this would be by including lower quality trees but this would require improvement of the technology to reduce costs, and very effective operational control.

2.8.4 *Logging costs*

It is well-established knowledge from experience in temperate and tropical forestry that well-planned and executed timber harvesting costs less than haphazard and unskilled logging. However, this fact has been frequently questioned by uninformed outsiders in discussions on sustainability. There is ample evidence that the badly planned and organized conventional selective logging by crawler tractor in the current fashion of roading and logging is more expensive with respect to direct and indirect costs than proper, sustainable selection harvesting and can be almost as expensive as helicopter logging. The reasons are that work performance is extremely poor and inefficient, machine wear and road maintenance are excessively costly, many of the internal costs are not even accounted for, and externalities are ignored. Lack of skills and poor management result in excessive soil movement and machine time used in road-making and log extraction. Poor standards of drainage, road and skidtrail construction waste time and labour, and cause excessive wear of machinery. Technically poor and badly maintained equipment and unskilled labour increase machine wear, operation time and timber waste in felling, extraction and transport. The rate of work-related accidents is extremely high, but hardly appears as a cost factor. Many of these points also applied in the past to the conventional practice of native *kuda-kuda* logging in Peatswamp forest (Bruenig, 1965c). Proper and orderly sustainable harvesting was intro- duced in the 1950s against initial opposition by most concessionnaires. The results were reduced costs, improved working conditions and safety and increased out-turn per felled tree, improving overall private and social profitability. Korsgaard (1985, p. 15) concluded from results of field work in Sarawak, 'there is no reason to tolerate harvesting operations that are wasteful or damaging'. He cited the practical experiment by Mattson Marn and Jonkers (1981) in Sarawak which proved that the following reductions of costs, waste and damage were possible by improvement of the conventional logging to the standard of RIL: skidding costs −25%, skidtrail area −22%, canopy opening −44%, loss of commercial residuals −33%, abandoned merchantable timber −48%, cost per cubic metre extracted timber −26% and working time per cubic metre −26.5%. Chua (1986) in an experimental trial in Sarawak also obtained a 30% reduction of damage and lower harvesting costs in RIL above 18.6 m³ ha⁻¹ out-turn. Putz and Pinard estimated that the introduction of a comprehensive RIL system causes the following additional costs, based on an out-turn of $80 \, \text{m}^3 \, \text{ha}^{-1}$ (expressed in 1993 US\$ ha⁻¹): \$50 for surveying and mapping of forest and of future crop trees, \$10 for planning and marking of skidtrails, \$35 for directional felling, \$40 for drainage of roads and skidtrails. The total additional cost would then be \$135 ha⁻¹ or \$1.70 m⁻³. These are not really 'additional' costs, but costs that have to be borne for orderly resource use (Putz and Pinard, 1993) and which are being ignored in conventional selective logging. These costs have to be weighed against the savings in extraction costs, increase in timber recovery and gains in productivity. Recent experimental trials in Africa and Amazonia corroborate the Malaysian experience, that sustainable harvesting is more profitable and cost-efficient than the conventional selective logging (Anon., 1995; Grammel, 1995; S. Schardt, ITW Berlin, personal communication 2 June 1995). Heli-logging, under otherwise equal conditions, is operationally 20–50% more expensive than the current crawler tractor logging and > 50% more expensive than RIL. Weighing against this are the silvicultural, ecological and environmental benefits and higher log prices, but socio-economic costs and revenue losses accrue for timber left unextracted.

2.8.5 *Logging and hydrology*

Hydrological catchment research in temperate forests has proved that traditional and adapted methods of thinning and opening of forest canopies have only a weak and ephemeral effect on the water balance. Also, in tropical rainforest catchments, the effects of logging are now well enough understood for compatible and efficient logging systems to be designed so that unnecessary damage to environment, climate and water bodies is avoided (Hamilton, 1983; Cassels *et al.*, 1985; Bruijnzeel, 1992). Stream water draining untouched tropical rainforest carries naturally substantial amounts of dissolved mineral and organic matter. High-intensity rains that saturate the soil cause overland flow and, on slopes, water pressure and flow in the soil and soil slumping, especially if the relief energy and rainfall intensities are high. As a result, large amounts of eroded soil are carried as particulate suspended and bed load sediment (Hamilton and Taylor, 1991; see also Chua, 1993; Lai, 1993; Nik and Yusop, 1986 for examples from Malaysia). Water infiltrability of undisturbed forest soil and of the soil in manual skidtrails are practically equal. Differences, if any, are not statistically significant. This is different on tractor trails. The weight of the tractors causes long-lasting and deep-reaching compaction (increases in soil bulk density) and reduced infiltrability and sorptivity, particularly in clayey soils (Malmer and Grip, 1992; Droste, 1995). This increases storm flow intensity, soil erosion, sediment load of water runoff and silting of water bodies, and lasting reduction of soil rootability and tree growth rates. Overland flow, erosion and river siltation are decreased in relation to the reduction of skid-trails and soil disturbance in road-making. Natural flow rates, and consequently the effects of logging, are extremely variable, depending on soil, topography, relief energy, rainfall intensity, vegetation structure and quality of logging. In Sabah, commercial selective logging in hilly MDF increased the suspended sediment yield fourfold after the logging road had been built across the head of the monitored catchment, fivefold after logging had proceeded to 37 m from the road edge, and 18-fold during the five months in which the remainder of the catchment area had been logged. A year after completion of logging, the largest monthly sediment yields were still 3.6-fold those of the undisturbed control catchment. The reduction indicates initiation of recovery, but the watercourses and narrow flood plains remained silted and logging trails continued to supply sediments to the drainage system (Douglas *et al.*, 1992). The time needed for the sediment load, after very cautious and careful RIL, to decline to pre-logging levels has been studied intensively in Queensland (Queensland Department of Forestry, 1983). The output from the forest stands is reduced rapidly and pre-logging levels are reached within 2 or 3 years, but roadside erosion continues unless drainage facilities are adequate and properly maintained. In constrast, the very heavy sediment load from conventional selective logging in steep hilly country in Central Sarawak is expected to decline markedly from a logged area only after 3–5 years but will remain high for the whole management unit. Large amounts of rotting slash, abandoned logs and dying trees release polyphenols (tannin), humic compounds and minerals (including iron), directly and through soil leaching, into the watercourses over a much longer period, possibly reaching a peak only after 10 years. The water quality of the discharge from a single water catchment area or from a whole management unit or regional working circle will never return to pre-logging levels (S. Anderl in Forest Department Sarawak, 1995). In addition to the damage done to the water bodies, the discharge of dissolved and suspended organic matter, nutrients, trace elements and soil is a loss to the ecosystems which affects productivity and sustainability.

2.8.6 Logging and nutrients

Research and experience in temperate forests show that removal of the relatively nutrient-poor log timber is only critical to the nutrient balance of the forest in whole-tree harvesting and short-rotation plantations. The proportion of timber actually extracted in RIL in SMS in tropical rainforest is in the order of 5–10% of the above-ground biomass. The nutrient losses are accordingly in the order of 3–8% of above-ground nutrient stocks. The amount removed can be restored by biological fixation and wet and dry atmospheric inputs within a felling cycle of 40–80 years, except perhaps for phosphorus, depending on soil conditions. The knowledge of nutrient budgets in tropical forests has been reviewed by Bruijnzeel (1991), using data from 25 sites. He found strong variation of nutrient fluxes in association with features of rooting, soil fertility, parent material and geomorphology. More, well-calibrated and standardized studies are needed for precise and accurate estimates. In the mean time simple comparisons of estimated stocks and removals can be used for tentative assessments of changes due to timber harvesting. A study of selection harvesting in a rainforest in Surinam gave the following relations: $23 \, m^3 \, ha^{-1}$ timber, equal to $17 \, t \, ha^{-1}$ phytomass, were extracted and $30.9 \, t \, ha^{-1}$ were left behind as slash (stemwood and stumps 15.6 t, crowns and leaves 15.3 t); $50 \, kg \, ha^{-1}$ nitrogen and $160 \, kg \, ha^{-1}$ other nutrients (Ca, K, Mg, P) were removed with the extracted logs. The nutrients removed are a tiny proportion of the nutrient content in the living phytomass, which are $2 \, t \, ha^{-1}$ nitrogen (removed 2.5%) and $5 \, t \, ha^{-1}$ other nutrients (3.2% removed). Replenishment of N from biological fixation and from precipitation would take less than 20 years (Jonkers, 1987), but longer for the other nutrients, depending on the intensity of inputs from various sources. Harvesting intensities of $40–60 \, m^3$ every 25 years as visualized in Malaysia would certainly exceed the rates of replenishment of any of the nutrients. Consequently, felling cycles have to be longer or artificial fertilization has to compensate for the losses.

The preservation and recapture of nutrients in the rooting sphere, and also P immobilization, in newly formed gaps is apparently very effective. The dryweight of live and dead pine roots and the rates of decay were assessed prior to, immediately after and several months after localized (experimental gap formation) and landscape-level (hurricane) disturbance in a subtropical wet forest in Puerto Rico (Silver and Vogt, 1993). The small-scale disturbance had little effect on the fine roots, which were only slightly diminished in 0.1-ha gaps. Decay of fine roots was slow, which reduced nutrient losses. The large-scale disturbance first caused an increase, but after 1–2 months fine roots declined as a result of accelerating root mortality. Experimental clear-felling without wood extraction of 0.5-ha plots in San Carlos de Rio Negro showed that nutrient losses in soil and stream water were very slight and ephemeral (Jordan *et al.*, 1980) probably due to rapidly developing regrowth and uninterrupted effective nutrient scavenging by the roots and microorganisms. The tentative conclusions are that the gaps formed and the log timber extracted in selection felling (RIL) do not impoverish the nutrient stock, provided safe minimum standards (0.1–0.3 ha gap area ha^{-1}, 2–5 trees felled, 30–60 m^3 extracted, > 40-year felling cycle) are maintained. According to Proctor (1992)

> It is important to realize that the analysis of felled trees will not give a guide to limiting nutrients and the trees will certainly have more nutrients in them than they actually require. Hence studies of nutrient removal by logging activities will tend to overestimate the impact of logging on forest production. Ideally, such studies should always be combined with an assessment of the actual nutrient requirements of trees but this is a difficult task.

The tendency to accumulate nutrients possibly beyond the actual needs of current phytomass growth appears to be contributing to the unexpectedly high nutrient levels in Peatswamp forest and KF and the small differences, if any, of nutrient stocks in the phytomass per ton along the site gradients of declining soil nutrient richness. It may be an important element of risk reduction in the tropical forest ecosystem to stockpile more species, more phytomass and more nutrients than are actually needed under 'normal' conditions.

2.8.7 *Logging and fauna*

Research into the effects of logging on fauna in recent years has consistently produced evidence that, after logging, certain specialized understorey bird and amphibian species may be prone to local extinction, but over larger tracts all species are likely to survive or are able to recolonize. Management must leave suitable areas unlogged as refuges in which survival, and from which recolonization, is possible. The area of the whole felling series will comprise untouched, freshly logged, young regrowth and mature forest areas and TPAs, which should provide adequate range and choice of habitat. Species which play a key role in seed dispersal, such as hornbills and imperial pigeons, can be accorded special protection by retaining habitats with nesting opportunities such as large hollow trees. Overmature primeval forest patches should be established as nature reserves, be demarcated and excluded from logging and protected as refuge, habitat and seed source. Stream embankments should be protected from interference by logging and roading for the sake of river-bank and habitat protection, and also for purposes of biological water purification. With these precautions the danger of even local extinction of birds and mammals is very small, possibly nil. The major problems concern the aquatic life in streams and rivers that are affected by increased turbidity, pollution and siltation, and the soil biology in areas affected directly by crawler tractors and excessive exposure. Both kinds of damage can be reduced by appropriate planning and engineering.

2.8.8 *Recovery of growing stock after logging*

Selective logging with blanket application of a minimum diameter limit as low as 50 or 60 cm in MDF, or even less in Peatswamp forest, removes immature trees which are in the 'great period of growth' with the highest current rates of basal area, volume and value increment. Basal area and volume increments of A-layer species growing uninhibited by competitors in gaps > 0.1 ha in primeval natural forest generally culminate in the diameter range 40/50–70/80 cm at a probable age of 70–100 years (Fig. 6.3), if the crown is well illuminated and intact. The culmination of basal area and volume growth in this diameter range is well known from Meranti and other A-layer Dipterocarps, and is described for three African species by Alder (1992) in an example of calculating sustainable yield. These trees should be preserved and not cut as in selective logging with diameter limits of 50 or 60 cm. Worse in practice, if a tree below the prescribed diameter limit is exceptionally well-shaped, straight and sound, loggers will rarely resist the temptation to cut it. A cursory look at any log landing will confirm this. The results of these practices are:

- The fastest-growing trees with the highest volume and value increment in the 'great period of growth' between 40/50 and 70/80 cm diameter are cut prematurely in ecologically excessive and economically unnecessary, heavy selective logging.
- The most superior élite trees for the next cycle are often cut illicitly below the prescribed minimum diameter limit, which worsens prospects of high PEP in the future.
- The growing-stock volume and value increments are consequently depressed below the natural level of NPP and PEP because the growing stock is depleted of the fastest growing élite tree population.
- The residual trees of lower average quality and lesser stature and growth rate do not compensate for the loss; the fastest growing, most well-shaped (trunk and crown) tree in a random tree population is generally genetically superior and retains its status throughout life; if it is cut and the inferior trees are left, the mean growth rate of the population will drop (Bruenig, 1971c, 1984b, 1986b).
- The prematurely cut élite trees will be replaced by the smaller trees in the C- and D-layers; these layers are naturally poorly stocked with

A-layer trees and require some time to adjust to exposure from drastic canopy opening and more time to reach their 'great period of growth'; consequently, felling cycles must be extended by 20–30 years or more (Bruenig, 1974).

- The residual growing stock resulting from overlogging is, therefore, of poor phenotypic and possibly poor genetic quality in the B-layer, and is far from the 'great period of fastest growth' in the C/D-layer; genetic upgrading from progeny of any left large, overmature giant trees will be effective after 100 years at the earliest.
- Consequently, future PEP declines severely to as low as 50% and less of the naturally possible volume and value increment.
- Excessive and poor roading and skidding, heavy soil erosion and compaction, and damage to the immature residual trees depresses NPP and PEP still further to between 25 and 50%.
- Consequently felling cycles of 25–50 years are not sustainable and 60–100 years are more realistic if the goal is to produce valuable high-grade timber of 80–120 cm diameter; on poor soils even longer production periods are required (Woell, 1989).

The first cycle of selective logging maximizes quick and easy cash flow and profit for the concessionnaire. The unnecessary private costs are the excessive costs of poor-quality logging. The social costs are the waste of merchantable timber, a drastically reduced PEP in volume and value, damage to water bodies and loss of amenity in the residual forest and in the landscape. The forest is downgraded by selective logging instead of upgraded by SMS. The process of ecological and economic degradation continues and becomes more acute with each subsequent logging. PEP will be lost completely when log sizes have declined below 30 cm diameter, which makes natural forest management uneconomic and non-competitive with plantations.

In Sarawak, the Ramin-bearing Mixed Peatswamp forest PC1 was placed under a sustainable harvesting and management system, offering easier and safer working conditions, in the 1950s. However, the blanket application of a diameter-limit system proved to be a mistake in the very patchy Ramin forest. The following example shows the problem. A 5.84-ha sample plot with 357 $m^3 ha^{-1}$ tree volume > 20 cm diameter in an ecotone of phasic communities PC1/PC2 was logged under the standard diameter-limit restriction in 1992. Ramin stocking and accordingly the felling intensity (28.4 trees ha^{-1} representing 42.4% of the basal area) were very high. This caused a canopy opening of 60–70%, which could be expected to be lethal for Ramin regeneration (Bruenig and Sander, 1983). In addition, 59.6% of the residual growing stock was severely damaged, 17.8% slightly damaged and only 8.2% undamaged (ITTO, 1990d). Almost 200 $m^3 ha^{-1}$ logging slash was left behind. Utilization of this slash would be unwise for ecological reasons even if industrial processing were possible (Eisemann, 1991). The recovery of the Peatswamp forest ecosystem after such heavy, but not unusual, disturbance will require more than a century for nutrients alone. The restoration of biomass and canopy structure will require considerably more than the 40 years prescribed as the felling cycle in the management plans. The average mean annual increment of biomass (Sections 1.13 and 6.7) in PC1 indicates 100 years as a more likely time horizon. The physiological and ecological effects of the drastic canopy opening and sudden exposure of the C- and D-layers, the high rate of logging damage to the residual growing stock and the naturally relatively low NPP potential of phasic communities PC1 and PC2 make recovery not only slow, but also a highly uncertain process.

The ecosystem processes of the repair of the damage and the restoration of the growing stock after natural destruction or heavy logging proceed with different speeds in the various compartments of the ecosystem. The fastest, within a range of several years, is the restoration of leaf area and mass, followed by NPP, litter production and soil protection. Zoomass recovery by recolonization, reproduction and regrowth is probably equally fast. The restoration of nutrients, tree phytomass and canopy structure after selective logging takes many decades and may require a century in MDF. Simple model calculations of growth and crown expansion show that the restoration of a reasonably structured canopy after the felling of seven trees, which caused 50% canopy opening, would take about 50 years on an above-average site (assuming a mean annual increment of B-layer trees of 1 cm diameter), longer on less productive sites. Clear-felling of the KF and Peatswamp forest may cause damage and change the soil to such a degree that recovery takes many centuries. On all sites and in all forest types, tree-species biodiversity in terms of species richness and dominance diversity takes the longest time to reach pre-destruction levels. This may be several decades under favourable circumstances or several centuries after very heavy impacts on very fragile, biologically sluggish soils (e.g. the Padang vegetation in Bako National Park, and Bana in San Carlos de Rio Negro; Bruenig, 1961a, 1965b; Bruenig *et al.*, 1979).

An example of a relatively speedy recovery on a very favourable site is the following report from a study of apparently habitually overlogged (removal of probably 50% of the biomass) MDF on the foothill of Gunung Lemaku, Sipitang, Sabah (annual rainfall 4000 mm). Ten years after logging leaf dry weight was 7 t ha^{-1} and leaf area index 6.7. These values are in the range of the average for untouched MDF. The biomass of living trees was only 251 t ha^{-1} dry-weight above ground (148 trunks > 20 cm diameter). Stumps accounted for 37 t ha^{-1} dry-weight. The original above-ground tree biomass was probably between 500 and 600 t (plus root biomass of 142 t). At least 250 t would then have been extracted or gone to waste 10 years ago, having been largely decomposed in the mean time. It would take at least another 30–40 years to restore the biomass stocking to the former level, or 40–50 years from the date of selection felling (Phillips, 1993). The results of a pilot study in the MDF of Peninsular Malaysia by Korsgaard show clearly that logging in recent years has reduced the growing stock and impaired BNP and PEP more than the obviously less destructive harvesting before 1980. It is also evident that few logged areas approach the stocking level of the virgin forest, and that there is a high variation in the stocking with increasing time since logging. Only a few of the areas harvested 20–40 years ago are reaching maturity for a second harvest. This observation is supported by the results of the third National Forest Inventory (Korsgaard, 1993a). The results of the properly executed SMS in Queensland show that these consequences could have been foreseen and definitely could have been avoided even with the benefit of larger net stumpage returns. This demonstrates clearly the advantages and practicability of RIL and indicates the feasibility of sustainable forest management and conservation in tropical rainforests (Queensland Department of Forestry, 1983; Vanclay, 1994).

2.8.9 *Timber harvesting and non-timber forest products*

Reduced impact harvesting of timber, including directional selection felling of marked trees and careful extraction by tractor (flat to gentle slopes), cable systems (moderate slopes) or helicopter (steep slopes), has little effect on the production of NTFP. The climbing rattan-producing palm species and sago-producing palm species benefit from moderate (felling gaps of 0.1 ha covering 20–30% of the canopy surface area) to very heavy (0.3–0.5 ha, 60–80%) canopy opening, while very light opening has not much effect (< 0.1 ha, 10–20%). Honey-bees are hardly affected because their favourite host trees, such as the huge A-layer *Koompassia excelsa* (Becc.) Taubert and *K. malaccensis* Maingay ex. Benth, are rarely harvested or damaged by felling of usually smaller neighbouring trees. Even shifting cultivators shy away from the extremely hard wood of the giant trunks and prefer to harvest honey. Fruit trees, such as engkabang (*Shorea* spp.), durian (*Durio* spp.) and *Artocarpus* spp., and ecological keystone species, such as strangler figs, are completely protected by law in Malaysia and elsewhere and must be neither felled nor damaged. Latex-producing species, such as *Dyera* spp. and *Palaquium* spp., as well as several resin- and incense-producing species, were in former times locally protected under management plans, but have lost their economic value except as timber producers. Proper selection felling, as part of a sustainable management system, is likely to have no, or positive, effects on NTFP. Their declining importance as an economic product in natural forests is due to exhaustion or the changes in the social and economic environment (Sections 2.2, 2.3 and 4.6), which favour production in planted forests and plantations (Sections 7.4 and 8.4).

2.9 *Tropical Rainforest and Global Climate*

The scale by which tropical rainforests, as a source or sink, could potentially influence the global climate is determined by the area of the tropical rainforests in proportion to the surface of the globe and the intensity or size of the state or flow variables of the forest as source or sink. The area (A_{rf}) as a percentage of the surface area of the globe (A_g) is approximately:

global surface (land and ocean) (A_g)	100%
tropical zone (land and ocean)	40%
tropical land surface	12%
tropical forest area	4%
tropical rainforest area (A_{rf})	2%

If C_{rf} is the mean concentration (e.g. watt cm^{-2} or $W\,cm^{-2}$) of a parameter in the tropical rainforests and C_g the global mean, the effectiveness (E_{rf}) of the influence of the tropical rainforest areas may then be defined as:

$$E_{rf} = \frac{A_{rf}}{(A_g - A_{rf})} \times \frac{C_{rf}}{C_g}$$

The effectiveness is then (Baumgartner and Bruenig, 1978):

whole tropical zone	$0.67 \times (C_t/C_g)$
tropical land surface	$0.14 \times (C_t/C_g)$
tropical forest area	$0.04 \times (C_{tf}/C_g)$
tropical rainforest area	$0.02 \times (C_{rf}/C_g)$

The scale of the influence of tropical forests on the radiation and energy budgets, the water balance and the chemical composition of the atmosphere of the globe may be calculated on this basis. The concentration ratios C_{rf}/C_g are > 1 for radiative emission, water vapour, carbon content and aerodynamic surface roughness, but < 1 for albedo and for some anthropogenic pollutants. The reason is that infrared emission, saturation vapour pressure and plant productivity and production are relatively high in the tropical rainforest, but albedo and the emission of pollutants and of some climate-effective gaseous and particulate substances are somewhat smaller in tropical rainforests than in other land cover types or outside the tropics.

The tropical zone covers 40% of the earth's surface but injects 58% of the water vapour into the global water cycle; of this 49% is from the oceans and 9% from the tropical land surfaces. Tropical forests occupy about one-third of tropical land surfaces, of which less than half are rainforests. Assuming a double rate of evapotranspiration for tropical rainforests, which is on the high side, their maximum influence on the global water cycle is about 2–3%. The effects of deforestation, reforestation, afforestation and timber harvesting on the global water balance must be assessed in relation to this figure. Deforestation will be followed by some form of regrowth which emits vapour again. Timber harvesting reduces evapotranspiration immediately, roughly at the relative scale of reduction of leafage, but increases transpiration by increasing aerodynamic roughness caused by gap formation. This, and leaf-flush of the residual canopy, increased growth in the understorey and regeneration on the ground, offset this reduction very quickly. Even in the impossible scenario of total instant destruction of all tropical rainforests and complete prevention of regrowth, the decrease of annual global evapotranspiration would be less than 2–3%, or less than the range of data noise due to measuring errors and natural variation. The same applies to the effects of deforestation on the global change of albedo (proportion of light reflected) from 13% over rainforest to 19% over grassland. The effects on the content of carbon dioxide, carbon monoxide and other trace gases of the atmosphere would be more substantial, particularly in the case of air-cleansing and photochemically active compounds. Even an unimaginable holocaust of all tropical rainforests, preventing any regrowth, assuming that 40–50% of the carbon dioxide is, at current rates, taken up in the oceans, would raise the atmospheric carbon content by a mere 10–15%. This is a tiny amount in relation to what is achieved by the burning of fossil fuels for the reduction of ore to metal, and for cement production, traffic, power generation and air-conditioning. Commercial timber harvested in the rainforests from PFE and from conversion land, and timber and wood collected for non-commercial uses, may perhaps amount to 800–1000 million m^3 annually. In addition, encroachment of deforestation in the rainforests may add another 2000 million m^3 woody matter that are not immediately balanced by concurrent regrowth. Assuming that all the fuelwood is burned and that 70% of the extracted timber waste is burned or rots, and 90% of the slash in the conversion and logging areas turns into carbon dioxide and only 10% to SOM (Fig. 1.12), 2500 million m^3 (1740 million t) would emit about 700–800 million t carbon annually into the atmosphere. This is a net discharge from forestry and deforestation in rainforest, which will continue until no more forest is converted to agriculture, timber increments equal yield in the PFE, all nature reserves and other TPA are constituted and safe, and planted forests have reached an equilibrium between growth, yield and decay.

Deforestation in all zones of the tropics for husbandry (mainly in America) and agriculture, with small contributions from urban growth and infrastructure, causes at present a net release of 1.6 ± 1.0 billion t C (Deutscher Bundestag, 1992, 1994a,b). The sources are burning and decomposition of above-ground biomass (11.7–18.0 kg m^{-2} total living biomass carbon in tropical forests) and release from the large pool of SOM, which in ferralsols is about 14.5 kg m^{-2}, in humic acrisols 17.8 kg m^{-2} and in podzols 14.2 kg m^{-2} (Sombroek *et al.*, 1993). To

compensate for this amount of release, about 700–800 million t C by
carbon-fixing from tree plantations would require immediate afforestation of about 125 million ha fertile land, which is unrealistic (Section
2.11).

2.10 Environmental Change and Forestry

Heske (unpublished lecture notes of 1931–1940, Chair of World Forestry,
University of Hamburg) postulated that modification, especially deforestation, of large tracts of rainforest, such as the Congo forests, would
have lasting climatic effects. It would interrupt the self-sustaining and
closed convectional water cycle of the rainforest–atmosphere ecosystem
and reduce annual rainfall on the site. As a result, the tropical rainforest would gradually disappear and never recover due to lack of
rainfall. The declining evapotranspiration from the rainforest would
weaken the circulatory regional water vapour–rainfall system and
subhumid, semi-arid and arid regions in the lee would receive less
water vapour and rainfall (Fig. 1.20). Since then, the focus of concern
has turned to the possibly more critical effects of global and inter-
regional changes of the atmosphere on the rainforests (Bruenig, 1991a).
Projections of future global climatic states are intrinsically uncertain
(Darmstadter *et al.*, 1994), but the general global trends are fairly clear
(IPCC, 1992) while regional resolution of predictive modelling is poor.
Assessments of the need and of how to pre-adapt forests at regional
and local scales are, therefore, essentially speculative. A precautionary
low-risk approach suggests strategies of pre-adapting the managed
forests by ecological and economical strategies that are feasible even
if the climate does not change, or if it changes in a different direction.
Generally, environmental changes are likely to have less effect on flora
and fauna that have undergone similar stresses in the not-so-distant
past (for an in-depth discussion on the complex subject of extinction
and survival see Lawton and May, 1995). We demonstrated earlier that
the tropical rainforest has survived considerable climatic fluctuations,
which permit us to consider the primeval forest as being pre-adapted
within this range.

In this situation of great uncertainty it would be most impracticable
to wait for predictions with certain probabilities. Action would then
be delayed until it is too late for effective reaction. The present most
likely changes during the next century in the equatorial belt,
particularly in the South-East Asia–Pacific region, are an increase of
temperature in the low atmosphere by about 2°C, continued rise of the
surface temperature of the sea, and consequently a higher energy
content of the atmosphere, which increases the chances and severities
of extreme and violent climatic events. Considering the history of
climate and vegetation during the Pleistocene and the fact that the

progress of global and national programmes to protect the atmosphere will remain slow, except in a few more enlightened countries, the most likely changes relevant for a precautionary strategy of sustainable and adaptive forestry in the rainforest biome are:

- increased moisture stress, alternating supersaturation or extremely intensive drying-out of soils and vegetation, with associated heavy loads of solar radiation and heat, more severe episodic droughts of long duration;
- warmer air and soil (Sombroek, 1990) temperatures, risk of loss of humus and nutrients and increase of pests and diseases, but possibly no substantial change of NPP, even with increased CO_2 levels;
- more energetic and violent convective, frontal and cyclonal air circulation with very powerful lightning, more severe gales and micro-bursts, more frequent intensive downpours;
- possible increase in ultraviolet radiation and concomitant diverse photochemical activities which may lower GPP rates;
- most likely an increase of atmospheric pollution and of its distribution, with either fertilizing and/or phytotoxic and soil-destabilizing effects.

These possible future changes will impact on fragmented tropical landscapes which are naturally vulnerable to erosion, floods, fire and outbreaks of pests and diseases and have been made even more fragile by human activities. There may also be inter-regional effects. The sporadic climatic 'butterfly' effects are among the better known phenomena of the unsuspected but far-reaching effects of local climatic events. Such effects may be triggered off by large-scale deforestation in such climatically active areas as the Malesian archipelago.

It may be concluded that the plant species of the rainforest are most likely sufficiently pre-adapted to survive the expected vagaries of climate change. The primeval natural forest will also survive the climate changes, possibly with some modification of biodiversity (Solbrig *et al.*, 1992) patterns. More critical could be the effects on the self-regulation and self-sustainability of modified and manipulated natural and planted forests. Precautionary measures must be taken to strengthen the protection of soil and water, increase wind resistance, reduce the susceptibility to drought damage and to bolster the capacities for self-regulation and recovery. Such measures will be discussed in Chapters 6 to 8.

2.11 Carbon Offset by Forestry

At present the notion is popular that forestry has the capability to reduce significantly, by carbon-fixing, the annual net addition of carbon to the world's atmosphere. This environmentally appealing idea has been promoted recently by scientists and is being commercially utilized by some power companies (Blakeney, 1993; Putz and Pinard, 1993) but its feasibility is doubted by others (Brown *et al.*, 1994). Forest preservation and conservation can mitigate carbon emissions caused by forest degradation and deforestation. Forestry can establish new sinks by planting new forests. Forest preservation and conservation is the most straightforward, simplest and most effective way to maintain the carbon sink at a high level, for example by naturalistic selection silviculture management and by establishment of TPA. Establishment of carbon sinks by the afforestation of treeless ground allows carbon to accumulate during the first production period (rotation), which in fast-growing plantation forests in the tropics is 10–20 years. After this, no more net carbon fixation occurs and new planting is necessary. All the biomass of the first generation must be permanently stored and not recycled if the programme is to be fully effective. Comparing the preservation of natural forest and compensatory afforestation strategies, to balance the carbon store of 350 t in 1 ha natural rainforest, an area of at least 35 ha of plantation forests must be established on bare land to fix the same amount in 1 year. In reality, the area annually reported as additionally planted in the tropics is less than 10% of the simultaneously deforested area (FAO, 1993). Tree planting, at most, currently only compensates for 0.3% (1/350) of the carbon released by deforestation and forest degradation in the tropics.

The proclaimed objective of compensatory carbon-offset tree planting, however, is a different one. It is to balance the carbon released by a specific new power plant which burns fossil fuel. This is a laudable objective in principle and within the limited scope of its purpose, but it does not contribute measurably to the reduction of net carbon emission into the atmosphere (Bruenig, 1971b; Heuveldop *et al.*, 1993; Kriebitsch *et al.*, 1993). To be successful afforestation demands fertile land, quality planting stock and skilled human resources. All of these are in high demand for other purposes. Consequently, the long-term opportunity costs of carbon-offset may be high, while the contribution to the protection of the atmosphere is small. Fertile land worldwide, especially in tropical countries, is becoming dangerously scarce as a result of erosion, degradation and population increase (Deutscher Bundestag, 1994b, pp. 167–173). Fertile land in the tropics, except perhaps in the Brazilian Cerrado, is needed for purposes other than timber growing (see Chapter 7) and the kind of timber grown is already in oversupply (Bruenig *et al.*, 1986a). Also not justifiable are schemes which compensate carbon emissions in a developed country by reducing logging damage and waste in logging and timber processing in tropical developing countries. These reductions are anyway an indispensable component of the on-going programmes towards sustainability, which furthermore could be achieved without external funding. More justifiable, effective and appropriate is to reduce the emissions from power generation, industry, households, traffic and land-use (forestry and agriculture) in the developed countries, and to use timber instead of energy-intensive substitutes (Table 2.1; Frühwald, 1994; ITW, 1994b) in all countries.

Sustainable Forestry in Rainforests 3

3.1 *Concept of Sustainable Forestry*

Haber, President of INTECOL states:

> 'Sustainability' seems to have become the foremost ecological catchword of the 20th century's last decade. But what is sustainability? Apparently it is a property natural ecological systems possess, otherwise life would not have persisted, and evolved, for more than three billion years. We can speak of 'self-sustainability' in analogy to self-organization and self-maintenance of natural systems. We consider these remarkable properties, which are not yet too well understood, effective enough to have ensured the survival of large-scale natural global changes of the past earth's history.
>
> (INTECOL Newsletter, vol. 21, no. 3-3, 1991)

A review of the subject and a useful discussion of sustainability from an economist's point of view is given by Toman (1992). The traditional ancient and modern definitions of sustainability in forestry are given in Appendix 1. Sustainability and its viability aspect at various levels of the ecosystem hierarchy concept are discussed by Grossmann (1979) and Grossmann and Watt (1992). The interpretation of sustainability is relative to the categories and dimensions of time, energy, matter, organization and space. This reference must be defined before sustainability can be described. Sustainability is a concept that has developed in the course of cultural evolution. It is not, and cannot be, a 'western' concept because there is no monomorphic 'western' culture and because the concept exists in all human societies from some early stage of settled agriculture onward. The differences are due to different levels of sophistication according to the stage of cultural evolution but not due to genetic differences between races. Its development in forestry happened to occur in Europe. Its present form is relevant and applicable with due adaptation to forestry in non-European forest countries. The major principles are:

1. The ethical principle that forests, as part of creation and natural heritage, are the responsibility of humankind and must be caringly managed and preserved, and its products prudently used with due consideration given to all living components of the forest and the landscape ecosystem because the human species is the only one endowed with the intellectual capability to recognize and bear this responsibility.

2. The fair balance between satisfying human needs and preserving the basic ecological properties of the ecosystem, such as viability, productivity, elasticity, tolerance and resilience (Hartig, 1820; Hunde-shagen, 1828).

3. The view that forests and forestry are integral parts of the national cultural heritage of nature, environment, economy, art and literature (von Carlowitz, 1713).

4. The ethical principle of intergenerational fairness, which imposes the moral constraint that the current manifold local, regional and national needs must be satisfied without compromising future options to choose and enjoy (V. Carlowitz, 1713; Heske, 1931b; Lubchenco *et al.*, 1991), considering that the world is constantly changing (Speidel, 1984).

The concept of sustainability has been the answer to the recognition that natural resources can be limited, overused and destroyed, which was realized even in classical Greek literature. Later milestones include the hypothesis of unifying basic natural principles operating in natural landscapes, especially in forest vegetation (A. v. Humboldt at the end of the eighteenth century), the emergence of ecology as a science (Haeckel in the late nineteenth century), the creation of the ecosystem concept (Tansley and Lindemann in the first half of the twentieth century) and of biocybernetics (Wiener in mid-twentieth century). The recent worldwide upsurge of public concern over resources and the environment added political relevance. The concept and practice of sustainability in forestry has developed with the evolution of the rural landscape (Fig. 2.4). The impacts of the early forest-dwelling gatherers, the forest-consuming shifting agriculturists and of forest-creaming loggers (phase I to II) were weak and the issue of sustainability was not recognized. Growing populations and more effective technologies began to exhaust and destabilize forest resources and the landscape in phase III. In response, the need to manage forest production, protection and amenity functions sustainably is recognized, and forestry is created.

3.2 *The Holistic Nature of Sustainability in Forestry*

Sustainability of ecosystems refers to the totality of components and processes of ecosystems and to its interactions with adjacent ecosystems of lower, equal or higher level. These levels are the natural bio-ecosystem (forests, lakes, rivers, fields, landscapes, regions to biosphere), technical–economic ecosystem (village to region and nation) and socio-political ecosystems (local to international). Therefore, any human interference at one level must be designed to retain sustainability, that is viability, health, functionality and efficiency, of all levels. Accordingly, forestry is concerned with the sustainability of the forest ecosystem with all its diverse individual components and processes, the landscape ecosystem of which the forests are an interacting component, and the cultural ecosystems of which forestry is part at local, regional, national and, finally, at global level. In forestry, the holistic approach implies the comprehensive lateral and vertical integration of local and regional management and of multisectoral development. Bottom-up and top-down strategies are not alternative approaches, but complementary and interdependent. Without such integration and combination, individual projects or management units may reach their targets, but results will only slowly spread to other systems, as efficient linkages are not provided for. Isolated projects will have little impact on general economic and social development.

3.3 *Unpredictability and Uncertainties*

Forest and landscape ecosystems are complex and dynamic. Individual processes may be biocybernetically regulated and reasonably predictable within a known probability distribution. The whole complex ecosystem of forest or landscape is neither biocybernetically regulated nor predictable. The multitude of interacting processes include catastrophic collapse, non-linear cause-and-effect relationships and fuzzy variables and states. This obscures probabilities and causes uncertainties that blur predictions. Finally, also the nature and quantity of impacts by external forces of the environment on the ecosystem are variable and unpredictable. Examples are episodic events of climatic extremes, emissions of volcanic minerals and, more recently, emissions of anthropogenic atmospheric pollutants from agriculture, traffic, industry and households, but also social upheavals and policy-instigated wars. Forestry is strongly affected by the vagaries and flippancies of the social and political environments, creating the largest exogenous element of uncertainty. The result of all this is that even with the most complete knowledge of the natural and cultural ecosystems we shall never be able to make long-term predictions with defined probabilities. If we cannot do this, sustainability cannot be expressed by narrowly quantified criteria and indicators (Section 10.2).

Climate change creates additional uncertainties (Sections 2.9 and 2.10), which are succinctly reviewed by Trefil (1991). He concludes that 'deciding what to do about the greenhouse effect is like deciding how much insurance to buy'. In contrast to decisions on insurance, in the case of forest ecosystems and Global Climate Change we cannot calculate probabilities and risks. As a result of wide planning horizons, manifold functions and dependence on unpredictable nature and society, both diverse and fickle, forestry has, inherently, to cope with high levels of various kinds of uncertainties. In such a case, a combination of compliance with safe minimum standards, prescribed by reasoned best judgement, and adoption of precautionary low-risk strategies in long-term and day-by-day operations seems to be the most practical, feasible and prudent approach towards sustainability. In forestry practice of medium and long-term planning, the best we can do is to choose the option which, in the fuzzy environment of uncertainty, is most likely to be the least wrong. The precautionary principles of least risk and safe minimum standards are, therefore, the best policy practice to secure sustainability of the tropical rainforest ecosystem. The corresponding strategy is to choose those alternatives that accord with, and make best use of, the basic principles of structure–function relationships in complex dynamic eco-systems. However, the most important, critical and crucial element of sustainability in forestry and in natural resources use and management generally is not the nature of the resource, the knowledge of it, the experience with it, nor the available technology and economic constraints, but the nature of basic human attitudes. The major uncertainties and obstacles to sustainability of tropical rainforest management and con-servation are the unpredictable, fickle and flip-flopping attitudes of the acquisitive human mind, supplemented by the intractabilities of the social and political environment which human nature creates (Sections 2.1–2.5).

3.4 History of Sustainable Forestry in Tropical Rainforests

Sustainable forestry excludes the function of forests as land reserved for eventual conversion to other uses. Forest areas which have been des-ignated for conversion to non-forestry uses relate to sustainability only in so far as the utilization of the forest growing stock before conversion must conform with the principles of sustainable economic development at the larger scale. Timber removal from such land must be synchronized with land and economic development. The timber of the felled trees must not be wastefully, but fully and properly utilized and the liquidated stump-page capital value must be reinvested in the national economy. The following text is concerned only with Permanent Forest Estate (PFE) and not with deforestation and encroachment, which are the concern of general politics and forest policing and not a component of sustainable forestry conservation and management in the strict sense of the terms

(Bruenig, 1989b,c,d; Amelung and Diehl, 1992). Forest management and silviculture in tropical Asia, and in Japan, China and Australia were initiated by European forestry thinking and foresters about the middle of the last century and later spread to tropical Africa and America. Neil (1981, pp. 37–122), on a suggestion by Dawkins, distinguishes five phases of development of tropical rainforest management (phases 1–5 below). I have added a future phase (6) to distinguish the current phase (5) as a period of restoration of traditional principles of sustainability with uncertain outcome, from the vision of holistic sustainability yet to come in phase 6.

3.4.1 Phase 1. Pre-management era before 1850

The originally purely acquisitive utilization of timber and non-timber products by local populations was gradually diversified by intentional tending, enrichment and conservation in forest, village and home gardens to produce for the growing domestic and export markets (Hesmer, 1966, 1970; Whitmore, 1990). Small-scale commercial timber exploitation creamed the forest of speciality timbers, such as incense wood, ornamental wood from forks and gnarled branches, high-class marine timbers and cabinet woods. Harvesting of timber and non-timber products was wasteful and usually destructive. Examples are the felling of the trees to collect fruits, edible eggs and animals, gutta percha, small pieces of ornamental or aromatic wood or to work a few boards or one dug-out from a large trunk.

3.4.2 Phase 2. Indo-Burma/Franco-German era, 1850–1900

Sustainable management procedures were introduced, mainly based on German traditions of multiple-use forestry and its organization. Initial attempts failed to transfer silvicultural technology directly from Germany. Commercial timber logging occurred, but was rare. In Ceylon, overcutting and growing stock depletion were caused by the increasing timber demand for the manufacture of tea boxes. Native forest creaming ('river-scratching') for timber and non-timber products continued to prevail. Forest departments began trials of experimental planting of native and exotic tree species for production and to rehabilitate degraded sites. Colonial forest services in Ceylon (Sri Lanka), Burma (Myanmar) and India introduced formal inventory and working plan procedures in response to an 'orgy of forest destruction' caused by the sudden increase of demand for land and timber by the tea industry, expanding agriculture and the growing timber trade (Trevor and Smythies, 1923). In Burma, teak became threatened by shifting agriculture (shag) and demands for ship building. In response, the German-born botanist at the University of Bonn, Dr Dietrich Brandis, later Sir Dietrich Brandis, was appointed to Burma as forester. In close cooperation with villagers, he established his famous Teak Selection System of yield control and silviculture integrated with *taungya*, the local system of rice cultivation, following the German tradition of *Waldfeldbau*. Later, Brandis was transferred to India to become Director General of Forests (Hesmer, 1975). A comprehensive forest policy was subsequently formulated for India and vigorously implemented, based on comprehensive German-type working plans (Schlich, 1989). Permanent forests were demarcated and gazetted. The first formal sustainable working plans were prepared and implemented between 1860 and 1865 by foresters trained in France and Germany. The quality of sustainable forest management and naturalistic silviculture in India became proverbial and strongly influenced development of natural forest management in Malaya, Thailand and Indonesia, and eventually also in Africa and America. The first planting on a large scale of Teak and Mahogany in Indonesia and Malaya in the second half of the nineteenth century followed Indian examples. The introduction of management systems in natural forests, beginning with the regeneration-orientated shelter wood system, were influenced by developments in India. Troup's (1928) *Silvicultural Systems* and (1921) *The Silviculture of Indian Trees* and Champion's (1936) *Preliminary Survey of Forest Types of India and Burma* were milestones and became classics of their kind. The latter was revised three decades later (Champion and Seth, 1968). In Africa, the focus was naturally on recultivation and French and German influences were strong in developing plantation forestry. Subsequently, the influence of German forestry declined in the wake of the two world wars, and has not recovered since. Germany became the world's largest donor of funds for tropical forestry (phase 4); but even then her intellectual impact remained weak, inconsistent and flimsy.

3.4.3 Phase 3. Malesian–African era, 1900–1960

Practical experience, scientific knowledge and forestry techniques were transferred from India and directly from Europe, mainly Germany, into Thailand, Malaya, Indonesia, and to Japan and China. The Malesian–African era showed the most vigorous adaptation and development of the art and practice of sustainable forestry, especially in silviculture, dendrology, utilization of non-timber products, inventory and management. The success of the introduction of the principles of sustainability in management practice was possible due to strong research support, political backing by governments and technological back-up from regionally or globally orientated academic institutions. Heske (1931a,b, 1937/38) issued early warnings against the emerging large-scale overuse and misuse of forest lands by shag in the tropical zone and by the timber industry in the boreal forest zone. He challenged German forestry and forestry science with the responsibility to set guidelines for implementing sustainability. Heske predicted that sustainability would become the priority problem for world forestry in the twentieth century; he was right! The international market for tropical timber remained tiny and restricted to speciality timbers of high value or durability. By 1960, practical experience with rainforest silviculture and management systems were well advanced towards the target of sustainability. Practised procedures were suitably codified in manuals (Francke, 1941, 1942; Walton *et al.*, 1952; Bruenig, 1957, 1958, 1961b, 1965a; Dawkins, 1958; Baur, 1962; Wyatt-Smith, 1963; Catinot, 1965). Comprehensive and vigorous ecological research was underway in some countries, such as Sabah, Sarawak and Malaya, to support forest management. Development in the rainforest zone in Africa lagged behind India and Malaya. The selective logging, mainly for the European market, was much less intensive and the population pressure in the wet equatorial climatic zone lower. It was as late as 1943 in Nigeria that it was realized that the old conventional selective logging system would, in the course of several felling cycles, lead to a gradual destruction of the forest (Blanford, 1948). Deforestation and forest overuse grew rampant in Central America. Some basic research, vigorous practical experimentation and thorough management planning were done in some areas such as Trinidad, Puerto Rico and Surinam. The Amazon basin remained outside forestry development throughout this period.

3.4.4 Phase 4. Pantropical exploitative era, 1960–1990

An unprecedented economic boom in the USA and Western Europe, followed by Japan, led to a vast expansion of the international pulp and timber trades. Tropical rainforests widened their share in the booming world market, adding commodity timber grades to the traditional speciality timbers of very high quality. Demand pressures and greed-propelled development concepts provoked unsustainable rates and destructive technologies of timber logging. Inequity of terms of trade in forestry kept prices low and quality requirements high, denying tropical countries the incentives needed to retain or regain forestry sustainability and social equality. In the producer countries, socio-economic conflicts, environmental damage and resource depletion accelerated. Ample cash flow from forest growing stock liquidation led to unsustainable spendthrift fiscal policies, apparent but unreal increases in gross national product (GNP) and capital flight. At the same time, the national natural resource accounts declined in forestry. Annual balances turned sharply negative, usually unnoticed by the public and governments. Habitual conflict and war mongering, hot and cold wars, general political disorientation, misdemeanour, scapegoating and scarecrow bashing, economic and fiscal mismanagement distracted governments and the people from the real problems and added to the forestry dilemma. Overuse and destructive abuse of human and natural forest resources precipitated social fiasco and political instability in many countries. International aid organizations, such as the Food and Agriculture Organization (FAO), favoured industrial forestry but were doubtful of natural forestry. The general trend was away from natural tropical forest research and management; tropical rainforest was considered a wasted asset (Bruenig, 1977a). Few countries with rainforests adhered to the principles of sustainable management. On the whole, forestry practice turned into uninhibited resource exploitation. Politics in many countries and in international organizations in the cold-war environment degraded to scrambles for power, prestige, profits and personal gains. Even in blessed and already advanced countries, such as Uganda and Nigeria, disaster struck and political instability resulted in wholesale forest destruction and management collapse (Rukuba, 1992). Another disaster struck forestry in the State of Queensland. The exemplary integrated rainforest forestry had to be abandoned in favour of exclusive nature preservation. These two countries had advanced site-specific management procedures (Dawkins, 1958; Baur, 1962; Queensland Department of Forestry, 1983) which could have continued to give badly missed and urgently needed guidance to others.

3.4.5 Phase 5. Restoration era 1990 to possibly 2000/2020

This period is a transitional phase of rethinking and restoration of the traditional principles of sustainable forestry in the tropical rainforests, and also in the world's forests generally. The activities of non-governmental organizations (NGOs) and UN agencies initiated change, especially the former International Union of Conservation of Nature, now the World Conservation Union (IUCN), the International Tropical Timber Organization (ITTO), and also FAO with the controversial Tropical Forest Action Plan (TFAP of FAO and World Bank). These activities created awareness, recognition and acknowledgement by the tropical and boreal forest countries that the critical state of their forest resource was threatening their national sustainable economic development. The Centre for International Forestry Research (CIFOR) was established in response to these global concerns with the mission 'to provide a global research partnership to enhance and sustain the contribution of forests to human well-being' (CIFOR, 1994). Finally, Heske's prediction became true: sustainability of forestry became the global priority of the century. It was a prominent issue of the United Nations Conference on Environment and Development (UNCED) in 1992 (Grayson, 1995). A global forest convention and certification of sustainability of forest production became priorities on the world agenda for protection of environment, resources and sustainable development. The preference of politicians and administrations for status quo ante, resistance by interest groups in the politico-commercial complex, and endless, esoteric and ignorant bickering over definitions and concepts between NGOs, governments and the trade, delayed progress towards sustainability but could not prevent it. The need for restoration of ethical principles and moral norms of conduct as preconditions of restoring sustainability was finally realized. Eventually, forestry and forest-based industries took the lead by means of a dualistic certification and trademarking scheme, initially focused on the European market (Chapter 10).

3.4.6 Phase 6. Approximating sustainable management and conservation

The achievement of the state of sustainability was targeted by ITTO in 1991 for the year 2000 but, even applying the conventional ITTO concept of sustainability, this is utopian. It is more realistic to set as target not a point, but a course towards sustainability. Implementing current principles of sustainable forestry in all the forests of the world will be a never-ending process which cannot achieve perfection and completion. Also, knowledge of the systems to be managed sustainably will never be perfect and complete. Sustainable forestry must constantly adjust to natural, economic, social, intellectual and environmental changes, without deviating from the underlying universal principles of sustaining system viability. Implementation will continue to be threatened by Global Climate Change, resource depletion, environmental pollution, persistent socio-political evils, socio-economic inequalities and political instabilities. If political and social morals and ethics are not restored soon, the principles of sustainable forestry will not be realized. The habitual preferences of present-day politicians, businessmen and bureaucrats for the profitable and comfortable *status quo ante* would then persist and continue to obstruct sustainable forestry and sustainable economic development worldwide.

3.5 History of Rainforest Silviculture and Management

Contrary to the misconception that colonial forestry services were generally charged with the responsibility to secure raw material supplies to the ruling country (Heske, 1942), priorities in reality were to ensure the supply of local needs. This applies at least in the British Empire, with the exception of naval timber from Burma (Myanmar), Canada and New England. In Malaya, forestry research and forestry practice at the turn of the century concerned chiefly mangrove protection and management to secure vital supplies of NTFP and few special timbers for the local market. The traditional native methods of harvesting and processing of NTFP were improved and new products developed in order to expand the range of products, improve returns and secure sustainability. By the 1930s, knowledge and its implementation in both areas were well advanced. Mangrove management was well established (Watson, 1928). The range of NTFP and methods to obtain them were exhaustively described by Burckill (1935) and little new knowledge or use has been added since. Silviculture and management of the Mixed Dipterocarp forest (MDF), originally adopting European systems, has since the 1930s developed a characteristic Malayan approach first in Malaya (Heske, 1932; Francke, 1941; Walton *et al.*, 1952), later in Sarawak (Lee, 1982). The development of a variety of silvicultural systems and procedures is described in the now classical *Manual of Malayan Silviculture* (Wyatt-Smith, 1963) and by Chin *et al.* (1995). A major change since 1963 has been the on-going large-scale conversion of the PFE, first in the lowlands, more recently also in the hills. The most productive management areas and valuable research plots were lost, and with them most of the examples of successful sustainable forest management. At the same time the forest-based timber industries expanded beyond the supply capacity of the shrinking PFE. The loss of the lowland areas of mainly Meranti–Keruing type of MDF forced the reorientation of silvicultural development in the peninsula to the hill forests. The immediate problem was to find ways to regenerate *Shorea curtisii* Hill Dipterocarp forest. The irregular fruiting with long intervals and failure of the seedlings to establish themselves on an episodically dry site are not unlike conditions in *Shorea albida* Peatswamp forests in Sarawak, in spite of the widely different edaphic conditions. The shift of focus of forestry to the hills has given forest hydrology and watershed protection and management a high rank in research and practice. The concurrent very changeable trends in forest plantation research and practice will be mentioned briefly in Chapter 7.

In Sarawak, similarly, forestry activities began at the end of the nineteenth century with improving the native technology of harvesting various kinds of latex, cutch, oil-seeds, rattan, Belian and sago. Silvicultural experiments and practice began in the early 1920s, mainly with enrichment plantings with local tree species to meet local demands for special timbers. Export-orientated silvicultural research and practice became topical only after the Second World War when sawn timber of Ramin (*Gonystylus bancanus* (Miq.) Kurz), previously a 'weed species', and logs and sawn timber of Jongkong (*Dactylocladus stenostachys* Oliv.) became major regular export commodities, followed by *Shorea albida*, the hard and heavy Alan as sawn timber and the medium light Alan bunga mainly as veneer. Mixed grades of peatswamp logs were dumped on the low-price East Asian market. Consequently, research concentrated initially on the improvement of harvesting and development of silvicultural systems for Mixed Ramin Peatswamp forests. Since the late 1960s, attention has been increasingly given to MDF with experiments to adapt selection silviculture to the conditions in the MDF in Sarawak (see Chapter 6). Plantation research and activities began in the early 1920s and followed the pattern set in Malaya, later Peninsular Malaysia of oscillating between native and exotic species preferences, with occasional 'wonder trees' making an ephemeral debut (see Chapter 7).

A sustainable selection-silviculture management system for Ramin (*G. bancanus*) Mixed Peatswamp forest has been developed and introduced step-by-step since the 1950s. A system of combined area–volume control and volume increment yield regulation on the basis of reliable minimum estimates (RME) (Dawkins, 1958) was introduced as a first step towards sustainability. Formally approved working plans were prescribed for all management units in the peatswamp PFE and felling plans outside. Attempts were made to integrate forestry with agriculture and the general regional development by means of formal Regional Management Plans (RMP), which covered all categories of forests and forest-based industries in a region. The first pilot RMP was prepared for the Balingian Peatswamp forests. The plan provided also for the conservation of representative ecosystems inside the PFE and for the gradual, synchronized harvesting on crownland prior to conversion. The plan gave long-term RME yield forecasts encompassing all management units and the conversion areas as baseline figures for economic development planning. By 1963 all Peatswamp forest areas in PFE and outside in Sarawak were covered by RMP. In 1960–1962 a comprehensive codification of procedures of forest inventories and of management planning and control was initiated in Sarawak after a review of the management procedures in use in the British Forestry Commission and the Commonwealth (Bruenig, 1961b, 1963, 1965a). The Sarawak Forest Manual (Forest Department Sarawak, 1959) prescribed full recognition of native customary rights in PFE establishment and management planning. In any management plan 'Rights, privileges should be cited in detail, and their effect on the plan noted. In addition to legal rights, it may be advisable to draw attention to purely local demands that must be satisfied e.g. fishing stakes from mangrove forest'. The dramatic expansion of logging in MDF in the 1970s and the problems it created, stimulated silvicultural research and practical developments, aided by an FAO project towards sustainable management systems, reinforced since 1993 by ITTO and bilateral cooperative projects.

Communal and village forestry has been traditionally a priority in the declared Sarawak Forest Policy and in forestry practice, but with variable success. Many communal forests were established in the 1930s and in the 1950s. The objective was to supply village needs, such as fuelwood, timber, fruits and other forest products, and at the same time to improve and protect environmental and ecological conditions and amenity in the landscape. A major problem was to sustain the interest of the villagers in managing their forests. The economic advancement of the state eventually made communal forestry less attractive economically. For example, the people of Kampong Bako (Bako Village) preferred to have their 870-ha communal forest, established when the park was constituted in 1957, revoked in 1994 and the area added to the Bako National Park, as the villagers feel that they derive more benefits from the park than from their communal forest.

In Sabah, research into logging damage, regeneration and growth after first cutting in MDF began early in the 1950s. In recent years emphasis has shifted as the primeval forest has been rapidly exhausted. The focus is now on the rehabilitation of the heavily overlogged, up to seven times re-entered and, consequently, badly degraded forests and soils. As in many rainforest countries, the question in Sabah is now not how to use the primeval forest wisely and prudently, but how to repair efficiently, at reasonable cost and low risk, the excessive damage done by overlogging. The solution is being sought in partnership with the German aid agency GTZ, which is developing sustainable management procedures and plans for the whole state, including transformation of the severely degraded Demaratok management unit into a guideline model management area.

The development of forestry in Indonesia probably began with planting of fruit trees, some species of Dipterocarps and Teak by villagers in Java some centuries ago. Governmental concern with forests originally centred on the protection and management of Teak in the seasonal forest areas, adapting the guidelines developed by Brandis in Burma (Myanmar). Extensive plantations of Teak and *Agathis loranthifolia* were established in Java. The rainforests of the outer islands remained outside the sphere of interest until well after the Second World War. As logging in Kalimantan expanded, a modified form of selection silviculture management, so-called Tebang Pileh Indonesia (TPI), was introduced (Directorat Jenderal Kehutanan, 1980). In the Philippines, the sustainable manual logging of the 1920s was replaced by destructive mechanized logging with unsuitable technology imported after the war from the USA. In response, the Forestry Bureau introduced a selection-silviculture management in 1958, the so-called Selection Improvement System (SIS), in Mindanao (Bureau of Forestry, 1965 and 1970; Uebelhör and Abalus 1991). However, the adverse social and political environment made effective application, and the protection of the forest against overlogging and encroachment by agriculture, impossible to this day. Papua New Guinea, the fourth country in the Malesian region with large tracts of rainforest, was literally overrun by logging companies from the region and East Asia in the 1980s and is experiencing at present a most destructively plundering timber bonanza. Particularly intractable problems in Papua New Guinea are the involvement of politicians in the timber business and native land ownership. Consequently, no serious attempt at sustainable management could be made until very recently (Barnett, 1989).

In Africa, forest management in the humid tropical forest zone began later than in Asia and followed a different course as a result of the different natural conditions and the different preferences of the colonial powers. The original deciduous forests had mostly been converted to savannahs in prehistoric times. Many of the more accessible evergreen rainforest on fertile ground had been modified by itinerant slash-and-burn agriculture. The replacement forest is a peculiar mixture of secondary and primeval elements. The emphasis of early colonial forest policies at the turn of the century, consequently, was generally on forest conservation, enrichment planting of impoverished or logged forests and on afforestation of deforested and usually much degraded lands, mostly in the deciduous forest biome. The evergreen rainforest received scarce commercial attention until after the Second World War, when large quantities of a few species of medium light hardwoods began to be exported in great quantities from West Africa as commodity timber to Europe. Overlogging caused local growing stock exhaustion and regeneration of the tall emergent trees was usually insufficient in these strongly successional forests (Francke, 1942), but little was done to remedy the situation. Exceptions were individual timber companies with large and relatively secure concessions which had an interest in maintaining sustainable standards of logging and management, especially in Congo, Zaire and Gabon. The problems of sustainability in African rainforests are very different from those in southeast Asia. The density of merchantable trees of commercial species is very low and the felling of one or two trees per hectare is insufficient to induce growth of the residual stand, which consists mostly of lesser known and hardly used but potentially commercial species. Regeneration of the present commercials is notoriously poor. Development of research and practice towards sustainability is urgently needed but is obstructed by the adversities of the political environment in most countries. Bilateral application-orientated research continued to some extent throughout the postwar period but application failed, again, mainly for political reasons. Very practicable silvicultural and appropriate codified management procedures had been prepared for evergreen forests in Uganda by Dawkins (1958, 1959, 1988), but their application was thwarted by political instabilities. It was not until 1993 that the political situation in Uganda improved sufficiently for the forest department to become effective in the field again, but much damage had been done in the mean time (Rukuba, 1992). The well-known Budongo forest in Uganda, a classic example of sustainable selection-silviculture management since 1930, and of ecological research in the 1940s (Eggeling, 1947), has luckily survived. The forest continues to be an important source of information on the effects of selection logging and silviculture on forest structure and the differential effects of forest management on the different species of wildlife, especially primates.

The dogma of maximizing uniformity for the sake of high volume production, which had been socially adapted and ethically fitting to the conditions in the eighteenth and nineteenth century in Europe, was transferred to West Africa and elsewhere in the tropics, and was tried with such silvicultural systems as the 'Uniformisation par l'Haute,' and the various forms of shelterwood systems. The uniformization concept proved impracticable for natural regeneration systems in equatorial rainforest (Francke, 1942; Baur, 1962; Catinot, 1965; Kio, 1983). Natural rainforest management has been practised with success in some large concessions of German companies in Gabon, Congo and Zaire. Small native licencees in the areas continued the practice of creaming, overlogging and too-short felling cycles (Osho, 1995). It is symptomatic that these concessions, unperturbed by bickering and haggling in international politics and attacks by non-governmental organizations, but with support by governments and the African Timber Organization (ATO), have in May 1995 declared that they will seek certification of sustainability (ITW press release, 16 May 1995) for the European timber market (Chapter 10).

In tropical America, orderly management in natural and planted forests began in Trinidad. The first formal working plan with a 25-ha annual coupe and 60 years rotation under a shelterwood system was implemented in Arena Forest Reserve during the 1930s and revived in the 1950s (Neil, 1981). Almost simultaneously, planned management of restored and natural forest was introduced in the evergreen forests in Mount Luquillo in Puerto Rico (Wadsworth, 1951/1952). Later, a management system to sustain timber yields in relatively poor natural forests was developed in Surinam (Graaf, 1982, 1986, 1991). The current situation in tropical America is reviewed by Figueroa et al. (1987). For the rest of tropical America, the social and political scene never was and still is not favourable for sustained management of forest and other resources. Simple and uncontrolled manual creaming or mechanized heavy selective overlogging of timber, and haphazard and unsustainable exploitation of extractives and other forest products prevail in most tropical American countries to this day. An exception is the Indio community-based integrated forest management and forest industry company Quintana Roo, Mexico (Bruenig and Poker, 1989). This company is also seeking certification mainly with a view to the European market.

3.6 Principles of Silvicultural Management

Almost 100 years of too little scientific research, but much application-orientated research (Eidmann, 1942) and experimentation and practical experience have accumulated some basic scientific knowledge and plenty of practical silvicultural experience. This sufficed to design if not the best, at least low-risk strategies and feasible tactical approaches for specific targets and conditions. The early tendency was to achieve uniformity by rigid silvicultural schedules and shelterwood-type systems, much under the influence of conventional European silviculture. The dogma of maximizing the functions of protection and production, adapted and ethically fitting to eighteenth and nineteenth century Europe, led to trials with silvicultural systems such as the African 'Uniformisation par l'Haute', the various forms of shelterwood sytems and the early Malayan Uniform System, which aimed at a more uniform structure of the growing stock. Trials, errors and failures, and better scientific knowledge and understanding of the natural and economic ecosystems gave tree-species mixture and biodiversity prominence again, first in Europe (Bruenig, 1984b, 1986a) and subsequently in tropical forests. The growing understanding of system dynamics helped to overcome dogmatic opposition to the introduction of the more rational, flexible and adaptable approaches of traditional, naturalistic silviculture.

The failures of the positivistic ideologies prevailing in the nineteenth century, that human beings could manipulate nature and humanity and overcome all natural obstacles, opened the way to more holistically conceived, system-orientated silvicultural management systems that neither force nor copy, but mimic nature. The structural features of the forest may mimic uniform pioneer forests embedded in a successional crop sequence or, at the other extreme, mimic the complexity and diversity of the mature primeval forest. Harvesting and natural regeneration will be fully integrated; ideally, no enrichment or improvement plantings are necessary. Single-tree or group selection silviculture limits itself to regulating mixture and competition by harvesting and a minimum of between-felling-cycle interventions. Success depends on the quality and proper execution of the first harvesting in the primeval forest (see Section 2.7 and Chapter 4). The silvicultural principles of mimicking pioneer forest and successional dynamics will be discussed in Chapters 5 and 6. Ways and principles of mimicking mature primeval forest ecosystems will be described in the following chapter.

4 Principles and Strategies of Sustainability

4.1 Hierarchy of Sustainability

The principle of sustainability applies in different forms to all levels of forestry from the global and national, through regional and local management unit levels to the forest ecosystem (forest stand). The various levels are interdependent and interacting. Sustainability cannot be realized by a forest stand, or concession or ownership unit, or by any nation, in isolation. The following brief review of principles at three levels includes the abstracted gist of the literature and of extensive annotated lists of criteria and indicators which several NGOs, particularly ITW and ITTO, have prepared in recent years. These annotated lists introduce nothing new and accord with the traditional principles, criteria and indicators of sustainability which have been used in forest management and conservation in central Europe over many centuries. During the past 100 years of trial and error they have proved their adaptability to forests and forestry in the tropics and in temperate forests outside Europe.

4.2 Principles at National Level

The chances for sustainability of forestry multiple-function (Fig. 4.1) are determined at national level. Success hinges on a resource policy which integrates forestry equitably with agriculture and economic development. Finally, success depends on the political will and strength to enforce it, and on the response of the general economic and social environment. National forest policy and forest law set the frame for defining the long-term goals and short-term targets for forestry and for allocating land and forests to various uses and ownership categories (Appendix 1). Major principles of integrating and optimizing are:

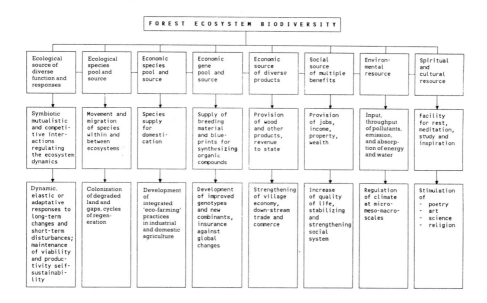

Fig. 4.1. Forests are a manifold source of diverse products, functions and services. Traditional European foresters accordingly developed the principle of sustainable multiple-use forestry very early in response to society's needs, material scarcities and environmental degradation. The top row shows the major source functions of forests, the middle row the multiple use functions provided by these sources, and the bottom row the processes to which the use functions contribute to holistic forestry.

1. Sustainability of forestry must be seen in the context of the whole national economy, the state of the forest resource must be continually monitored by forest inventories and comprehensive natural forest resource accounting, and socio-economic compatibility analysis.

2. Allocation of goals, costs (fees, taxes, etc.) and benefits to forest categories (Appendix 1) must fairly consider and be compatible with the diversity of needs and potentials among the various public and private ownership categories and the multiple interests of society as a whole.

3. Development of the type and capacity of the forest-based industries must accord with the development of sustainable yield of the permanent forest estate (PFE) and the schedule of production from conversion-forest lands.

4. Forest stand, management and policy structure and functions must be tolerant to possible environmental changes.

5. Designs of sustainability of national forestry must recognize the impacts of Global Change, in particular the dependence on the unpredictable international economic trends and the uncertain course of global environmental change.

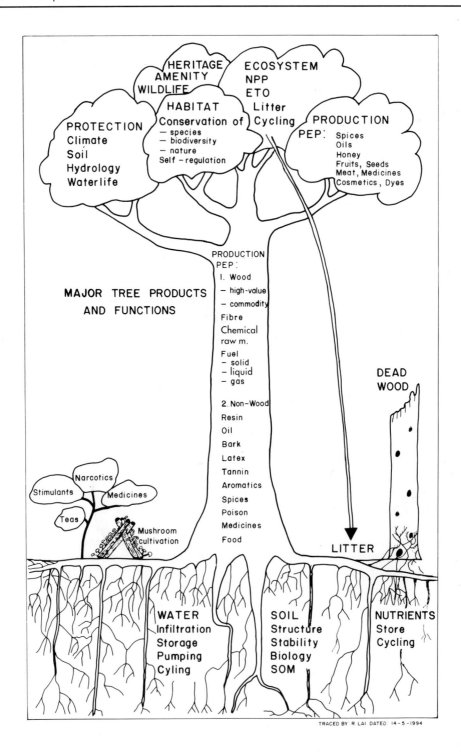

Fig. 4.2. (*opposite*) The many different products and ecosystem functions and services of tropical trees and forests. All product functions were at some time important and of essential survival value to forest-dwellers and local rural and urban people. Growing populations and industrialization increased the value of wood as raw material and energy source while the various other production functions decreased with increasing income levels. With growing affluence the relative values of the social, spiritual, habitat, amenity, recreational and environmental functions increase. The impending climate change will give recreational and environmental functions added value.

4.3 *Principles at Regional and Management Unit Level*

At regional and management unit level, the national goals are translated to medium-term targets that are primarily pursued by optimum allocation of sites and systems of conservation and management. Instruments are regional forest development or management plans and local management or working plans in the PFE, and felling plans in conversion forests. The objective to secure natural and cultural sustainability in forestry necessitates matching resource use with the potential of the ecosystem types, their site conditions and the social and cultural environment. The essential core principles of planning and control are: sufficient, reliable and integrated multisectoral database for objective and balanced assessments (Bruenig, 1993b); social and environmental system analyses, cost/ benefit and sensitivity analyses as mandatory components of all land and forest use planning; optimal ecological, economic and social site matching; long-term security of forest and land tenure and direct liability of holders of titles or concessions to stimulate motivation and acceptance of responsibility; vertical integration between production in the forest and consumption in the downstream processing and manufacturing industries; balanced and socially fair and practicable shares for government, industry and communities in forest ownership, privileges, authority and benefits/costs.

4.4 *Principles at Forest-stand Level*

The goals and targets set for the forest management unit determine the principles for forest-stand treatment. They prescribe the kind and amount of production and non-production functions (Fig. 4.2). From these derive the structural features of the forest which are modified by harvesting and manipulated by silviculture. The following principles are generally relevant: within the limits set by the goals, the target for stand manipulation should be to achieve the goal with as much diversity and species richness as site conditions permit; management should aim at forest structures which keep the rainforest ecosystem as robust, elastic, versatile, adaptable, resistant, resilient and tolerant as possible; in natural forest, canopy opening should be kept within the limits of natural gap formation; stand and soil damage must be minimized; felling cycles must be sufficiently long and tree marking so designed that a selection forest canopy structure and a self-regulating stand table are maintained without, or with very little, silvicultural manipulation; production of timber should aim for high quality and versatility; a high degree of self-regulation and self-sustainability must be aimed at by appropriate forest structure (canopy) and by single-tree management favouring species with wide ranges of tolerance and high resistance to climatic extremes and Global Climate Change. The basic principle is to mimic nature as closely as possible to make profitable use of natural ecosystem dynamics and adaptability, and to reduce risks and costs (Appendix 2). Management systems must be multi-use orientated, economically viable and socially acceptable, which means high biological and economic diversity, site-adapted and site-protecting canopy structure and low-risk, low-investment strategies. Adequate ecological conditions in and on the soil and a balance between canopy configuration and carrying capacity of the site are the keystones for sustainability of tropical forest management and conservation. Compliance with management and felling plans must be secured by positive (subsidies, security of tenure, supportive programmes of training and marketing) and negative (fines, loss of tenure) incentives. If compliance with the principles of good forestry is secured, management of the natural tropical rainforest ecosystem, thanks to its ecological properties, is even easier and sustainability less difficult to secure than in temperate or boreal forests (Chapters 1 and 2), provided the simple and basic principles of technically appropriate harvesting technologies are carefully and skilfully applied. However, the reality of tropical forestry rarely meets these conditions (Prabhu *et al.*, 1993). Therefore, enforcement guided by testable safe minimum standards is the most crucial requirement for achieving sustainability at the present time. Scaling such standards and defining critical loads are and will remain a puzzle in the fuzzy and dynamic systems concerned (Chapter 10).

4.5 Timber Management and Conservation

A properly planned and executed selection silvicultural management system (Section 6.5) creates a wide range of habitat conditions to ensure that the majority of the animals and plants of the primeval forest can survive. Some animal species avoid harvesting areas and migrate into a refuge, sometimes into secondary forest, but eventually will return. Some species which are rare or absent in primeval forest may increase in or invade logged areas. The major threat to large animals is not forestry but hunting for meat, medicines or sport. According to research in Sabah (Lambert, 1990) nectarivorous and most frugivorous bird species are more tolerant of logging than insectivores, and are more abundant in forest 8 years after logging than before. Capture rates in logged forest were nearly three times higher than in primeval forest, which may be due to greater bird density or to wider ranging activity. Studies by IUCN (1988 and following issues) have confirmed practical field experience that species extinction by timber harvesting alone is unlikely to happen beyond rare and localized events if vulnerable and specialized habitats are destroyed, such as stream beds or ridge tops. Heywood and Stuart (1992) came to the conclusion that there is presently little actual evidence for the reality of the mass extinction rates that some authors derive from theoretical models (see also Lawton and May, 1995). Forestry experience is that integration of conservation and productive and protective forest management does not come automatically, but can be achieved (Sections 5.3, 6.4, 6.5 and 7.5). In Sarawak, Sabal Forest Reserve, RP 146, selective logging in Mixed Dipterocarp forest (MDF) and creaming in Kerangas forest (KF) has not reduced but more likely increased species richness in a forest, which is among the richest, if not the richest, forest in Malaysia (Droste, 1995; Bruenig *et al.*, 1995).

Experimental comparison of tree-species richness in 20 1-ha research plots in Liberia between primeval natural forest and modified selection-harvested forests with and without silvicultural manipulation has shown that 15–20 years after intervention no statistically significant differences in tree-species richness and spectra exists between them. Species losses equalled additions of ecologically and economically equivalent tree species, except for one recorded net reduction of one species that occurred in a primeval control plot. The dominance diversity shifted slightly but remained within the range of natural variations in the course of normal regeneration dynamics and successional medium- to large-scale gap dynamics (Bruenig and Poker, 1991; Poker, 1992). Regrowth after clear-felling in San Carlos de Rio Negro (Heuveldop and Neumann, 1980) and in Gogol, Papua New Guinea (Lamb, 1990) showed surprisingly little subsequent floristic change (Sections 2.8 and 6.4). In about one to two centuries, the physiognomic structure and therefore the habitat properties of the original forest in the clear-felled areas in San Carlos will most likely be restored. In contrast, areas subsequently converted to eucalypt monocultures in Gogol, Papua New Guinea, changed the ecosystem drastically and species richness and diversity are lost (Johns, R.J., 1992) for the sake of a highly risky endeavour to supply chips to Japan (Lamb, 1990) at low prices but high social cost.

Sustainable natural forest management for timber production alters the structure of the primeval forest by harvesting all or part of the hump of mature and overmature large trees > 80 cm diameter (Fig. 6.3). Examples of the large giants and tracts of the original primeval forest should be preserved for a number of reasons. Foremost is the ethical imperative to preserve heritage, but also the practical reason to provide habitat, refuge and seed source. The backbone of management-integrated conservation is totally protected core areas that are supplemented by smaller reserved strips and plots in the production forest area. This scheme is very similar to the IUCN concept of core totally protected areas and managed buffer zones (Oldfield, 1988) and the UNESCO concept for biospheres that combine and integrate conservation and development (Draft Statutes of the World Network of Biosphere Reserves, UNESCO, 24 March 1995).

4.6 Sustainable Alternative: Non-timber Forest Products?

The rainforest has a great and well-known potential for growing a wealth of many types of plant and animal produce (Burkill, 1935; Nepstad and Schwartzman, 1992; Godoy and Bawa, 1993; Godoy *et al.*, 1993; Hall and Bawa, 1993; Saulei and Aruga, 1993). It has become fashionable to advertise the alternative use of tropical rainforest as an 'extractive resource' of diverse non-timber forest products (NTFP), which could sustainably yield more value than timber and offer a viable alternative use. The history of multiple NTFP utilization in Europe and in advanced tropical countries, such as Malaysia and Brunei, does not support this view. In Sarawak, NTFP (dammar, gutta percha, india rubber, birds' nests, rattan, camphor, bees wax) and Belian timber contributed, in 1876, Sarawak $12,588 or 7% of the government revenue and Sar.$156,731 or 21.3% to the export value (in 1870 Sar.$213,555 or 27.8%) (John, 1879). Jelutong and cutch made a large contribution from 1910 to 1970. Beyond this, only rattan maintained its position, while all other NTFP faded out (Table 4.1). A recent inventory of the tree species and NTFP in 1-ha plots in logged and primeval MDF at three locations in the Apo Kayan and in the lower Mahakam valley, East Kalimantan, and a survey of markets in Samarinda yielded the following results (van Valkenburg and Ketner, 1994): 121 taxa were recorded, of which 40 were quality timbers, 19 low quality, 54 firewood only, 8

Table 4.1. Export of timber and non-timber produce (NTFP) from Sarawak. Timber in 1000 m³ round-timber equivalents, non-timber products in tonnes. The ratio of recorded free-on-board value of timber compared with non-timber products in 1991 was million Malaysian Ringgit MR3525: 18 million (0.5%) in spite of an illipe nut harvest.

Produce	1870	1900	1910	1920	1991
Timber (1000 m³)	0	0	0	5	17,018
Gutta percha (t)	c. 500(?)	476	101	24	0
Gutta jangkar (t)	0	0	387	71	0
Jelutong (chicle) (t)	0	24	6363	5227	0
Damar (resin) (t)	c. 30(?)	298	123	679	0
Rattan (t)	c. 10(?)	2458	1238	1054	1858
Cutch (t)	0	0	0	3640	0
Nipah sugar (t)	0	292	173	577	1

Sources: Senada (1977); Forest Department Sarawak, Annual Report (1991).

multipurpose trees (fruit, exudate, bark); 13 taxa produced edible fruit. Logged and primeval forest did not differ. The diversity of NTFP actually offered on the markets was rather poor and did not accord with expectations raised in the literature. There was a trend 'capture to culture', e.g. rattan (64 taxa recorded, 19 traded under 13 trade names) was planted by villagers in logged forests nearby.

Even in such poverty-stricken regions as Amazonia, the long-term trend of social and economic evolution towards improved living conditions will make the collection of NTFP or 'extractives' in natural rainforest less attractive. Forest reserves for NFTP capture in Amazonia are the so-called and fashionable *reservas extractivismas*. They are 'instrumental in protecting marginalized groups of rubber-tappers' and are sustainable as long as underdevelopment, economic stagnation, unemployment and low wages persist. However, in its current form, 'extractivism does not represent a satisfactory alternative for the long-term future' (Clüsener-Godt and Sachs, 1994). The trend is clearly away from capture towards culture by domestication and cultivation (see also Section 2.2). The same applies in principle to the shift from commercial hunting to game breeding. Hunting for sport, however, has good prospects if cropping is effectively controlled. However, heavy hunting pressure can affect forest regeneration and species distribution. An example from French Guyana is described by Maury-Lechon (1991), but where the critical thresholds lie is uncertain. NTFP are not a priori more sustainable in natural rainforests than timber. On the contrary, they are threatened by overexploitation and abuse as much as timber is threatened by wasteful overlogging. In addition, they are socially more endangered than timber and depend on poverty (Parnwell, 1993; Taylor *et al.*, 1994; Parnwell and Taylor, 1995; see also Section 2.2).

5 Sarawak Forestry: Tortuous Road towards Sustainability

5.1 History of Forestry in Sarawak

Contents and relevance of sustainability are particularly variable in a multiracial, multicultural and dynamic country such as Sarawak. Goals and targets must be compatible with different ethnic groups and conditions in different regions. Sustainability will change before it can be achieved. Sustainability can only be approximated but never finally secured because criteria, social priorities and technical targets change too fast. Progress has constantly to adapt to the dynamics of the complex natural, economic and social system. The history of forest use and forestry in Sarawak is no exception. It has, with ups and downs, advanced through six periods of forest use and forestry development. The prehistoric conditions in the area are described in a number of articles in the *Sarawak Museum Journal*. The history of forestry in Sarawak, until the formation of Malaysia, has been described by Smythies (1963) and Porrit (1993). Details on forest policy, forest classification, silviculture, inventory and forest management during the colonial period are given by Bruenig (1957, 1958, 1963, 1965a,c). The current situation is documented by the Forest Department Sarawak (1990a).

5.1.1 Phase I. Early phase: occupation and cultural evolution, pre-1841

This was a poorly documented period of cave-based settlements 50,000 years ago and subsequent invasions by various people of different geographic and ethnic origins, of original land occupation and continued migration and incessant warfare, of alternating interference by foreign sovereigns, towards the end by the Sultan of Brunei. Traditional native forest exploitation was mainly for heavy, durable hardwoods, incense wood, rattan, resins and wildlife products, mostly for domestic consump-

154

tion but also for export to Europe and China since time immemorial. Beads I found in 1958 in burial caves in the area of Mulu National Park are probably relics of a barter trade between the Mediterranean and Borneo about 3000 years ago.

5.1.2 *Phase II. The Raj of Sarawak: pre-Forest Department, 1841–1918*

In this period, Sawawak was first Brunei dependent, then an independent state as sovereign Raj. From 1888 the country was under British protection. Pre-Forest Department forest utilization began to be regulated by the administration with emphasis on non-timber forest products. The barter trade between forest dwellers and local traders was conducted on supervised market days to protect the interests of the collectors. Native harvesting techniques were improved and trial plantings established to secure sustainability. Generally, this was a period of gradual pacification and consolidation which continued well into the twentieth century. Progress was essentially slow, but economic development was steady and in accord with the cultural diversity, needs, capacity and wishes of the multi-ethnic society.

5.1.3 *Phase III. The Raj of Sarawak: Forest Department, 1919–1941*

The Forest Department was established by the Third Rajah in 1919 and a first statement of forestry policy was issued in 1924, which emphasized the social functions of forestry. Durable speciality timbers became internationally marketable. The Forest Department introduced further technical improvements of the customarily careless and mostly destructive native techniques of collecting chicle, cutch, gutta percha, mangrove poles and timber. Forest reserves and communal forest, and experimental plantations with native multiple-use tree species, were established. Early firewood plantations were never utilized because firewood went out of use before the plantations matured. Attempts to bring some legal order and equality into the process of land occupation, to contain shifting cultivation by law and by agricultural improvements had limited success. The first simple timber cruising by rough-and-ready methods assessed merchantable growing stocks.

5.1.4 *Phase IV. The Raj of Sarawak: Japanese occupation, 1941–1945*

During the Japanese occupation, sustainable forestry was suspended and activities reduced to exploitation of accessible forests for timber to supply war-time needs. There was an increase of shifting cultivation in remote refuge areas.

5.1.5 *Phase V. British Colonial Government: Forest Department reinstated,* *1946–1963*

Highlights of the activities of the re-established Forest Department under the British Colonial Government were: formulation and implementation of conservation and sustainability concepts, restatement of Forest Policy (1954), vigorous expansion of the permanent forest estate (PFE), establishment of totally protected areas (TPA), revival of communal forestry initiatives; first statistically designed forest inventories including remote sensing and data on soil, site, minor vegetation and trees > 10 cm diameter to felling diameter limit, first steps to integrate multipurpose and multisectoral forest surveys and management planning, initially in the Peatswamp forests, at the management unit, region and state levels. All Peatswamp areas of PC1 to 3 (Section 6.7) were covered by five regional management plans (RMP) by 1963. A first provisional RMP was drafted for the Baram Mixed Dipterocarp forest (MDF) (Forest Department Sarawak Annual Report 1963), which provided for large permanent working circles and TPA (rugged country, forest nomad territories). First properly designed inventories in MDF included defect and timber quality subsampling. The growing stock and terrain survey at Bukit Raya, Rejang valley in 1961 was followed by the first orderly planned and sustainable logging in 1962. Forest Department inventory staff went to the Solomon Islands as experts to aid an inventory programme. Negotiations were begun with the Food and Agriculture Organization (FAO) in 1961–1962 for a comprehensive MDF inventory for the preparation of RMPs and management plans in MDF. Botanical and ecological research was intensified and institutionalized to provide management with information for the planning of utilization and silviculture. The government instituted ethnological, anthropological and agricultural research for the long-term planning of integrated and sustainable regional and local development. The first national park in Borneo, Bako National Park, was established in 1956 and the Batang Ai and Lanjak–Entimau areas were explored in 1954–1955 and 1990 (Section 5.2). Towards the end of the period, Japan and the USA had exhausted accessible timber stocks in the Philippines and began to open timber sources elsewhere in Southeast Asia. Japan gradually began timber acquisition in Sarawak to supply large amounts of raw timber, pressuring communal forests (Porrit, 1993).

5.1.6 *Phase VI. Sarawak Forest Department in Malaysia since 1963*

Malaya, Sabah and Sarawak federated to form Malaysia. Forestry remained a state matter, but harmonization in the federation was being developed. The Forest Department faced an unprecedented logging and export boom, increasingly from MDF (Fig. 5.1), for which it was unprepared. The logging companies introduced forms of mechanized logging which caused excessive timber waste, heavy damage to site and growing

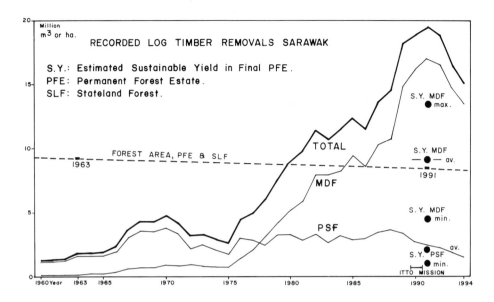

Fig. 5.1. Recorded removals of logs in million cubic metres from the Permanent Forest Estate (PFE) and Stateland Forest (SLF) areas in Sarawak. MDF, Mixed Dipterocarp forest; PSF, Peatswamp forests, including PC1 mixed ramin PSF, PC2 Alan batu association and PC3 Alan bunga consociation. Forest area in million hectares of PFE and stateland forest. S.Y., sustainable yield calculated on different assumptions concerning the condition of the forests after first logging in the productive portion of the PFE in MDF and PSF. The actually felled volumes are higher than the extracted and recorded volumes (for explanations see text). November 1989–March 1990: period of field surveys by the ITTO Mission. Data obtained from the annual reports and statistics of the Forest Department Sarawak, 1960–1994.

stock and rapid area consumption. The rates of forest licensing and of the progress of logging far exceeded capacities of the Forest Department and the sustainable yield capacity of the forests (Fig. 5.1) in contradiction to the primary aim of the Forest Policy 1924 and the primary objects (1) (a) and (b) of Forest Policy 1954 (Section 5.2). The Malaysia–FAO forest inventory and management project was eventually implemented but deviated from the comprehensive integrated design proposed in 1962. Subsequently, hill forest logging expanded so rapidly and in such a manner that the government thought fit to invite the International Tropical Timber Organization (ITTO) to review the state of forestry in Sarawak. The ITTO Mission visited Sarawak in 1989–1990 (ITTO, 1990a). The report subsequently led to restructuring of the Forest Department, strengthening of enforcement capability and to the implementation of measures, including bilateral and multilateral international cooperation, to regain sustainability by the year 2000. In the 1990s, the forest-based industry has expanded and diversified downstream timber processing. At the same time government reduced the annual cut (Fig. 5.1) and the export of round logs in accordance with ITTO recommendations.

5.2 *The Forest Policy of Sarawak*

A forest policy provides the framework of preferential political goals and demands within which the goals of forestry are defined considering the present situation, actual problems, forseeable future problems and the condition of uncertainty. A statement of policy was approved by the Sarawak Government in January 1924, which reads (somewhat abridged) as follows (Smythies, 1963, p. 238):

> The primary aim is to insure the permanent maintenance of a sufficient area of forest to supply all the needs of the inhabitants. Some countries possess such extensive forests that they can afford to export a proportion of their produce. Today Sarawak is one of the latter but owing to wasteful methods of cultivation and primitive methods of extraction, timber is already becoming scarce in some districts. This, however, is a local scarcity only which can in time be corrected. At present it may be said that all timber produced in the country is consumed within it and that an export trade is non-existent. There is, however, a large export of minor (i.e. non-timber) forest produce such as damar, jelutong, rubber, gutta percha, canes, illipe nuts and other products, only insignificant quantities of which are consumed within the country. The primary aim is to maintain and increase the export of minor forest products. The secondary aim is to encourage the export of timber which is surplus to internal requirements. The area of reserved forests in Sarawak is still considerably less than one per cent of the country and reservation must continue for a long time to come.

This statement reflects the overall policy of the government to maintain as much as possible the cultural diversity and identities of the people in the multiracial society and to proceed cautiously with export-orientated economic development. This basic approach was affirmed in phase V, 1946–1963, by the renewed statement of forest policy in 1954, which requires the Forest Department to pursue the following goals:

> (1) Reserve permanently for the benefit of the present and future inhabitants of the country forest land sufficient
>
>   (a) for the assurance of the sound climatic and physical condition of the country; and safeguarding of soil fertility, and of supplies of water for domestic and industrial use, irrigation and general agricultural purposes; and the prevention of damage by flooding and erosion to rivers and to agricultural land;
>
>   (b) for the supply in perpetuity and at moderate prices of all forms of forest produce that can be economically produced within the country and that are required by the people for agriculture, domestic and industrial purposes under a fully developed national economy.
>
> (2) Manage the productive forests of the Permanent Forest Estate

with the object of obtaining the highest possible revenue compatible with the principle of sustained yield and with the primary objects set out above.

(3) Promote, as far as may be practicable, the thorough and economical utilization of forest products on land not included in the Permanent Forest Estate, prior to the alienation of such land.

(4) Foster, as far as may be compatible with the prior claims of local demand, a profitable export trade in forest produce.

(5) Foster by education and publicity the value of forest among the public and to provide for vocational training in the field of forestry and forestry industry.

(6) Ensure a regular and uninterrupted source of funds for the advancement of forestry in the State.

(Forest Department Sarawak Annual Reports 1954 and 1991)

The unwritten, simple and conservative practical forest policy of the second half of the nineteenth century and the two written forest policy statements of 1924 and 1954 provided an adequate base and the terms of reference for forest legislation and for the implementation of sustainable forestry with the primary objective to meet the needs of the country's people, government, environment and nature. During parts of phase VI adverse external forces interfered with the process towards sustainability and successfully, but temporarily, diverted development away from sustainability and back to earlier phases of capture and acquisition, but at a much grander scale. This process is being gradually reverted to the principles of traditional Sarawak forest policy since the ITTO Mission, when Sarawak reconfirmed its commitment to the policy statement of 1954.

5.3 History of Conservation

The traditional use of soils and forests by wandering forest-dwellers and by migrating pioneer shifting cultivators is not conservation orientated (Sections 2.1 and 2.2). Their forest utilization threatened even vital plant resources. Beccari (1904, p. 158) mentioned the Upas or Ipoh tree (*Antiaris toxicaria* (Pers.) Lesch.) as 'another forest species utilized by man, which will probably in time disappear altogether as a wild tree'. But he also found Upas trees planted by Dayaks 'in the forest and amidst fruit trees', a kind of *ex-situ* gene preservation. John (1862, p. 32) commented on the local overuse of sago as indigenous people feared to penetrate far into the forests. Similarly, Jelutong (*Dyera polyphylla* and *D. costulata*), Gutta percha (*Palaquium gutta*) and Gaharu (*Aetoxylon sympetalum*) are known to have become locally extinct as a result of overuse and misuse by local communities. Conserving activities by indigenous people have been confined to the preservation of abodes of spirits and graveyards, usually ridge crests and hilltops with unproductive skeletal soils and haunted alluvials rich in disease-carrying vectors. The first independent government of Sarawak was preoccupied with the establishment and preservation of peace and the development of trade throughout the nineteenth century. The general need for conservation and forest protection, however, was already realized. Beccari (1904), during his second visit to Sarawak in 1878, observed 'with the system of rice-cultivation in Borneo, any extension would lead to a corresponding destruction of forests'. As a countermeasure he suggested the establishment of small agricultural stations in different localities for culture experiments and for gene preservation in something like botanical gardens; also, a considerable extent of primeval forest should be preserved in proximity to the capital, Kuching, and kept in its natural condition, somewhat on the lines of national parks in the USA. Beccari proposed to establish the Matang hills as a national park

> where if the forests were cleared the soil would in all probability after a short time become unproductive and get overgrown with lallang grass … the destruction of forests diminishes the frequency and abundance of rain … [and] causes a notable increase of temperature around the capital.
>
> (Beccari, 1904, p. 370)

This sounds very much abreast with present-day environmental and conservation problems and concepts (see Sections 2.2, 2.9 and 2.10).

The history of an official policy of conservation of nature, plant and animal species in Sarawak begins with the establishment of the Forest Department under Mr J.P. Mead on transfer from Malaya as Conservator of Forests in 1919. Subsequent statements of Forest Policy (1924) and legislation (Forest Rules, 1919; Sarawak Reservation Order, 1920; Forest Order, 1934) initiated nature and species protection and regulated forest resource utilization. The Forest Ordinance and Forest Rules, 1954, the Statement of Government's Forest Policy by the Governor-in-Council, 1954, the National Parks Ordinance, 1958, and the Report on Fauna and Flora of the Special Select Committee of the Sarawak Legislature, 1988, provided the legal instruments required by a modern policy of integrated and comprehensive, environmentally favourable land and forest utilization and conservation. A good postwar aerial photo reconnaissance guided the selection of priority areas for conservation. The first national park, Bako National Park, was established in 1956–1957 with the enthusiastic help of the villagers who have been participating successfully since. More than 50 animal species or groups are now totally protected and over 50 partly protected, and over 50 rare, endangered or keystone plant species or groups are protected, including 23 high-forest and mangrove tree species.

The statements on conservation in *Forestry in Sarawak, Malaysia* (Forest Department Sarawak, 1990a) reads:

Sarawak is keenly aware of the need for forest conservation, not only to ensure a sustained supply of timber but also to maintain environmental stability, to provide sanctuary for wildlife, and to serve as an invaluable storehouse of genetic resources useful for the improvement of our indigenous tree species, agricultural crops and livestock. Conservation deals with the biological interaction of plants, animals and microorganisms and the physical elements of the environment, i.e. water and soil, to which they are intimately bound. These natural assets are permanently renewable if wisely conserved. Improvement in economic status and human welfare cannot be sustained unless the conservation of these living resources is specifically drawn into the process of development. Conservation based development is not only sensible but also economically sound. The relationship between development and conservation is one of mutual support.

This policy and concept definitely puts the onus on forest management to be not only compatible but also integrated and symbiotic with species and nature conservation. This is a sensible and practicable concept, but its successful implementation depends on the political and social environment, which now again are favourable. In Sarawak approximately 8% of the land or 10% of the designated PFE area are, or will be, protected as a National Park or Wildlife Sanctuary. This is more than double the 3.8% of the total surface area of the Brazilian hydrographic Amazon basin that has been designated protected forest. It is about equal to the 10% protected in the Ecuadorian Amazon region (Fundacion Natura, 1991), but less than the 17% or 19 million ha of 110 million ha PFE that are TPA in Indonesia (Collins *et al.*, 1991). According to the Ministry of Forestry Indonesia (1993), nature reserves and national parks and protection forests occupy 49 million ha. In 1994 Sarawak, Malaysia and Kalimantan, Indonesia, amalgamated the Lanjak–Entimau Wildlife Sanctuary (177,000 ha), the Batang Ai National Park (24,000 ha) in Sarawak and the Bentuang–Karimun Nature Reserve (600,000 ha) in Indonesia into a transboundary biodiversity conservation area (801,000 ha), the first of its kind in the region (Fig. 1.16).

5.4 Timber Removals 1960–1994

From before the Second World War to 1960, logging in MDF was by poorly planned trial-and-error experiments (Smythies, 1963) which usually failed to meet yield and profit expectations. After Malaysia was formed, the recorded timber removals from MDF gradually increased, but there was no corresponding increase of management, or at least sustained-yield regulation, in PFE and stateland forest. As much was cut as the markets demanded, especially in Japan (Poritt, 1993; Fig. 5.1).

Recorded removals increased sharply during the 1970s. The recorded fellings in the PFE reached the most likely level of sustainable yield in 1987, after which the cut rose steeply above any of the estimated sustainable yield levels, calculated as permanent yield for the area of *c.* 4–4.5 million ha productive and intact PFE and as temporary yield for the 2–2.5 million ha stateland or conversion forest. The PFE could have produced, with properly applied reduced impact logging (RIL) under selection silviculture management system (SMS) and full utilization and accounting, at least 12 million m^3 sustainable yield during a 60-year first felling cycle on 4 million ha with an increment of 3 m^3 ha^{-1}. To this must be added the reduction of primeval growing stock to average SMS level on a 75,000 ha annual cut, which would have amounted to 2–3 million m^3 per year. The clearing of 2.5 million ha stateland forest for agricultural development, if spread over 50 years, would yield another 2.5–3 million m^3, depending on the actual land requirements during the conversion period. The total would have been a sustainable cut of at least 17 million m^3 plus thinnings for 50–60 years, thereafter 13.5 million m^3 plus thinnings. In reality, recorded removals nominally remained below these levels, but the actual resource consumption by far exceeded them. The ITTO Mission argued that a decline of sustainable yield to below 4.5 million m^3 was now unavoidable. Responsible for this dramatic decline of productivity are the logging of the 50–80 cm diameter élite candidate trees, the heavy damage to the residual growing stock and to the soil, the rapid logging progress and a very short (25 years) felling cycle. This loss is not due to deforestation (broken line: decline of forest area of PFE and stateland forest, in Fig. 5.1).

The recorded timber removals from Peatswamp forest (PSF) responded less dramatically to the tempting lures of a demanding market (Fig. 5.1). By 1963 harvesting in all PSF areas in Sarawak were regulated by five RMP, most Peatswamp forest areas in PFE were under working plans and stateland PSF under felling plans (Bruenig, 1961b; Forest Department Sarawak Annual Report, 1963). These plans prevented excessive over-logging. The sustainable allowable annual cut could have been about 2 million m^3, but the blanket application of excessively low diameter limits for harvesting, the rate of legal and illegal overlogging and the short felling cycle in Ramin PSF, PC1, and clear-felling in Alan bunga PSF IPC 2 and 3, reduced the future sustainable yield to less than 50% of the possible sustainable primary economic productivity which is still better than the 20–25% in MDF. In future the sustainable yield from PFE in Peatswamp forest will be less than 1 million m^3.

5.5 *The Present State of the Forest Resource*

The Sarawak Government recognized in the nineteenth century that the forest resource of Sarawak was overused but underutilized, and that the forest lands were threatened by little productive, but highly exacting migratory shifting cultivation. Appropriate legislative action during the past 75 years has provided adequate legal conditions for sustainable and social multiple-use forestry and effective forest protection. A large body of research and survey data on the ecology, growing stock and silviculture of all forest types in Sarawak and on the human environment is available (Bruenig, 1957, 1974; Anderson, 1961a; Wyatt-Smith, 1963; Ashton, 1964; Lee and Lai, 1977; Primack *et al.*, 1987; also the reports by Sarawak Museum, Borneo Research Council and Sarawak Forest Department) and provides an adequate basis for sustainable forestry. However, application of knowledge and enforcement of law and policy have fallen short of needs especially in MDF (ITTO, 1990a). This contradiction between adequate policies, laws and scientific knowledge and decline of the forest resource in Sarawak is generally faced by rainforest countries with active logging. The total land area of Sarawak is 12.5 million ha. The forested area in 1993 was 8.4 million ha of which at least 4.5–5 million ha have been modified by selective logging (Table 5.1). The PFE was 4.7 million ha but eventually should be at least 6 million ha, of which a substantial part, perhaps 30–35%, will be forest on unfavourable or fragile, erodible soils, steep slopes and inaccessible sites that are unsuitable for timber production and should not be logged. The area of totally protected national parks in 1993 was 0.1 million ha and of wildlife sanctuaries 0.2 million ha. A further 0.5 million ha of natural forest at least are designated as TPA, but establishment is nowadays exceedingly slow and encroachment is a serious threat to the designated areas. Approximately 2 million ha of stateland forest are presently being selectively logged to prepare the land for

Table 5.1. Areas in 1000 ha of Mixed Dipterocarp and Kerangas forest (MDF/KF), Peatswamp forest (PSF) and Mangrove forest in Sarawak. The sizes of the permanent forest estate and of the totally protected areas are planned to be substantially enlarged.

	MDF/KF	PSF	Mangrove
Permanent forest estate (1991)			
Area	3699	762	37
Logged	1073	575	37
Annually logged	74	22	1
State land (1991)			
Area	3324	448	+
Logged	2091	328	+
Annually logged	110	15	+
Totally protected area (1991)			
National park	104	+	+
Wildlife sanctuary	175	+	+

Source: Sarawak Forest Department (1991).

transfer to agriculture, but part may eventually not be needed. This area may finally become PFE and require silvicultural rehabilitation. About 6000 ha of tree plantations exist, and more are planned, mostly on suitable parts of the 3.4 million ha of abandoned traditional shifting cultivation land. It is hoped that the forestry industry may invest in tree planting.

A feature of the conventional selective logging by tractors is that it underutilizes the felled trees but overlogs the forest by misuse of timber, overuse of area, and abuse of growing stock and soil. Due to this and illegal logging, logging progress in terms of area and growing stock has been excessively fast. The government, realizing the situation in 1990 (ITTO, 1990a), began to reduce the annual cut in the PFE and strengthened the Forest Department's capacity of enforcement. The priority problems which have to be solved by the Forest Department are:

1. Implementation of a rapid change from wasteful selective logging to profitable selection RIL and SMS.
2. Rapid assessment for long-term growth and yield planning, based on existing and newly collected reliable data of site, growing stock, growth potential and commercial increment in the PFE, especially in the logged areas, and adjustment of length of felling cycle to crop and site conditions.
3. Restoration of productivity and sustainability by flexible silvicultural procedures and holistically integrated management systems for the overlogged PFE (Sections 6.5–6.8).
4. Assessment of the state, growth potential and suitability for inclusion in PFE and recultivation of overlogged stateland forest and of the secondary forests on all land categories (Chapters 7 and 8).
5. Establishment of an integrated data and information system comprising research, forest management and forest policy administration.

The immediate objective must be to mitigate the current decline of economic productivity of the PFE. The long-term objective must be to restore productivity to its original levels. The potential to compensate the decline by plantations is limited by problems of legal availability of land, high risk of investment and poorness of soils. The example of decline in Fig. 10.1 is based on real-life data in Malesia and illustrates the threat of impending collapse of economic forestry if determined and effective action is not taken very soon. This rapid action is not utopian in Sarawak. ITTO (1990a) had commended the high standard of the management system and procedures in Sarawak. The mission also stated that the management in the Ramin PSF was an outstanding example, rare in the tropical rainforests, of sustainability, unfortunately impaired by a shortening of the felling cycle from 60 to 40 years and threatened by pressures to shorten cutting cycles even further. The MDF management system was technically well advanced but fell short in enforcement. As a result, increment and sustainable yield in the productive area in the PFE will decline drastically, unless action is taken effectively and efficiently, without delay. Volume- and value-adjusted documentation of removals and exports, illegal logging and low timber fees have caused substantial loss of revenue both in Sabah and Sarawak. With full and proper collection by the state treasuries, 'the current revenue from timber exploitation could have been obtained from about one third of today's logging area' (Uebelhör and Heyde, 1993). If, in addition, RIL and proper utilization standards had been enforced, the logged area could have been even further reduced to one-quarter of the actually logged area without loss of profits and revenues. This smaller area would have been properly harvested, and would hardly need silvicultural treatment before the first timber stand improvement fellings are due. The whole PFE could have been managed sustainably at low cost and risk levels, with realistically long rotations and felling cycles, maintaining high levels of growing stock, biodiversity, productivity and volume and value yields. The problem in Sarawak, as in Sabah and Kalimantan, now is to restore productivity, protective functions and social value of the excessively large area of overlogged forest, while at the same time timber yields and revenues decline, plantations cannot compensate and pressure for premature re-logging will increase, especially if log exports continue at a high level (Marsh, 1991).

To restore social values of the forest estate, Taylor suggests revival of the old permit system to utilize forest resources at small scale outside the concession areas. Permits would be issued to small businesses, villagers, families and individuals and managed with assistance from the forest service which would have to be restructured and decentralized for this purpose (Taylor *et al.*, 1994). Such a scheme, of course, would require rigorous enforcement of rules and regulations, fairness and security of land and forest tenure, and must be developed in full consultation with the rural communities and effective and efficient corporate management infrastructure. The most crucial points are the development of flexible marketing strategies and securing profitable markets. The concept cannot be imposed on them autocratically from above or outside. More practicable would be a transfer of a greater share of responsibilities for sustainability of conservation and management to the concession holders in return for assured long-term tenure on condition of compliance.

6 Naturalistic Natural Rainforest Management in Sarawak

6.1 Targets and Principles

State forestry in Sarawak, as elsewhere in the world, is legally charged with the responsibility to implement the national forest policy in compliance with the constitution and the special laws of the country (Chapter 5). The forest policy defines the general goals of forestry in its social context from which the specific targets of public or private forest management are derived. The targets of management are formulated in terms of products and service functions. Timber production in rainforest began with speciality timber of high quality and value but changed to the present commodity timber preference. In future, preference will most likely again be on high-quality timber, to which the production of commodity timber can be coupled as a secondary priority. Consequently, management should concentrate on growing a few large and tall élite trees which leave growing space for economically and ecologically important biodiversity (Bruenig, 1970b, 1984b,c). This target of management determines the technical objectives and procedures of silviculture and integrated harvesting. Targets and objectives must accord with the complex dynamic nature of the forest and cultural ecosystems, and accordingly with high degrees of uncertainty and risk. This applies in particular to non-timber forest products (NTFP). The commercial prospects for the production of NTFP from natural rainforest are sometimes assumed to be bright (Peters *et al.*, 1989; Posey and Balee, 1989; Veevers-Carter, 1991; Godoy and Bawa, 1993; Godoy *et al.*, 1993) but above a certain threshold of standard of living and quality of life, simple 'capture' NTFP collecting in primeval or modified forest becomes economically unattractive and socially unacceptable (Section 4.6; Anonymous, 1994a; Clüsener-Godt and Sachs, 1994).

166   The goal of naturalistic silvicultural management is to establish forest

Fig. 6.1. MDF after first cut under a sustainable Selection Silviculture Management System. Sarawak, 1993.

stands (Fig. 6.1) and forest landscapes (management units) which are robust, possess a high degree of resilience and tolerance against stress, strain, shock and natural environmental and socio-economic change, and which are capable of reorganization if collapse (destruction) should occur. Forestry experience is that the floristic and spatial forest structure (Figs 6.1, 6.2 and 6.8) influence the manner in which natural and man-made perturbations spread and are buffered in forest stands and in forest landscapes. Species-rich, diverse, multifunctional and structurally complex and dynamic forests with a balanced relation of structural elements (phytomass, SOM) and processes (organic matter and nutrient cycles) buffer impacts, repair damage and adapt to new situations more efficiently and easily than uniform, simple-structured forests which are narrowly focused on maximizing one particular type of product. The principle of naturalistic silviculture is to mimic the self-regulating and self-sustaining mechanisms of nature in order to use natural dynamic forces and mechanisms to achieve goals and targets cheaply and safely (Fig. 6.1). Harvesting is the most important silvicultural operation. It manipulates forest canopies and thereby guides growth and regeneration

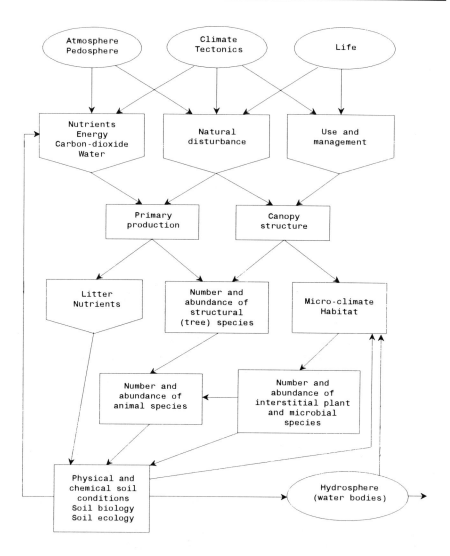

Fig. 6.2. Biodiversity is a key functional element, but not a practical, easily measurable indicator, among the manifold elements of self-sustainability in forest ecosystems. Its preservation is crucial for viability, tolerance and resilience of tropical rainforest ecosystems and their sustainable management. Biodiversity at gene and species levels depends on the canopy structure and the productivity of the structure-building tree species. Structure and production are influenced by physical and chemical growth factors, natural disturbances and anthropogenic interference, which in turn are determined by natural and cultural environmental conditions. The arrows indicate the major functional direction of the interdependencies and flows of effects through the forest ecosystem from environmental compartments above to hydrology below. The interactions in reality go both ways and are more complex than could be depicted in this drawing, which emphasizes aspects of particular relevance to conservation and management. Forestry can influence PEP and biodiversity effectively by modifying and manipulating the structure of the A, B, C and D layers of the canopy.

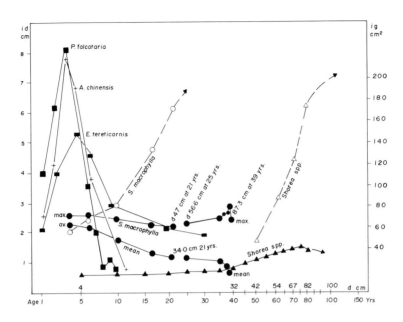

Fig. 6.3. The current annual increment of trunk diameter id with age (logarithmic scale) of (1) pioneer species in short-rotation plantations (Chapter 8): *Paraserianthes falcataria*, ■, Mitchell, 1963; *Anthocephalus chinensis*, +, Fox, 1968; *Eucalyptus tereticornis*, ■, Seth, 1968 in Bruenig, 1971c; (2) mature forest species planted in secondary forest: *Shorea macrophylla*, RP 76-A1, planted 1973 on udult ultisol, Figs 7.3 and 7.5, ● from ages 4 to 21, upper line is fastest growing trees, diameter 47 cm at 21 years, lower line is plot mean, diameter = 10.9 cm 6 years and 34.0 cm 21 years; *Shorea macrophylla*, TP 7c, planted 1936 on flat alluvial, Fig. 7.1, ● from age 25, 1961, to 39, 1975, no more measurements since, upper fast line is diameter 56.6 cm at 25 years, 87.3 cm at 39 years, lower slow line is stand mean diameter, indicating decline when there is in fact still increase (fast line), but need for thinning; (3) mature forest A-layer *Shorea* species growing in primeval Mixed Dipterocarp forest (MDF) in Mindanao (Reyes, 1958). Canopy opening by proper harvesting and TSI 10–15 years later would increase diameter-growth in the 30–80 cm range considerably. The differences between fastest tree and the mean in (2) indicates the great potential of silviculture for improving productivity (breeding, selection and release thinning, early fertilizing). The basal area increment ig culminates much later than diameter increment. A diameter limit below the great period of growth of diameter and basal area kills the goose which lays the golden eggs. (From Bruenig, 1971c, fig. 212.2–1; RP 76-A1 and TP 7c, Forest Department Sarawak, W. Then.)

to the desired targets. Harvesting creates canopy gaps to which growth responds in two ways: horizontally by expansion of crown and, correspondingly, stem diameter in the B-layer, but relatively less among the big and mature trees in the A-layer whose crowns in overmature trees may even shrink; vertically by stimulating height growth among regeneration in the C- and D-layers. The reactions continue in both cases until the canopy is closed again. The time needed depends on the initial gap size and the growth rates in both directions (Bruenig, 1973b; Bruenig *et al.*, 1979). A 25–30 m wide canopy gap in the A/B-layer could be closed by

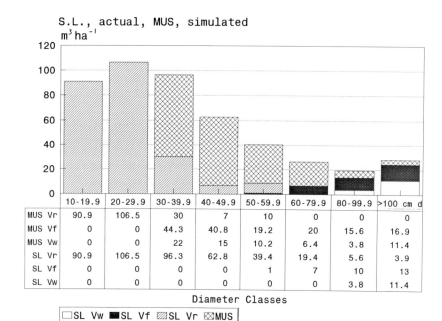

S.L., actual, MUS, simulated
m³ ha⁻¹

	10-19.9	20-29.9	30-39.9	40-49.9	50-59.9	60-79.9	80-99.9	>100 cm d
MUS Vr	90.9	106.5	30	7	10	0	0	0
MUS Vf	0	0	44.3	40.8	19.2	20	15.6	16.9
MUS Vw	0	0	22	15	10.2	6.4	3.8	11.4
SL Vr	90.9	106.5	96.3	62.8	39.4	19.4	5.6	3.9
SL Vf	0	0	0	0	1	7	10	13
SL Vw	0	0	0	0	0	0	3.8	11.4

Diameter Classes

☐SL Vw ■SL Vf ▨SL Vr ⊠MUS

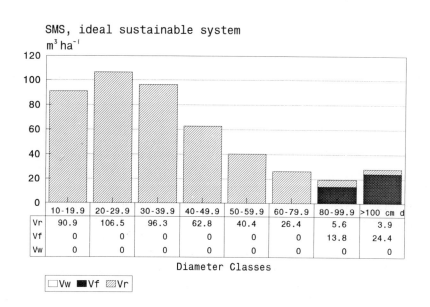

SMS, ideal sustainable system
m³ ha⁻¹

	10-19.9	20-29.9	30-39.9	40-49.9	50-59.9	60-79.9	80-99.9	>100 cm d
Vr	90.9	106.5	96.3	62.8	40.4	26.4	5.6	3.9
Vf	0	0	0	0	0	0	13.8	24.4
Vw	0	0	0	0	0	0	0	0

Diameter Classes

☐Vw ■Vf ▨Vr

growth in both directions within 40–60 years, after which new canopy opening by harvesting would be feasible. Harvesting removes phytomass from the site (Section 2.8). The principle of sustainable naturalistic management is to maintain continuously enough living and dead phytomass on the site to protect the soil, maintain SOM, and stimulate biological activity and nutrient cycling. Analysis of data from 212

Fig. 6.4. (*opposite*) Above-ground stem volume per hectare of all trees > 10 cm diameter in 2-ha *Dryobalanops beccarii* Mixed Dipterocarp forest in RP 146, Sabal Forest Reserve, reconstructed condition before a low-key selective logging in 1978. The volume has been calculated as length of bole (ground to crown) times basal area and form factor of 0.5. The total tree-stem volume is 424.6 m^3 ha^{-1}. The present above-ground plant (tree) biomass is 468 t ha^{-1} dry matter. A light to medium selection felling takes only a small portion of the 'structure trees', except in the diameter classes > 80 cm and leaves moribund giants. Malayan Uniform System (MUS) would have removed almost all merchantable growing stock above 50 cm diameter, and some below. SMS (bottom) combined with RIL is ideal for sustainable high-value production at low cost and low risk. SL, blank: merchantable timber wasted. Vf, dark shaded: timber extracted in a light selection logging. MUS, cross-hatched: timber that would have been harvested under MUS in addition to SL and Vw. Much of the smaller sized residuals > 50 or 10 cm diameter would be removed by poison-girdling (pg). SL, hatched: residual stand after MUS. Vw, blank: merchantable log timber on the ground in 1990, rejected for unknown reasons and not extracted in 1978 (15.2 m^3 ha^{-1}). Simulated MUS includes shaded and double hatched. Vr, light hatched: residual volume of standing growing stock volume as enumerated in 1990 (424.6 m^3 ha^{-1}).

permanent sample plots in rainforest in North Queensland gave no evidence that repeated cautious selection harvesting under SMS in Queensland would statistically significantly reduce diameter increments below the site potential for tree growth (Vanclay, 1990). This indicates that the removal of timber did not damage the essential functions of the forest ecosystem. It is the art of the naturalistic forester in all forests of the world to strike the right economic and ecological balance between the amount and kinds of timber removal and canopy opening, and the amount and kinds of living and dead phytomass which are maintained on the site for soil protection and maintaining self-regulatory functions and growth.

6.2 Potential and Actual Economic Production

Ecologically and economically sound silviculture has been developed by foresters largely as a response to destructive natural impacts, economic pressures and societal demands (Bruenig, 1971c, 1973a, 1984b,c, 1986b). The similarities, differences and linkages between natural–ecological and cultural–economic ecosystems are succintly reviewed by Perrings (1995) stressing the well-known uncertainties and unpredictabilities of the interactions within and between these systems, which are largely non-linear and constantly changing. As a result, the future economic production (increment which can be harvested as economic yield) cannot be predicted. The extreme heterogeneity and dynamics of tropical rainforests at spatial scales from groups of associated trees to continents, and the rapid cultural changes in the tropics, make prediction even more uncertain.

The Mixed Dipterocarp forest (MDF), in comparison to African and

American rainforest, has a taller stature, higher density of biovolume of large trees, higher proportion of commercial species, higher merchantable volume stockings. High incidences of defects are the inevitable result of the high diversity of very active life forms, but economically are an enigma. Locally, especially in West Borneo, the incidence of stem hollowness is very high but varies between sites, forest types and at random (Panzer, 1976). Age is another enigma. Small trees in the B- and C-layers are not necessarily younger than large trees in the A-layer and trees of equal size are not necessarily of the same age. Age information is essential for growth and yield estimation, but is impossible to determine in practical management (Borman and Berlyn, 1981). Consequently, response to release and exposure is difficult to predict (Dawkins, 1963; Poker, 1989, 1992). Different natural stand structures and the economic objective of maximum profit have led to different selective logging strategies, which have created different ecosystem structures after harvesting (Bruenig, 1967a). The general tendency in the first cut in primeval forest has been towards higher diameter limits for harvesting in the Philippines and lower diameter limits in Peninsular Malaysia. The result is wide variations of volume yields between about 20 and 200 m^3, and of the condition of the residual growing stock (Figs 2.12–2.15 and 6.4).

The commercially exceptionally attractive Asian MDF is very much more at risk from overlogging and wasteful utilization than most rainforests in Africa and America. The area of rainforest and the annually logged area in Asia are relatively small, but the initial growing stock and the recorded removals of 33 m^3 ha^{-1} are much larger than in Africa (14 m^3 ha^{-1}) or America (8 m^3 ha^{-1}) (FAO, 1993, especially table 4; Baaden, 1994). Management towards sustainability with such removal rates should, in theory, be very easy, but practice is different. Waste and illegal removals have to be added to arrive at the actual felling rates, which may be 50–100% higher. Fiscal and private financial interests and political forces favour overlogging; enforcement of sustainability often meets strong resistance and is practically ineffective and often impossible. 'A management system is as good as its silviculture' (Appanah and Weinland, 1992) means as good as the degree of productivity it maintains after the first and subsequent harvests in relation to the site potential. Usually, as there is no integration between logging and the silvicultural management system, this degree declines to a level far below the potential of forest and site.

6.3 Growth and Sustained Yield Potential of SMS in MDF

The potential net primary production (NPP) and primary economic production (PEP) are determined by the natural potential of the site for plant growth. Within this potential, the actual NPP and PEP are determined by the condition of the forest stand, modified by timber and NTFP harvesting and manipulated by silviculture. In zonal forests, such as the MDF, the upper ceiling of the potential PEP is 6.5–10.0 $m^3 ha^{-1} a^{-1}$ on above-average sites, normally not available for forestry, and 3–5 $m^3 ha^{-1} a^{-1}$ on the sites to which forestry usually is relegated (Section 1.13). It is very likely that at least 50% of this will be thinnings to release the high-quality élite trees from growth inhibition by competition. This leaves 1.5 $m^3 ha^{-1} a^{-1}$ of high-quality logs from trees of 80–100 cm diameter in yield class 3 $m^3 ha^{-1} a^{-1}$, and 2.5 $m^3 ha^{-1} a^{-1}$ of 80–120 cm diameter in yield class 5 $m^3 ha^{-1} a^{-1}$. The 50% intermediate 'by-product' from thinnings will be utility timber of diameters between 30 and 60 cm for domestic outlets. Woell measured higher annual increment rates than 5 $m^3 ha^{-1}$ in long-term experimental plots in Dipterocarp forests in the Philippines. Selected élite (final-crop) trees, about 70 ha^{-1} of light and medium weight Dipterocarps, produced a mean annual increment between 2.1 m^3 (droughty sites or high altitude, felling cycle up to 105 years) to 8.4 m^3 (favourable above-average lowland sites, felling cycle 50–70 years). Release of the élite trees from competition by thinning stimulated diameter increment by one-fifth and volume growth by about one-quarter. Consequently the felling cycles for the same target diameter could be shortened by 20%. Woell assumed a high initial density of élite trees (70 N ha^{-1}), which will lead to crowding beyond 50 cm diameter (crown diameters *c.* 12–15 m). Timber stand improvement (TSI) thinning is then indispensable to maintain high rates of growth of high-quality élite trees (compare Sections 7.4–7.6). There was no evidence that integrated selection felling and silvicultural management would lead to losses of timber species among the population > 10 cm diameter (Woell, 1989). This agrees with findings by Poker (1989) for Grebo National Park, Liberia, where the only loss of one tree species occurred in one undisturbed control plot. The French–Indonesian silvicultural project STREK reported 9.3 m^3 mean annual peeler quality timber increment under SMS in MDF. The annual diameter increment of Dipterocarps is reported to lie between 0.8 and 1.5 cm in TPI (Tebang Pileh Indonesia)–SMS areas, which is about twice the growth in silviculturally untended stands (MOF–CIRAD, 1994).

The potentials and limitations of rainforest productivity and the problems of crowding have been critically and exhaustively discussed by Dawkins (1959, 1963). Later, Dawkins (1988) concluded that, with a few exceptions (mineral-rich soils, higher elevation), sustainable mean annual timber yields in tropical rainforests lie below 8 $m^3 ha^{-1}$, but mostly below 5 $m^3 ha^{-1}$. Pure natural regeneration systems have, in practice, seldom exceeded 2 $m^3 ha^{-1}$. Higher yields are commonly, but misleadingly, quoted from small areas, few trees, irrelevant sites, short periods, single instances, and even misprints. Kio (1983) concluded from his earlier research that the frequently quoted rate of 2–3 $m^3 ha^{-1}$ annual increment in natural silviculture may be increased to 10 $m^3 ha^{-1}$ by cautious state-of-the-art logging and improvement of silvicultural techniques. Bruenig

(1966) assumed sustainable annual timber increments between 2.4 and 7 m³ ha⁻¹ in naturally regenerated lowland Mixed Dipterocarp forest. Graaf (1982) considered natural silviculture with a cautious girth-limit selection system and low-cost silvicultural treatment economically viable even at the very low rate of 1 m³ ha⁻¹ of annual commercial volume increment, which he regarded realistic at a 20-year cutting cycle on poor, acid sand in the Mapane forest in Surinam. In Sabal Forest Reserve, RP 146, the mean annual stemwood increment over bark 1978–1990 on the 7 ha in *Dryobalanops beccarii* MDF, which were illicitly relatively lightly selectively cut and logged with light tractors (Fig. 6.4, SL alternative) in 1978 most probably, in response to canopy opening was 10 m³ ha⁻¹. If logging had happened later, increment would be proportionally higher. Aerial photography in 1982 already shows the logging tracks (Droste, 1995; Tables 6.1 and 6.2). The adjacent creamed *Agathis borneensis* Kerangas forest (KF) on medium deep, humus podzol well supplied with water probably would have had almost the same increment of the residuals if not almost all élite candidate *A. borneensis* had been felled. The *Casuarina nobilis* KF would have at most half the post-harvest increment if it had been logged, judging from the decline of canopy height and complexity

Table 6.1. Growing stock 1963 and 1990, and the volume of timber cut in 1978 (14 trees ha⁻¹ felled, 30% below diameter limit, timber volume estimated from stump diameter, and logs left behind). The enumerated standing tree volume > 10 cm in 1963 was 391 m³ ha⁻¹; after illicit selective logging in 1978 the viable residual tree stand probably was 218 m³ ha⁻¹. The enumerated growing stock in 1990 was 359.3 m³ ha⁻¹. The mean annual increment 1978–1990 was approximately 10 (maximum 11–12) m³ ha⁻¹ tree stem volume. Full restoration of the original quality of the growing stock, with seven trees > 80 cm diameter, will at least take another 30–40 years (42–52 years after logging). Regeneration < 10 cm diameter to 0.3 m height in felling gaps and skidtrails in 1990 was ample (61,950 N ha⁻¹, 43% Dipterocarps). RP 146, Sabal Forest Reserve, Sarawak.

	1963			1978	1990			
Size class (cm diameter)	No. ha⁻¹	No. dipterocarps	Dipterocarps (%)	Timber cut (m³ ha⁻¹)	No. ha⁻¹	No. dipterocarps	Dipterocarps (%)	V (m³ ha⁻¹)
10–20	260	68	26.2	0	506	127	25.1	26
20–30	82	39	47.6	0	110	39	5.5	107
30–40	54	25	46.3	0	38	13	34.2	95
40–50	14	11	78.6	0	21	10	47.6	63
50–60	7	4	57.1	1	11	5	45.5	39
60–70	7	7	100	7	5	4	80.0	19
70–80	0	0	0	10	3	2	66.7	10
> 80	7	7	100	28	0	0	0	0
Total	431	161	37.4	46	694	200	28.8	359

Source: Bruenig *et al.* (1992); Droste (1995) (see Fig. 2.15).

Table 6.2. Deadwood compartment in Kerangas forest and Mixed Dipterocarp forest in RP 146, Sabal Forest Reserve, in 1989–1990. The tree densities are enumerated data, the volumes of standing trees from $0.5g \cdot h$. The tree biomass (phytomass, PM) equals volume × weight conversion factor ($v \cdot cf$). Deadwood > 10 cm diameter on the ground has been enumerated and mapped. KF, *Agathis*-bearing Kerangas forest; MDF, *Dryobalanops beccari*-bearing transitional KF–Mixed Dipterocarp forest. Biomass figures in brackets are totals including high stumps and crown, and logs left behind by the loggers.

Category	KF, partly creamed	MDF, selectively logged 1978
Standing living trees		
Trees (N ha^{-1})	688	696
Volume (m^3 ha^{-1})	407	417
Biomass (t ha^{-1})	306(447)	313(459)
Standing dead trees		
Trees (N ha^{-1})	84	64
Volume (m^3 ha^{-1})	21	31
Biomass (t ha^{-1})	14	24
Deadwood on the ground		
Trees (N ha^{-1})	272	336
Volume (m^3 ha^{-1})	97	123
Biomass (t ha^{-1})	63	80
Abandoned merchantable logs		
Volume (m^3 ha^{-1})	–	17.5 (38%)[a]
Biomass (t ha^{-1})	–	11.4

[a]The proportion of cut and cross-cut merchantable timber (logs) of felled trees which for unknown reasons were not extracted in 1978.

from *Dry. beccarii*-MDF to *A. borneensis*-KF and *C. nobilis*-KF (Fig. 1.10) (Bruenig, 1966, 1974). Unavoidable harvesting losses, timber defects, market-orientated grading to avoid rejection of logs and management failures reduce the actually merchantable yield by at least 50%.

Stem diameter increment is a good indicator of the length of time needed between harvests to grow log timber of certain size and grade. A long-term study, Research Plot RP90, Forest Department Sarawak, was established in 1975 in a 300 ha logged MDF in Niah Forest Reserve, south of Miri (Fig. 1.16). The plot was measured six times between 1977 and 1990 by the standard procedure prescribed by Synnott (1979). During the 13-year observation period, the mean annual diameter increment for trees > 10.0 cm diameter was 0.39–0.50 cm in thinned subplots and 0.23 cm in the control subplots. In the thinned subplots, Dipterocarps grew annually by 0.71 cm, non-Dipterocarps 0.37 cm in diameter. The *Shorea* spp., Meranti group had a higher growth rate (0.77 cm) than other Dipterocarps (0.58 cm). Merchantable basal area increment was 0.16 m^2 ha^{-1} a^{-1} in the control plots and 0.41–0.86 m^2 ha^{-1} a^{-1} > 30.0 cm diameter, in the thinned plots. The merchantable volume increment was approximately $2.0 \, m^3 \, ha^{-1} \, a^{-1} > 45 \, cm$ diameter (Chai *et al.*, 1994). On average, diameter increments of immature A-layer trees in zonal rainforest lie between 0.3 and 0.6 cm, but between 0.5 and > 1.0 cm in Dipterocarps (Reyes, 1958; Dawkins, 1959, 1963; Carvalho and Lopez, 1989; Chiew and Garcia, 1989; Rai, 1989; Thang and Yong, 1989; Fig. 6.3). The performance of prudently selected and released elite trees exceeds these values by 25% and up to 100% (Dawkins, 1958; Poker, 1992), so that means of sample plots are of little value to planning. Diameter increment is useful as an indicator of volume and value increment only if related to basal area increment. It can be very misleading if used alone to optimize growing schedules and times of harvesting. A diameter increment of 7 mm in a 40 cm diameter tree produces $44 \, cm^2$ or 3.5% tree basal area increment. Half this diameter increment (3.5 mm) in an 80 cm diameter tree produces about the same ($45 \, cm^2$), which is only 0.9% basal area increment, but relatively more value increment. If the diameter of the same 80 cm diameter tree grows at 1.3 cm (Fig. 6.3), the basal area increases by $165 \, cm^2$ or 3.3%.

The conclusion is that basal area increment should be accelerated by thinning as much as possible and feasible, especially during the great period of growth between 40 and 80 cm diameter, and certainly trees in this diameter range should not be harvested. Accordingly, liberation thinning was tested, recommended (Hutchinson, 1981; also Jonkers, 1982) and applied on 35,000 ha in Sarawak, but later abandoned. The opening of the canopy by selective logging was considered sufficient to stimulate growth adequately to produce an economic crop equal to the original forest within 25 years. This is probably true, but not many good trees are left to grow after conventional selective logging and those few need to be released from any inhibition. Hutchinson rightly stressed the importance of illumination of the crowns of the leading desirables (selected élite trees). This conforms with earlier results of Dawkins (1963) and later findings of Woell (1989) and Poker (1992). In contrast, Primack *et al.* (1987) found in an analysis of plots established between 1935 and 1955 that release (improvement, TSI) fellings did not increase mean diameter increments in comparison with trees growing in primeval forest, though variation in site characteristics, initial size distribution and species composition may obscure any silvicultural effects. Primack *et al.* found long-term patterns of annual diameter increment dominated by a striking but unexplained 3–6 year cycle of rapid growth followed by slow growth. Years of slow growth are known to be correlated with drought, flower and fruit production and this may have affected these rhythms. Primack *et al.* also detected in their data an unexplained decline in growth rates during the 20 years covered, which is not seen in any other datasets.

In the shifting dynamic mosaic of species-rich tree-association groups in a variable climate and on a heterogeneous soil, the growth of trees with or without harvesting and the effects of TSI are not easy to detect and predict. Yield tables and stand table projections require the lumping of tree-association mosaics into one population in order to calculate means which are neither statistically (Bruenig, 1973d) nor biologically (phasic mosaic dynamics, Poker, 1992) sufficiently homogeneous. Therefore, while the effects of TSI are not predictable with certainty, evidence (Table 1.10) suffices for a precautionary low-risk strategy in favour of application of TSI. The costs should wherever feasible be recovered from sale of the fellings for board and utility timber manufacture. Pre-logging climber cutting is also feasible and generally advocated. It is helpful in harvesting, reduces damage and is therefore economically worthwhile. Less simple is the answer to the question of sufficiency of gap size and growing space to promote the growth of selected élite trees and stimulate the growth of regeneration. It is also hazardous to estimate how many élite tree candidate trees in the 30–80 cm diameter classes are required as residuals in the first cut to ensure future sustainable supply of crop trees over successive felling cycles (Alder, 1992; Osho, 1995). Even more intractable is the prediction of the reaction of a rich mixture of species to canopy opening. The various species will react differently not only to increased exposure, but also to climatic, especially rainfall, and nutrient input fluctuations. A pattern of reactions in a certain period of observation may be completely different in another period with different climatic conditions or nutrient inputs. Previous experience is therefore of little use for predicting the future.

Long-term silvicultural experiments in better-than-average Philippine MDF showed that small gaps below 0.1 ha did not induce demonstrable growth of small trees and regeneration of the light-demanding emergent Dipterocarp species (Woell, 1989). Gaps had to be at least 0.1–0.2 ha, to stimulate significantly growth of fast-growing Dipterocarps. These findings accord with observations elsewhere (Stocker, 1985). The fall of one A-layer tree (diameter 80–120 cm, height 45–60 m, vertical crown sectional area 20×25 m = 500 m^2) causes one A-layer gap of about 400–600 m^2. The area below the canopy gap will be occupied by C- and D-layer trees. The falling tree will create another gap in the impact area. The size is very variable but averages around 500 m^2. If five trees on 1 ha are skilfully harvested, with at least two crowns falling on the same spot, the total gap area would theoretically be 0.2–0.3 (A-layer) + 0.2 (ground) = 0.4–0.5 ha (Fig. 6.5). The 50% gap area exceeds what would be needed for adequate growth stimulation. Damage by the falling trees to the residual stand and regeneration will be heavy. A cutting limit of four trees would appear to be a safer minimum threshold standard. With reduced impact logging (RIL), this could yield approximately 50–60 m^3, cause 30–40% gap area and 4–5% skidtrail surface, which would be tolerable if felling cycles are > 40 years. The basic structural features of the canopy will be maintained, but the aerodynamic roughness, heterogeneity of the canopy and the amount of rain and light reaching the ground will increase. Additional precautionary limitations are necessary to assure minimum damage and rapid restoration of the protection and habitat functions in

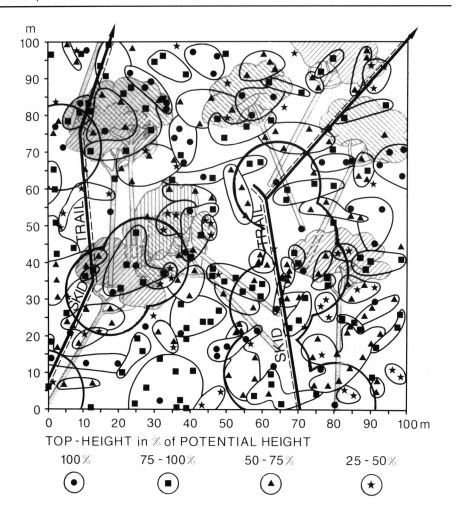

m

100

90

80

70

60

50

40

30

20

10

0

0   10   20   30   40   50   60   70   80   90   100 m

TOP - HEIGHT in % of POTENTIAL HEIGHT

100 %	75 - 100 %	50 - 75 %	25 - 50 %
⊙	▣	▲	★

Fig. 6.5. A combination of the dynamic mosaic model (Poker, 1992, 1995) and simulated selection harvesting assuming pre-felling liana-cutting and directional felling. The simulation can be used for an assessment of the scale of gap formation and possible damage under different types and intensities of harvesting. In the example the Grebo forest mosaic (Figs 1.1 and 1.23) has been used to stimulate the harvest of 10 trees of Mixed Dipterocarp forest size which contain 100–120 m^3 merchantable timber, but only 75 m^3 logs are extracted. The skidtrails cover 4.5% (exclusive of 'criss-crossing'). Removal causes 20% gaps in the A/B-canopy layer. The falling trees cause 15.3% gap by the crown impact (Fig. 1.25) and 2.3% by the stem. The net area of gaps on the ground and in the canopy is 37.6%. The total disturbed area is 37.6 + 4.5 = 42.1% (excluding criss-crossing). While still debatable, this appears to be beyond the threshold of the minimum safe standard of sustainability (ITW, 1994a,b,c,). (From Bruenig and Poker, 1989, 1991.)

sensitive areas of the management unit, such as watersheds or wetlands. Simple and inexpensive measures, which will affect PEP by at most 5%, are the exclusion within the production forest of logging from river-banks, stream-course strips, narrow ridges and ridge tops. This would create a patchwork of primeval stands with very high habitat and amenity value, which can provide seed, protection of water, soil and site, refuge and migration routes for animals. Similarly, the retention of overmature giant trees as habitat and genebank at a density of one tree per 1–3 ha will reduce PEP of the next harvest maybe by 1–3% at most, but will have a beneficial effect on regeneration, gene preservation and ecosystem self-regulation in the long run. The problem is that the feller is a financially tightly squeezed sub-subcontractor who is left to himself in the forest without advice and instruction except to produce logs. This he does as fast as he can. Sustainability is no concern of his.

6.4 Unsuitable Systems

The conventional simple 'selective logging with 50–60 cm' diameter (girth)-limit and too-short (15–30 years) felling cycles, and clear-felling of rainforests over large tracts remove the fast-growing and value-producing trees in the 40–80 cm diameter classes and destroy the architectural and organizational structure of the ecosystem and the PEP potential (Figs 6.3 and 10.1, Alder, 1992; Chiew and Garcia, 1989, fig. 4). Both change essential functions and the habitat value (Figs 6.6–6.8) and are not sustainable management systems. Harvesting a very large proportion of the growing stock in a uniform system (Fig. 6.4, the shaded and cross-hatched portions) produces high immediate cash flow and maintains high rates of volume (phytomass) production (Dawkins, 1959, 1963), but low rates of value and quality production (Bruenig, 1967a) and fails in habitat, soil and water protection. Andulau Forest Reserve, Brunei Darussalam, Compartment 2, was very heavily logged and, in addition, all trees > 5 cm diameter were poison-girdled according to the Malayan Uniform System (MUS) in 1956. Criss-crossing crawler tractors and poison-girdling caused very heavy gully and sheet erosion. The poison-girdling also removed the subcanopy matrix, which is essential to diversify the light climate and the growing space so that the Dipterocarp regeneration differentiates itself naturally while growing through this matrix. This fundamental interdependence between vegetation structure, microclimate and the radiative energy/plant/leaf/growth regulation applies globally. Modification of crown illumination by manipulation of the light climate in the vegetation is the most effective and artful operation in growing valuable timber with healthy trees at little cost and for hardly any direct expenses. In Andulau, the canopy became, predictably, dense and uniform, and was very badly overcrowded in 1993, 37 years after MUS treatment. The diameter growth was severely suppressed, by at least 20–30% but possibly > 50% below potential. Release operations were long overdue (Fig. 6.9). The crop did not meet the goals of high value production, habitat diversity and effective soil and water protection. If there had been fewer Dipterocarp seedlings and saplings on the ground in 1956, the development might have been less

ECOLOGICAL-SOCIAL FUNCTION INDICATORS
(humid and wetter tropical rainforest)

1. Natural Pristine-Primary and Secondary Successional Forest

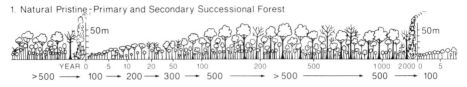

2. Edaphic Gradient in Pristine-Primary Forest (excess-optimum-pessimum)

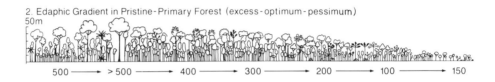

3. Anthropogenic Gradient Pristine to Deforested Land/Intensive Agriculture

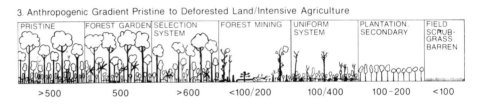

I. Aerodynamic Roughness Estimator "z_0"

Fig. 6.6. The change of rainforest stature and canopy structure and the correlated change of aerodynamic roughness and actual evapotranspiration (Fig. 6.7) which are sustainability indicators for ecological–social functions in three cases: 1. Natural succession after catastrophe such as shag or clear-felling in the zonal Mixed Dipterocarp forest formation on deep sandy–loamy–clay soil. The time scale is logarithmic. 2. Vegetation change with declining suitability for tree growth of soil conditions from relatively nutrient-rich and supersaturated alluvial through deep, well-drained upland soil (zonal formation with tallest stature) to very unfavourable shallow, badly drained, nutrient-poor and alternately saturated and parched soils (see Figs 1.11 and 1.15). 3. Human interference by modification (logging), manipulation (silviculture) and replacement (secondary forest, plantations, agricultural crops, degraded land) (see Chapter 2). z_0, aerodynamic roughness estimator (calculated from canopy parameters) (Bruenig, 1970a). The change of structure and stature is associated with corresponding changes of the aerodynamic roughness expressed by the estimated (or measured) parameter z_0 as indicator of architectural canopy complexity, habitat diversity, tree-species richness and dominance diversity patterns (evenness of mixture) (Bruenig, 1969b).

unfavourable, but this is uncertain. On the other hand, if there had been no regeneration on the ground at the time of felling, the residual forest might today be economically unproductive. The alternative, in this case to postpone felling until seeds fall and regeneration is present, is unrealistic. Mass fruiting of Dipterocarps in MDF, KF and PFS occurs regionally simultaneously at irregular intervals. Timber harvesting would have to be restricted to the 5–7 years after fruiting during which regeneration, on average, survives; this is impractical. Also impractical are regeneration

ECOLOGICAL-SOCIAL FUNCTION INDICATORS
(humid and wetter tropical rainforest)

1. Natural Pristine-Primary and Secondary Successional Forest

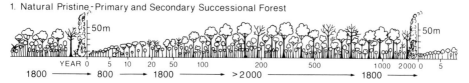

2. Edaphic Gradient in Pristine-Primary Forest (excess-optimum-pessimum)

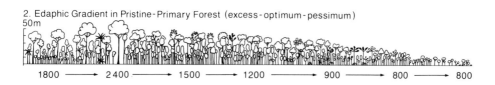

3. Anthropogenic Gradient Pristine to Deforested Land/Intensive Agriculture

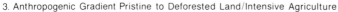

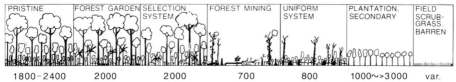

II. Actual Evapo-Transpiration ET_0

Fig. 6.7. Vegetation as in Fig. 6.6, change of the calculated (Bruenig, 1966, 1971a) and measured (San Carlos de Rio Negro, Heuveldop, 1978, 1980) rates of actual evapotranspiration, assuming unrestricted water availability and excluding drought conditions, along the vegetational gradients of canopy structure and soil conditions (Figs 1.11 and 1.15).

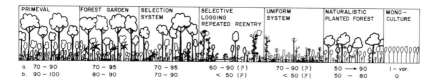

Fig. 6.8. The effect of forestry on tree-species richness and diversity. (a) Tree-species richness expressed as number of tree species in 100 individuals > 1 cm diameter and (b) evenness of tree-species mixture as percentage of maximum possible for the given species richness according to the McIntosh index (Figs 1.21 and 1.22), in relation to different systems of forestry interference (see Figs 2.12–2.14, 6.4 and 6.6). The canopy complexity and height of the naturalistic planted forest can equal the primeval forest or forest garden if given sufficient time (60–100) (Figs 7.7 and 8.1).

release operations prior to felling, which have to start far ahead of logging and must be synchronized with the rhythm of fruiting; this is difficult enough in temperate forests with highly developed access systems, but impracticable in most tropical rainforests. If there is no regeneration and

Fig. 6.9. MDF, exploited and poison-girdled under MUS in 1956/7, grew into a congested, overcrowded tree stand by 1994 (background) which urgently requires costly tending to release future-crop trees. Logging and heavy poison-girdling caused erosion which buried fertile alluvial land and killed the forest, which adds to the high social costs of the system (foreground). Andulau F.R., Brunei Darussalam, 1994.

MUS-type of harvesting is done, the regrowth may even stagnate, as parts of the Kelantan storm forests show. Another problem is the poison-girdling of a large number of trees, which may have undesirable side effects. Firstly, it brings human beings in contact with poisons and, secondly, it brings poison into the environment. Especially dangerous in both respects is sodium arsenite, which is absorbed and fixed in acrisols/ultisols in the form of organometal complexes or adsorbed on clay minerals. Any subsequent clearing and burning causes defixation and leaching of the poison into the water bodies (Titin and Glauner, 1993).

Extremely heavy forms of interference, such as the classical MUS or the Modified Malaysian Uniform System (MMUS) with a higher diameter limit (about 10 cm) for poisoning, as well as clear-felling, may cause conflict with legally recognized native customary rights of forest usufruct. The clear-felling in the Madang project in Papua New Guinea for chip production for Japan (Davidson, 1983; Nectoux and Kuroda, 1989; Johns, R.J., 1992; Sections 2.8 and 4.5) diminished shade-loving medicinal plants, valuable animal species and other NTFP, and traditionally used timbers. It also eliminated canopy-dwelling animals and plants, and river fauna and flora. However, some vertebrate game species, lianas and climbers, such as rattan, and fruit-producing, light-demanding secondary forest species increased. On balance, the habitat value for hunters and gatherers declined and customarily used timbers disappeared at least for the next 50–100 years (see also Section 8.3). The alternative option of village participation is not without complexities, vagaries and contradictions. The attitudes of the participants often suddenly and unexpectedly change (Songan, 1993). Management systems which depend on and are exclusively designed for active participation carry a high risk of failure unless they are based on a thorough socio-economic and cultural system analysis and confer substantial benefits. In conclusion, MUS and clear-felling may in certain forest types produce the highest immediate cash flow (Bruenig, 1967a), but are not suitable options for sustainable management of MDF (see also Wyatt-Smith, 1963) even if local people agree and participate. SMS, which harvests only marked trees above 80 cm diameter by RIL but leaves derelict giant trees standing (Fig. 2.12, large tree on left) thus retaining a near-natural diverse habitat structure, supplemented by totally protected areas (TPA), is more likely to reach the specific goals of the forest policy and the general goals of social forestry.

6.5 Technique and Standards of SMS in MDF

The tropical rainforest is naturally very robust. Even in extreme cases of severe forest ecosystem degradation through misuse and abuse by loggers, such as in Demaratok forest in Sabah, there is evidence that the MDF ecosystem is capable of recovery (Sections 2.3 and 2.4). The storm forest of Kelantan and very old, descriptively documented (John, 1862) secondary forests in the Melinau–Tutoh triangle in Gunung Mulu National Park, Sarawak, support this view. On the other hand, the stagnation after fatal defoliation of the whole canopy of *Shorea albida* in the Peatswamp forest and the stagnating Padang scrub types after Gambir cultivation in Bako National Park are obviously examples of transgression of the critical threshold for recovery. They are a warning against clear-felling and tree monocultures on fragile sites (Sections 6.7, 6.8 and 7.4). Such documented cases of 'ecosystem collapse' may warn, but give no direct indication where the levels of safe standards for sustainable modification and manipulation may be. If nature does not provide indicators and definite guidelines, we ourselves must judge and decide. Features such as diversity of taxonomic and physical structure of the forest, complexity and vertical integration of the canopy, amount and kind of litter production and humus formation are key elements of robustness, self-regulation and self-sustainability. Management must be designed so that these key elements are not adversely affected, but possibly improved. Key indicators are canopy structure, illumination, ground vegetation and site-specific indicator species (*Weiserpflanzen*). The social balance between costs of silviculture, conservation, protection and management, and the value of the multiple benefits will indicate roughly where thresholds of modification and manipulation may be reached.

The intact primeval forest can serve ideally as a prototype of successful survival. Primeval zonal equatorial rainforest, outside the cyclone belts, has natural gap sizes in the A-layer that range from 0.02 to 0.1 ha (single tree mortality) and up to about 0.3–0.6 ha if several large trees are killed by lightning or thrown by wind. Rare gale-force wind bursts may cause windthrow on several hectares to hundreds of hectares. Selective logging increases the number and size of gaps drastically, selection felling in SMS less so. Well-operated SMS does not reduce complexity and diversity; it may even increase species richness and diversity and maintain the complexity of habitat structure. A threat of decline arises possibly with successive felling cycles. There may be unidirectional changes in floristic and architectural structure (Weidelt and Banaag, 1982), if felling causes drastic structural changes in the A- and B-layers of the canopy at each successive felling. In some forests, such as the successional African rainforests and moist forests, the phasic communities in the Malesian Peatswamp forests (Section 6.7) and also the Malaysian *Shorea curtisii* hill forests, natural regeneration of the present emergent canopy species may be sparse or absent. Heavy felling may then cause a drastic floristic change. Inducement of natural regeneration is difficult, as it requires the formation of large gaps at the right time, which is unpredictable in most years. Planting the desirable species may be ecologically possible, but economically, is not feasible.

A good SMS combines harvesting with invigorating the natural dynamics of the forest. It promotes, if done well, natural regeneration and growth of trees of all sizes from seedlings to trees in the B-layer of the canopy, and to a lesser degree also in the A-layer, with the least possible change of canopy structure as a precaution against disrupting nutrients, water and organic matter cycles. Planting into logged forest is an admission of careless overlogging, skidding and roading or of ignorance. It was argued in Sections 2.7 and 2.8 that appropriate selection of trees and proper directional felling, and an extraction system which is adapted to the site, soil and forest condition, are the keys to success. Harvesting must pay and the growth of the B-layer residuals must be stimulated. This sets the minimum requirement for change of canopy structure. The upper limit is set by the need to sustain productivity, protection and habitat functions. As a compromise, harvesting should remove not more than about three to four A-layer trees, 10–20% of the volume and biomass, and not more than 25–40% of the canopy (A, B and C). Trees should be felled so that the crowns do not fall into gullies and water courses, where the decomposing organic matter contaminates the water. Good workmanship assures that skidding does not require the devastating turning of logs sideways (Section 2.8). High standards of engineering in harvesting and roading, minimizing earth movement and optimizing drainage and gradients, reduce water runoff, soil erosion and loss of organic matter. The target should be an upper limit of increase of sediment and water discharge of 5–7% (2–3% of the rainfall) from the whole area of the management unit (working circle). Standards for water quality and maximum critical loads of contamination have yet to be found.

The requirement to stimulate growth of the residual growing stock is difficult to monitor. The average growth rates in yield tables and stand-table projection models, if used at all, do not represent the actual, very fluctuating growth processes and are of limited usefulness as a yardstick and for forecasting. It is the future-crop élite trees which are of interest to timber management, not the average progress of stand means. Defects, especially the high frequency of hollowness in apparently old trees, also affect volume and value yield predictions (Vanclay, 1991c) as much as market fluctuations. Consequently, there is always a strong element of uncertainty, which creates the need for more sophisticated growing stock and system analysis and dynamic growth simulation (Korsgaard, 1988a,b; Vanclay, 1989, 1991a–c, 1994; Bossel and Bruenig, 1992; Ong and Kleine, 1995). The most desirable upper-canopy tree species are light responsive and require, for good growth responses, a well-illuminated crown from seedling stage onward. This requires heavy canopy opening, but suggested canopy opening above 50% (Ajik and Bernard in Putz, 1993) will cause excessive damage and soil exposure and, if regeneration is successful, would eventually lead to a more uniform system such as the MUS. At the other extreme in practice, forest field staff and loggers prefer to mark the prescribed number of trees for retention as residuals among the more abundant trees in the 20–30 cm diameter classes, which makes work easier, and as far away as possible from the area where the canopy will be opened and felling damage can be expected. The result is that 'the trees with the least probability to reach the target diameter of 60 cm dbh represent the bulk of the residuals' (Uebelhör *et al.*, 1990), that is, they are small trees oppressed or suppressed by a dense B/A canopy.

In Sarawak and elsewhere, the following shortcomings, risks and uncertainties are common even when SMS is prescribed and adopted:

- The length of felling cycle and minimum harvesting diameter are set too low, or undersized trees are illicitly cut; consequently the potentially most vigorous trees in the 30–80 cm diameter classes are removed. Remedy: proper system-analytical growth and yield estimation and simulation; flexible diameter limits and felling cycles to suit the local situation in the forest.
- The forest will become more patchy and heterogeneous if tree marking and felling are not properly done and enforced; patches with heavy stocking of merchantable trees are practically clear-felled while overmature or poorly stocked patches remain untouched; this may, ecologically, even have advantage, but is undesirable from an economic point of view (Appanah and Weinland, 1992). Remedy: strict enforcement of tree marking and logging supervision, TSI.

- Excessive damage is done by falling crowns. Remedy: directional felling; if possible, reduction of felling intensity, pre-felling climber cutting (this will not eliminate climbers and not endanger key food sources) and above all training of the fellers.
- Excessive or unnecessary silvicultural treatment which reduces habitat value and biodiversity and is expensive, is routinely applied. Remedy: ecological and economic cost–benefit analysis to define optimum level of interference and apply silvicultural treatment only if definitely necessary.
- Necessary liberation thinning is not applied (Uebelhör et al., 1990). Remedy: model simulation of individual and stand tree growth and yield to guide decision; efficient planning and monitoring (Chin et al., 1995).
- Excessively heavy felling as a result of erroneous increment assessment or failure to comply with yield prescription. Remedy: reassessment of increment by more sophisticated approaches, continuous forest inventory, effective enforcement.
- Excessive damage to soil and growing stock by unskilled work of fellers, tractor drivers and road builders, skidtrails and roads are without effective drainage, soils erode, water bodies are polluted and turbid. Remedy: survey, problem analysis, correction of causal factors, upgrading of engineering and harvesting, effective enforcement by intensive training and motivating concessionnaires and contractors from management to labour levels.

Enforcement can be secured in two ways. One way is by control of private operations by the public forest service (negative incentive linking compliance to penalties) who also does all silvicultural operations. The other way is to transfer the responsibility for compliance in harvesting and silviculture to a private company under a long-term tenure contract (positive incentive linking compliance to benefits). The latter arrangement, which restricts the governmental action to supervision and monitoring, is usually more efficient and effective. In Sarawak, the first approach is practised, regulated by the Sarawak Forest Management Code (1961), Sarawak Forest Department Logging Block Inspection Rules (1960) and Instructions for the Inspection of Logging Areas (1982). The prescriptions of the code and rules include the traditional subjects: engineering works (roads, drainage, log ramps), survey of forest site, soil, growing stock, rare and protected species, legal situation and protection of boundaries, water catchment and nature reserves. The management plan prescribes volume yield and felling areas for the current cycle, including preharvesting mapping and postharvesting inspection of coupes and blocks and any silvicultural treatment. The mapping of coupes and blocks includes demarcation of boundaries, position and drainage of roads and skidtrails, position of all trees to be harvested and residuals (élite trees, seed bearers, habitat trees, protected tree species and native-use tree species). After completion of harvesting operations in a block, compliance with RIL prescriptions is checked and the block closed to re-logging. Especially during the height of the boom, logging prescriptions were not observed because the concessionnaires had no incentive and no under-standing, and were not enforced because the public forest service was neither motivated nor adequately empowered. Silvicultural prescriptions for girdling, planting, release operations and liberation thinning (Hutchinson, 1981; Jonkers, 1982; Chai, 1984; Sagal, 1991) suffer from uncertainty about the feasibility of silvicultural treatment. SMS, if properly applied, enhances some of the values of the forest as a source of NTFP for the local people, such as game, fruit, honey, cane (rattan), latex, special kinds of traditional timber, spices and medicinal products, but transgressions by the loggers are habitual. Unique ecosystem types, stream banks and nature reserves (virgin jungle reserves) are or should be mapped and maintained as TPA to serve as documents, refugia, seed sources and genetic reservoirs. Logging on slopes steeper than 25° is now generally prohibited, but enforcement of this prescription, as of management plans generally, is difficult and falls short of needs in practice (ITTO, 1990a).

The most impressive example of successful application of SMS at large scale and practical level was the Queensland Selection System. Small-scale examples at experimental level are the Puerto Rico Selection System, CELOS Selection System in Surinam, selective harvesting-cum-silvicultural management systems as described among others by Bruenig (1957, 1973a), Lee (1979) and Lee and Lai (1977) in PSF and in some cases in MDF in Sarawak, and the few documented instances of still existing SMS in surviving patches of lowland MDF in Peninsular Malaysia (Francke, 1941). In all cases the first condition is that timber harvesting is done properly. Successful prewar and postwar examples in the lowlands of tropical Asia and Africa (Francke, 1942) generally have disappeared or are disappearing as a result of shifting cultivation, agricultural development and overlogging. The establishment of a regional documentation and integrated network of statistically adequately designed research and monitoring areas is, therefore, an urgent need to quantify the effects of management (Korsgaard, 1993c,d), to test experimentally a wide range of felling grades and types, and to study the physio-ecological effects of various types of canopy opening.

Rainforest protection and management is, with the exception of the above-mentioned and now mostly historic examples, inadequate; upgrading is overdue in the whole rainforest area. Particularly urgent is the phasing-out of non-sustainable selective overlogging and its replacement by SMS and RIL harvesting. A rapid transformation is foremost in the interests of the forest-owning countries and the loggers themselves. It is technically possible, socio-economically and financially profitable in the short and long run, and socially not only desirable but absolutely imperative. It requires initial expenditure in the form of investments into organization, infrastructure, training and equipment, which are quickly paid back and produce additional profits after 2 or 3 years, in PSF even earlier. Improved quality harvesting and utilization and efficient sustainable forest management has in all trials in Malesia, Africa and America proved more profitable than selective logging. The introduction of RIL in Peatswamp forests PC1 in the 1950s in Sarawak, against initial protest and opposition by the loggers who feared reduced profits, quickly produced higher net stumpage values and more timber from fewer trees felled. The same has been demonstrated to hold true in MDF in Sabah and MDF and PSF in Indonesia. The introduction of professionally skilled management and the creation of a motivated, skilled permanent labour force create visible expenditures in the company budget, but it reduces the manifold invisible losses caused by the habitual inefficiency and under-capitalization of the cash flow-orientated logging industry. The Malaysian–German Forestry Research Project for Sustainable MDF – Management in Sabah estimated the following costs for their SMS: RM34 ha^{-1} (DM19) for medium-term (20 years) management planning, including integrated inventory, or RM1.70 (DM0.95) ha^{-1} a^{-1} (Kleine and Heuveldop, 1993). Total management costs, including silviculture and protection, for the Indonesian version of SMS, TPI or TPTI (Tebang Pilih Indonesia) according to the Direktorat Jenderal Indonesia (1980) costs about DM40 m^{-3} for production (silviculture, management, protection), plus about DM80 for harvesting, roading and overheads (Sutisna, 1994). Naturalistic SMS is ecologically and environmentally superior to MUS or Tropical Shelterwood System (TSS). It is technically feasible, economically attractive and socially far more acceptable than other natural forest or plantation systems and selective logging, which is no system, in the wet tropics. Current forest development and land-use policies and programmes in the Malesian rainforests have to be reviewed in this light.

6.6 SMS for Fragile Oligotrophic Upland Soils in KF

6.6.1 Site conditions

The structure and the physical and chemical conditions of the fragile oligotrophic soils in Malesian KF, as well as in Amazonian Caatinga, are wholly or almost entirely dependent on the biological activities in the humus above and in SOM within the soils, and on atmospheric inputs (Sections 1.3 and 1.7). The organic soils on limestone karst are excluded because they are not suitable for sustained timber management. The oligotrophic Kerangas and Kerapah soils are extremely acidic, have a precarious nutrient status and are hydrologically unfavourable for plant growth (Section 1.3). The only effective absorptive complexes for nutrients and water, and sinks and sources for nutrients in most KF and Kerapah (KrF) forests, and in all Peatswamp forests, are the decaying standing trees, coarse and fine litter, the raw humus on the soil surface and the organic matter in soil. The clay component is dense, impermeable; the e.c.e.c. of the simple clay minerals is predominantly occupied by aluminium ions (Kehlenbeck, 1994; also Section 1.7). Kehlenbeck concludes from field and laboratory data on soils in RP 146 in Sabal Forest Reserve that the physical properties of permeability to water, penetrability to roots and water-holding capacity of the analysed humic spodosols, aquic spodosols and humult ultisols are ecologically more important and relevant than the nutrients in the mineral soil component. This accords with Bruenig's earlier view that these ecosystems wholly depend on nutrient replenishment from the air and from water inflow (Bruenig, 1966, 1986a). The sandy humus podzols have an aggregate structure in the Ah and Bh horizons which is cemented by SOM. The Ae is weakly structured and easily crumbled. The A horizons in the sandy podzols are not as susceptible to compaction as the clayey MDF soils. White-clay and skeletal (rocky) soils in KF are already very dense; pressure from vehicles does not increase density much further. Regeneration on skidtrails on Kerangas sands is rapid as long as organic matter and mineral soil are not completely removed. The extreme, small-scale heterogeneity of the KF soils (Bruenig, 1966) with respect to rooting depth and hydrological conditions and the high susceptibility of the low-absorptive but SOM-rich soils to drought, fire and nutrient loss are an enigma to management.

6.6.2 *Kerangas and Kerapah vegetation*

The floristic structure, and the sclerophyllous, xeromorphic physiognomy and low, aerodynamically smooth canopy of KF are as distinctly heterogeneous (Sections 1.5, 1.6 and 1.7) as the soils. This may indicate direct adaptations to the adversities of the site conditions (Bruenig, 1970a). Illumination in the C- and D-layers is favourable for regeneration in a small-gap or shelterwood regime, except in the more lush forest types on soils with more favourable hydrology (Figs 1.10, 1.14 and 1.18), which require larger canopy gaps. The variation of the levels of chronic stresses and the occurrence of episodic severe strains, especially drought (Bruenig and Muellerstael, 1987), between sites is reflected by the variation of structure and physiognomy of natural KF. Pioneer fast-growing, short-lived tree species with high water and nutrient demands do not occur in KF (Browne, 1952; Bruenig, 1966, 1974). The potentials of Kerangas and Kerapah vegetation for growth and repair after disturbance are low. Bruijnzeel *et al.* (1993) hypothesized that a possible cause of stunting in a cloud forest at 870 m a.s.l. on Gunung Silam, Sabah, may be caused by a high carboxylation resistance which results in low photosynthetic rates and transpiration. As nitrogen concentrations in the leaves or litter are not lower than in non-stunted forest at 680 m a.s.l., it is suggested that the high concentrations of phenols in the leaves and litter, on leaching into the soil, could interfere with ion uptake and cell division in the fine roots and with photosynthesis and transpiration in the leaves. Non-hydrolysable tannins or polyphenols have long been suspected not only of masking the N-containing proteins in the litter but also of interfering with photosynthesis and respiration in the leaves (Bruenig, 1966, 1970a), which was corroborated by evidence for interaction between organic compounds in peat soils and photosynthesis and respiration in Ramin leaves (Bruenig and Sander, 1983). Further problems on these very acid soils may be poor accessibility of phosphorus to the roots, but there is no clear picture of leaf nutrient contents in relation to KF and MDF (Sections 1.3 and 1.7; Riswan, 1991). The most important impediments for management are drought related: the high fire risk and the uncertainty arising from drought-induced mortality among seedlings and saplings (Becker and Wong, 1993). The success of regeneration after seedfall on these soils is the result of many interacting factors, such as the root–shoot ratio of seedlings and saplings when drought occurs, and the length of drought conditions. Illumination in KF is not a critical factor for regeneration, except in the richer KF types (Fig. 1.18, Yevaro forest).

6.6.3 Biological and economic productivity potentials

The ecological research in Kerangas forest was a spin-off of a biomass reconnaissance survey in 1954–1955 which was discontinued when it became obvious that the biomass stocking and the primary productivity are generally low (Section 1.13). Both are very low in KF on shallow clay, shallow lithosol and shallow humus podzol soils, KrF and the floristically closely related Peatswamp forest, PC4, 5 and 6. All these forest types are economically and ecologically unsuitable for any kind of production forestry but are locally rich in extractives and aromatic woods. Destructive collecting of NTFP by natives over centuries has eliminated this resource in most accessible areas. The only accurate timber growth data from mature natural KF are from two research plots. One is *A. borneensis* KF on medium deep humus podzol (MHP) in the former 20-ha long-term ecological research area in the Nyabau block of Similajau Forest Reserve north of Bintulu. The second is the coastal *Dryobalanops rappa* KF in RP 21 on groundwater humus podzol (GHP) in Anduki Forest Reserve in Brunei. The mean annual diameter increment was 0.4 cm in Nyabau and 0.5–0.9 cm in RP 21 of trees of 20–90 cm diameter. Thinning of the dense *D. rappa* stand increased diameter increment by 30–50% (Bruenig, 1974, p. 137). These annual diameter increments represent absolute maxima in KF on optimal sites with favourable hydrology. They correspond to an annual harvestable volume increment of selected final-crop élite trees in the 50–80 cm diameter range of 0.08–0.15 m^3 log timber per tree. The rotation or production period would be in the order of 100–120 years. If 50 candidates for the final timber crop are well spaced in the diameter range 20–90 cm, five to seven trees could be harvested every 20–40 years in the 60–90 cm diameter class, allowing 50% for unavoidable mortality, other

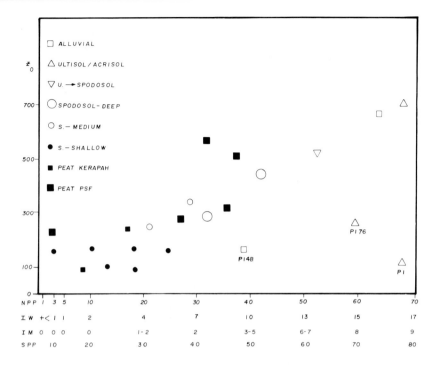

Fig. 6.10. Association between soil type, aerodynamic roughness of the forest canopy, the net annual primary production above ground of potentially usable wood in metric tonnes dry weight per hectare and the tree-species richness in research plots in Sarawak, Malaysia and Territorio Amazonas, Venezuela. SPP, species richness expressed as number of tree species above 1.5 or 5 cm diameter in 100 tree individuals in the natural forest stands only. The productivity (IW and IM) has been collated from published growth and increment research on ultisol–acrisol soils, alluvials and PC1 peatswamp soils, and assessed from canopy heights in PC2–4 and in Kerangas forests. The plantation (P1) increments are measured (RP 4B and 76) or collated from literature. IW, annual production of stemwood above ground in t^T ha^{-1}; IM, idealistic annual production of merchantable sawlog wood in t ha^{-1}; PI, planted forest research plots 4B (*Shorea splendida* (de Vries) Ashton) and 76-3A1 (*Shorea macrophylla* (de Vries) Ashton) and a high-yield monoculture (the SPP line does not apply to PL stands).

losses and calamities. These trees would yield 20–40 m^3 merchantable timber in the final harvest or a mean net annual increment (m.a.i.) between a minimum of 0.7 and a maximum of 2 m^3 ha^{-1}. Intermediate release thinnings could possibly add the same amount. This yield is so low that it would require exceptionally favourable site conditions to make sustained timber management economically attractive. Growth on poorer KF sites is even lower. The annual diameter increment of 0.9–1.0 mm calculated by Heuveldop and Neumann (1980) for the A-layer trees in the Caatinga high forest in the International Amazon Ecosystem Study at San Carlos de Rio Negro (Section 1.13) indicates very long production periods and very low volume yields. The growth rates vary widely between sites, but on average KF sites are considerably lower than in MDF (Section 1.13).

Potential commercial timber increments in KF probably range from zero on the poorest soils to a maximum of 5 m^3 ha^{-1} a^{-1} intermediate and final fellings on the more favourable sites and soils (Fig. 6.10). The very best sites in KF may produce diameter and volume increments that come close to average MDF rates (Sections 1.13 and 6.3, Table 1.6). These generally low growth rates in KF on poor to average soils are possibly a combined effect of slow release of N and P from slowly decomposing tannin-rich litter and slowly mineralizing SOM, and of the chronic stresses and episodic strains on trees and soil life from the extreme aseasonal fluctuations of water availability (Sections 1.2 and 1.7, Table 1.5).

6.6.4 Sustainable management

The low productivity renders KF and KrF suitable for commercial timber production only on the most favourable, medium deep to deep humus podzols well supplied with water, groundwater podzols and sandy humult ultisols on almost flat to gently sloping and easily accessible terrain. Such forests could be sustainably managed under an SMS or group-shelterwood management system. Biodiversity and canopy structure must remain adapted and not exceed the long-term carrying of the site. As in MDF, low-impact logging (RIL) and full utilization of felled trees are mandatory. The scope for silvicultural treatment (thinning, fertilizing) is limited by the sensitivities of the ecosystem (leaching, drought, fire) and by the low PEP. Only the hydrologically more favourable sites and valuable élite trees of such species as Bindang (*A. borneensis*), Sempilor (*Darydium* spp.) and a range of high-quality broadleaf tree species deserve silviculture and management. KF contains a variety of species which produce durable, high-quality timbers, such as members of the genera *Shorea*, *Dipterocarpus*, *Dryobalanops*, *Vatica* and *Anisoptera* in the Dipterocarps, and *Calophyllum*, Guttiferae. Some species produce cabinet woods, such as *Sindora* spp. (Sepetir), some *Anacardiaceae* (Rengas) and *Sapotaceae* (Nyatoh). A wide range of speciality timbers like Gaharu or incense wood (*Aetoxylon sympetalum* (Steen. ex Domke) Airy Shaw) have become rare through overuse but could be reintroduced if the formation of aromatic or ornamental heartwood could be induced successfully (Bruenig and Droste, 1995).

Principles of naturalistic silviculture in KF are to mimic the structural features of the primeval forest on the site and to retain the site-specific adaptations and species richness and diversity. The latter may be increased by reducing dominance cautiously to secure adequate ecological conditions. Dense patches of gregarious species could be opened by harvesting to favour more evenly mixed natural regeneration, provided this does not reduce adaptation. Enrichment planting is ecologically very risky and economically not feasible except perhaps for such valuable species as Bindang, Sepetir, Nyatoh and Gaharu on very favourable sites. In all types of KF, canopy uniformization, or making the canopy too rough and complex, clear-cutting and conversion to plantations are highly risky, non-sustainable practices. Reclamation of already degraded KrF ecosystems by artificial means is futile. Recovery should be left to natural succession and will take a long time, usually centuries (e.g. Bako National Park; Bruenig, 1961a, 1965b, 1966). The extreme heterogeneity of conditions of the generally marginal KF sites makes it essential to base any planting decisions on a very careful and thorough survey of site and soil, assessment of the hydrological conditions and a thorough analysis of ecological (drought, fire, acidification) and economic (cost/benefit) risk factors. The replacement of logged KF by plantations with hardy, site-adapted native tree species is possible, but ecologically risky and econom-ically not feasible. The same applies to enrichment planting into logged KF. Major limiting factors restricting silviculture are the high risk of failure due to chronic moisture deficiency, episodic drought, the low NPP above ground and the prohibitively low PEP. Dead organic matter above the soil, the microorganisms associated with it and SOM in the soil are the crucial elements for the efficient conservation and cycling of nutrients and water, and for the regulation of water storage and discharge from the KF soils with very strong and fast lateral water movement. Humus management demands RIL with low-yield levels to reduce disturbance of the humus layer and the canopy. As in MDF, modification of the canopy should remain within the limits of natural gap formation, which in KF differs strongly between sites. Harvesting should remove not more than three trees per hectare at a time. The gap area caused by harvesting should not exceed 20–30% of the canopy and ground surface. In comparison to MDF, KF suffers a much greater risk of fire during drought, and of soil degradation by exposure and disturbance. Long-distance cable systems would be more suitable in KF than tractors.

The most appropriate use which is fully adapted to the conditions in KF is recreation and nature education in the periphery of towns. KF offers excellent opportunities for tourism and eco-education in national parks (Bruenig, 1966, 1965b, 1974). Accessible KF are the most endangered forest ecosystem in Borneo. They are damaged by native overuse, and destroyed by commercial logging, fire, conversion to plantations and urban devel-opment. The fragility of KF vegetation and soils, and the adversity to plant growth of tropical podzols and related soils and the constraints this imposes on their use in agriculture and forestry are well documented (Dames, 1956, 1962; Bruenig and Klinge, 1969; Klinge, 1969), but are rarely observed in agricultural and urban development and in forestry practice.

6.6.5 Conclusion

The KF and KrF ecosystems cope with exacting site conditions in a number of ways which are a source of information for forest crop design under more favourable conditions. They possess a more favourable light climate in the C- and D-layers for A- and B-layer tree species to regenerate than MDF or PSF, PC1. The forest interior is brighter, drier and more breezy than in MDF, which creates high risks of fire and of drought damage. NPP and PEP are generally lower than in MDF, and the risks of drought-induced mortality and growth checks are very high. Timber production is only feasible in the taller and richer KF types on groundwater, giant and medium deep podzols and sandy humult ultisols (Bruenig, 1969c). In these forest types low-intensity SMS is feasible if the area and the growing stock are large enough. The main value of KF and KrF is for recreation, nature or eco-tourism, education in ecology and natural sciences, research and nature conservation. These highly adapted and resilient forests are an invaluable source of scientific information on the performance of very complex to relatively simple, more and less dynamic ecosystems under stress and strain, which is much more difficult and costly to study in MDF. KF and KrF may be included in landscape designs as part of networks of nature reserves and as corridors for the migration of wildlife through agricultural landscapes, and as green areas for recreation and wildlife habitat in urban areas. They are unique and valuable components for designing landscapes with high amenity value. Both KF and KrF are unsuitable for intensive plantation forestry and agricultural tree plantations. Only the most favourable sites are suitable for naturalistic silvicultural management. Groundwater podzols and humid grey–white leached clay soils may, under very favourable site conditions, be suitable for conversion to intensive, irrigation water-fed field agriculture if alternative land is not available and agricultural expansion is absolutely essential from a national economic point of view (Fig. 11.6).

6.7 SMS for Fragile Oligotrophic Peatswamp Soils

6.7.1 Peatswamp soils

The Peatswamp forest soils of Sarawak were described by Anderson (1961a, 1964, 1983) who commented on the diversity of the geomorphic features of the peat and the underlying substrate, which may be mangrove clay, sand–clay mixtures of beach vegetation or any of the Kerangas or Kerapah terrace soils (Bruenig, 1969b). The peat soils fully depend on rain and fog-drip as sources of moisture and atmospheric inputs of nutrients. There is no clear pattern of relationship between nutrient and trace element contents of the peat and forest type distribution. The presence of physio-ecologically active organic compounds (Bruenig and Sander, 1983) imposes restrictions on silviculture and poses a potential risk of failure to forest management. Other soil-borne risk factors are episodic drought, chronic SOM mineralization and subsidence of the peat after opening, removal or replacement of the canopy, due to warming of the soil and deficient litter production. Similar forest peats, except for the absence of *S. albida* peat domes, in West Malaysia have been described by Hahn-Schilling (1994), in Kalimantan by Driessen and Rochimah (1976) and in southeast Asia generally by Andriesse (1974).

6.7.2 *Peatswamp forest vegetation*

The vegetation of the Peatswamp forests in Malesia have been described by Anderson (1961a, 1964, 1983), Hahn-Schilling (1994) and Bruenig (1989f). The Peatswamp flora is related to KF and KrF, but poorer and less diverse. In spite of a pronounced tendency to patch-pattern, the forest stands within the phasic communities are more homogeneous within forest types than in MDF or KF, but differ distinctly between the six phasic communities described by Anderson (1961a, 1964, 1983) and Anderson and Muller (1975). The catena PC1 to PC6 (Figs 1.3, 1.4 and 1.11), according to Anderson and Muller (1975), represent a successional series, well documented by pollen profiles, from PC1 to finally PC6. In contrast, based on above-ground field surveys in Brunei, Yamada (1995) suggests two reverse and different sequences. PC1, especially Ramin, would in both series invade and replace Alan forest. If this were true, silvicultural strategies would have to be reconsidered, but the two sequences are difficult to reconcile with the data of Anderson and Muller (see also Sections 1.4 and 1.11) and the poor physiological performance of Ramin in Alan forest peat (Bruenig and Sander, 1983). Of these six PC series, only PC1, Ramin PSF, PC2, Alan Peatswamp forest (Alan association) and PC3, Alan bunga Peatswamp forest (Alan bunga consociation) are ecologically and economically suitable for sustainable management. PC2 and 3 have particularly severe ecological constraints. The fragility of these ecosystem types and problems with natural and artificial regeneration make sustainability a very distant goal.

The initial phasic community, PC1, Ramin (*Gonystylus bancanus* (Miq.) Kurz) Mixed Peatswamp forest is by far the most widespread PSF type. It occupies extensive tracts of newly forming peatswamp in the coastal plains. In older and rising peats it forms more or less narrow belts at the outer fringe of the raised peat dome. It is structurally the most complex, species-rich and dynamic phase of the Peatswamp forests. PC1 covers about 80% of the peatswamp area in Sarawak and is economically important as a source of Ramin timber. Ramin is common and locally gregarious in PC1 and rare in PC2, but also very widespread as a characteristic species on a variety of soils in KF and KrF (Newbery, 1991). It occurs in commercially attractive quantities only in Peatswamp PC1 where it occupies an intermediate position in the canopy between the A- and B-layers.

The next two phasic Peatswamp forest communities are dominated by Alan (*S. albida* Sym.). The phasic community PC2, *S. albida*–*G. bancanus*–*Stemonurus secundiflorus* Association or Alan Forest, usually forms a narrow band behind PC1 at the flank of the peat dome, but may in some flat, not dome-shaped, peatswamp areas cover large tracts. In PC2, Ramin is suppressed first into the B-layer, then into the C-layer and finally in PC3 completely wiped out by the vigorous, biochemical apparently aggressive (Bruenig and Sander, 1983) and expanding Alan. Gaps of 0.1–0.3 ha, attributed to lightning and wind, are conspicuous and common in the aerodynamically very rough canopy of PC2 (Bruenig, 1973a; Figs 1.3 and 1.4). These gaps, as far as we know, do not benefit the regeneration of Ramin, but large-area defoliation of Alan by an un-identified hairy caterpillar *ulat bulu* in the PC2 and 3 ecotones (see below) is claimed by loggers to stimulate diameter growth of residual Ramin trees in the B- and C-layers. The next stage of succession, PC3, *S. albida* consociation or Alan bunga forests, has a dense, single-species, uniform and aerodynamically moderately rough to smooth canopy of very tall (to 70 m) and slender Alan bunga (height/diameter ratios of 50 to 90) (Figs 1.11 and 6.5). Lightning gaps are smaller and crown breakage more common than in PC2. Gaps close fairly rapidly by crown expansion from the edges; however, gaps may be enlarged by windthrow. Episodically, drought conditions are likely to occur. Large-scale forest destruction by the caterpillar *ulat bulu* occurred between 1946 and the 1970s (Anderson, 1961b, 1964; Bruenig, 1973a) (Fig. 1.4). The amount of damage from *ulat bulu*, lightning and storms is large enough to affect sustainability of management. The obvious failure of Alan in the defoliated and wind-thrown areas to regenerate, and the very slow growth of the exposed previous understorey trees, mainly *Lauraceae, Fagaceae, Sapotaceae* and *Euphorbiaceae* (Ngui, 1986) speak against the clear-felling as currently practised in Sarawak.

The next phasic communities PC4, 5 and 6 to the centre of the peat dome are sclerophyllous forest, woodland and scrub respectively. They are species poor, low in stature and biomass, and probably extremely slow growing. Floristically and physiognomically they are more closely related to KF and KrF and some Montane forests than to PC1 (Bruenig, 1966). The unfavourable growth conditions and low stocking in PC4–6 in the centre of the peat dome render these forests non-commercial. The fragility of the vegetation and the uniqueness of flora and ecosystem, and the singular scientific and heritage value make PC4–6 natural TPA. Logging in PC3 and 4 threatens the important hydrological functions of the peat dome. The utilization of the unique primeval growing stock of PC4 and 5 for board manufacture, suggested in the 1960s, would certainly irreparably destroy the ecosystem.

6.7.3 Sustainable management of Mixed Peatswamp forest

Management in Peatswamp forest began in the 1920s in PC1 with the introduction of efficient, non-damaging tapping techniques and the preparation of regular tapping plans for the tapping of Jelutong (*Dyera polyphylla* (Miq.) Ashton in litt.), including improvement planting in special Jelutong working circles. Chicle from Jelutong latex was a major export commodity for the buoyant world market of the late 1930s and the postwar period (Table 4.1). During the boom, tapping by unskilled local natives was careless and damaged the cambium, termites entered and the trees decayed and very quickly collapsed. In Brunei at Sungai Ayam in 1958 (author's secondment as State Forest Officer Brunei), tappers working in an extractive reserve had to walk for 2 or 3 hours to get to the first surviving productive tree. A long day produced only 3 gallons of latex, which gave about 10 katies (approximately 6 kg) of chicle. Jelutong chicle sold for 37 ringgit per pikul (100 katies = 60 kg) at the bazaar. So the tappers could only earn 3.70 ringgit gross a day for very hard work and long working hours. After deduction of the expense for tools and chemicals, at most 2 ringgit were left, hardly enough to feed a family even then. Jelutong tapping ceased to be attractive with declining yields, rising costs and higher standard of living, and finally fizzled out during the 1970s and 1980s. During the 1930s and 1940s, Ramin was poison-girdled in Sarawak to favour Jelutong in PC1. After the war the first experimental Ramin exports to Australia met success and after 1950 Ramin became the most favoured export timber and Jelutong subsequently a 'weed', but recently Jelutong regained a position as peeler wood. Ramin felling and extraction by light rail and *kuda-kuda* (Section 2.7) began in most early licence areas in chaotic fashion and was extremely wasteful of forest area, timber and manpower. There were no data for the calculation of sustainable yield and length of felling cycles. The response of residual Ramin trees to release was erratic and sometimes negative. Seedling regeneration of Ramin was patchily distributed and reacted negatively to the sudden full exposure caused by removal of all trees above 40 or 50 cm diameter. Seedlings usually grew chlorotic and eventually died (Bruenig and Sander, 1983).

During the 1950s integrated long-term regional management plans and local working or felling plans were introduced. Prescriptions included rail-track survey, block demarcation, skidtrail (*kuda-kuda*) marking, directional felling, full utilization, post-logging damage and waste assessment (Fig. 6.11, Section 5.1; Bruenig, 1963, 1965a,c). The 60 years felling cycle, harvesting and yield (allowable annual cut) prescriptions were made palatable to the industry by creating the awareness that they were losing money by the customary non-system and would make bigger profits if they adopted RIL and proper management. The annual and periodic (10 + 10 years) working plans were enforced effectively by negative (fines) and positive (promise of long-term tenure) incentives, and close cooperative supervision of work in the forest (Fig. 2.16). A more intractable problem of yield regulation was, and still is, the question of proportional sustainable yield levels for the various species, mainly Ramin, Jongkong (*Daclylocladus stenostachys* Oliv.), Meranti (*Shorea* spp.) and Sepetir (*Copaifera palustris* (Sym.)

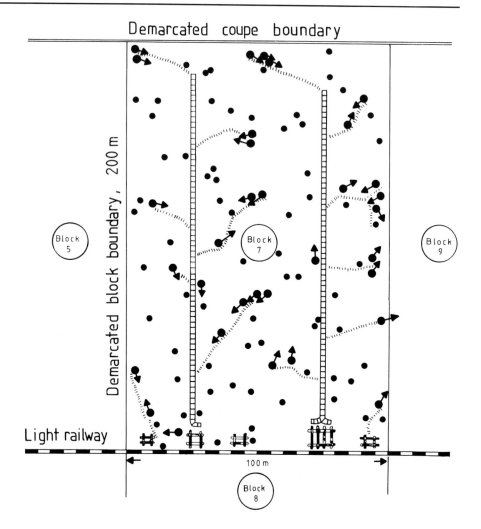

Fig. 6.11. Well-planned and carefully executed harvesting is essential for sustainability in all forests. The example shows the layout of a 2-ha block in Ramin-bearing Mixed Peatswamp forest in Sarawak. In the 1950s, well planned and orderly harvesting replaced the customary combination of snaking light rails and haphazard, tortuously chaotic and, for the workers, extremely dangerous native practices of *baring* and *kuda-kuda*. Dot, undersized commercial tree. Dot with arrow, tree above harvesting size with direction of felling. Only a selected and marked number of these trees will be felled in SMS, but all will be taken in selective logging. The block size is 100 × 200 m or 2 ha. The noticeably clumped distribution of the commercial tree species and merchantable sizes is typical for PC1 and requires silvicultural management systems focusing on individual trees and not on stand averages.

de Wit). The problem was complicated by the successionally very dynamic conditions of soil and vegetation in the Peatswamp forest. Alternatives were to retain the random natural species spectrum by artificial regeneration, or leave development of species proportions to the response of natural succession to harvesting. Insistence on certain mixtures of species and a certain proportion of Ramin would run against the natural dynamics of the ecosystem and involve high costs for silviculture with uncertain success. Broad production goals with a wide range of acceptable species in different mixtures would keep costs and ecological and economic risks low, but worried some conventional foresters. The forest growing stock would, in future, remain the usual mixture of fast-growing, more light-demanding species, but the slow-growing, shade-tolerant Ramin would be more subordinate and less gregarious. For PC1, practice was well supported by intensive research (Lee, 1979, 1991; Chai, 1986; Hadisuparto, 1993). Patchy clear-felling resulted from the initially introduced simple diameter-limit felling prescription and, like in MDF, created a very undesirable, excessively patchy stand structure and an unsuitable microclimate for Ramin regeneration, which may suffer from sun-scorch and die back from excessive respiration (Bruenig and Sander, 1983). There is a possibility that a new seedling population establishes itself from residual Ramin seed trees after the crowns have expanded and the canopy has closed again, perhaps 20–30 years after logging. Research into this question is needed. The problem could have been avoided by the simple remedy of single-tree selection and tree marking for directional felling.

Bryan (1974), evaluating the results of research plots on growth and effects of silvicultural treatment in Peatswamp forests in Sarawak (Clarke, 1964), came to the following tentative conclusions and recommendations for the PC1 that are still valid today:

- Small-sized regeneration is usually adequate for all species, including Ramin, but response to canopy opening by selective logging varies between and within species; planting is neither necessary nor recommended.
- Trees in the intermediate size-classes require several years after logging to recover and respond; silvicultural treatment to release élite trees and stimulate growth should, therefore, be applied 10–15 years after harvesting, followed by a second thinning 5–10 years later (also Chai, 1986, 1991).
- A higher diameter limit of felling (50 cm) produces a higher sustainable long-term increment than a lower (40 cm) diameter limit (which in principle accords with the conditions in MDF, Sections 6.3 and 6.5, and KF, Section 6.6, Table 1.6, Fig. 6.3); a diameter limit of 60 cm would probably yield even better results.
- The period between harvests (felling cycle of a formal management system) should be between 40 and 60 years, depending on species composition and silvicultural treatments, but not less.

6.7.5 *Sustainable management of* S. albida *Peatswamp forest*

The unique *S. albida* Peatswamp forests of Sarawak and Brunei are possibly the most threatened rainforest formation and ecosystem globally and may, in Sarawak, disappear within a decade. These forests are unique in the world as living documents of forest conditions in the Tertiary, which produced brown coal and amber. In the equatorial tropics they are, outside the mangrove, a unique example of a natural monoculture. Scientifically, they are invaluable as a source of ecological and silvicultural knowledge on the functioning and risks of pure monocultures. Nationally, they are a renewable but extremely fragile natural resource of great economic potential for sustainable timber production if judiciously and scientifically managed. Globally, they are an important sink of carbon. Overlogging by rapidly progressing clear-felling creates large tracts of stagnant scrub and pole stands, which become a source of carbon dioxide and methane on subsiding peat. Clear-felling started in the late 1950s in the Saribas Peatswamp forest, when the wasted-asset theorem was popular (Sections 2.7 and 3.4), and has since spread to the Rejang and Baram regions. Regeneration in Alan forests is a scientific puzzle and in practice a random gamble with unknown odds. Experimental planting (Lai, 1976, 1978) has shown the ecological risks and uncertainties involved. Bryan (1974) made the following comments on the two phasic communities with commercial potential, PC2 and 3.

PC2, Alan batu Peatswamp forest

The abundance and composition of regeneration in the PC2, Alan batu forest, is similar to that in PC1, except for the presence of Alan. Repeated harvesting of the (invasive) very large Alan batu trees could eventually restore the previous PC1 conditions or alternatively accelerate development into PC3, or deflect succession into something quite different.

PC3, Alan bunga forest

The observations and opinions on Alan bunga regeneration are inconsistent, contradictory and inconclusive. As in PC2, even ample stocking of regeneration is no guarantee of successful performance after logging. 'Research studies and diagnostic surveys of cutover Alan bunga forest lack any examples of Alan taking over dominance of stands unassisted by treatments', but there is evidence of an abundance of reproduction of other species and some evidence of possibly density-related growth checks in all species except Alan (Bryan, 1974, p. 59). Planting with or without prior burning is an option that requires a much improved database before decisions could be taken (Lai, 1976, 1978). The same lack of data exists on growth, yield, diameter limits and harvesting cycle. The possibilities for sustainable management by a shelterwood or shelterwood-strip system need to be explored. Research into the present state of the *ulat bulu* areas would give useful information on the natural trends. The most urgent action would be to stop the simple and clearly unsuitable clear-felling before it is too late and the whole PC3 and 4 are converted to stagnant, excessively dense scrubs and pole stands which enrich the atmosphere with carbon and methane.

6.7.6 Conclusion

The sustainable management of PSF PC1 poses few problems if the initial mistake of the 1950s, to apply a simple girth-limit system uniformly, and the later mistake to reduce the felling cycle for Ramin to 40 years are corrected and tree marking introduced. The greatest threat to sustainability is the pressure to lower the already too low diameter limit further and to allow premature re-entry as the virgin Ramin growing stock becomes exhausted. The main problem in PC1 is the effective enforcement of a management system which needs improving, but has generally proved suitable. In contrast, management in PC2 and 3 lacks the necessary physio-ecological and ecological–silvicultural knowledge base. The extraordinary uniformity of the architectural structure of the forests does not reflect age but may be attributable to the rapid responses of canopy trees and regeneration to additional light, water and nutrients available in small to medium-sized gaps (Anderson, 1961a; Bruenig, 1973b, 1989c). A uniformly even-sized, but obviously uneven-aged single-species crop, with a strongly fluctuating population of seedlings and saplings and unknown growth dynamics, is difficult to manage by any silvicultural system with natural regeneration, while planting, though possible, is expensive (Lai, 1976). This state of uncertainty and the inability to estimate probabilities for future trends and states are due partly to the inherent indeterminant nature of the forest ecosystems, but also due to simple human ignorance. Absolutely certain, however, is that utilization of the growing stock by clear-felling is, under such conditions and on absolute forest sites, premature and extremely hazardous and contrary to the principles of sustainability and to the interests of the national economy and coming generations.

6.8 Overcoming the Enigma of Uncertainty

Forest management and conservation in the rainforests universally has to cope with numerous uncertainties due to the indeterminacy of the natural and cultural systems and due to greed, indolence and ignorance of the human actors. The effects can be mitigated by adopting the simple principle to choose among possible alternative options, the one which needs the least amount of information and promises to be the least risky and wrong. Ecologically, this principle is met by retaining as much as possible of the basic floristic and physical structure and the vitality, tolerance and resilience of the natural primeval forest by integration of naturalistic silviculture and RIL harvesting. Economically, this principle is met if the product target is the most versatile, elastic and valuable product which can be grown at low cost and risk in a healthy, robust and self-sustainable crop. In SMS, it means growing high-quality timber with élite trees in the 40–80 cm diameter range to produce a target diameter of 60–100 cm (PSF, KF and low-yield MDF) or 80–120 cm (high-yield MDF). This strategy is most likely to be least wrong because it requires the least amount of information on future states of the natural, economic and social environment and the least amount of investment in silviculture and management, and it mimics natural ecosystems which have survived successfully during 100 million years of tribulations (Sections 1.2 and 1.4).

Under conditions of great uncertainty, as in forestry, it is perilous to maximize performance to meet narrowly defined goals and targets. Rather, goals and targets must be broad and flexible, and production designs broadly optimized according to the biocybernetic principles of system design (Appendix 2). Forestry measures must be supplemented by an efficient, highly productive, low-cost, high-quality downstream timber industry which reduces the uncertainties of the market. Sustainability in forestry depends on sustainable and rewarding home and export markets. In 1993, about 50% of timber exports from Peninsular Malaysia went to ASEAN markets, 23% to East Asia, and 17% to the EU. The proportions of the declared money proceeds from these exports were, respectively, 26%, 27% and 35%. During the same period, sawn timber from Sarawak sold per cubic metre for RM373 to China, RM591 to Japan, RM1033 to Germany and RM1047 to The Netherlands (largely on transit to Germany) (Maskayu, 1993). The consequences for strategic planning of sustainable forestry and forest-based industry are self-evident. The puzzling problem of uncertainties can best be met by low-cost, low-risk naturalistic silviculture in complex, dynamic forests with high degrees of self-sustainability which supply large and high-quality timber for the production of versatile, high-quality value-added timber products in the downstream industry to be sold to sophisticated high-price markets abroad, such as the EU, and at home.

Restoration of Degraded Ecosystems

7

Degradation of forests and fragmentation of landscapes culminate in landscape development phase IV (Fig. 2.4) when the effects of exploitative, predatory resource plundering in agriculture and forestry become obvious and are physically felt by the people. Income disparities, poverty at the base and excessive wealth at the top of the society cause discontent and tension which, in modern democratic societies, can no longer be subdued and ignored. In the countryside overuse, misuse and abuse have created a variety of modified forests, secondary vegetations and barren land (Figs 2.4, 6.7–6.9, 7.1 and 11.5). Forestry has been relegated to an administrative fiscal agency without real power and is now being pressured by the public to protect and rehabilitate the forest resource. Eventually, in phase V, action is taken. The degraded ecosystems are being rehabilitated, wherever possible, by natural succession. If natural succession cannot achieve the target, planting may be necessary. In areas with a long history of settled agriculture, multipurpose planting of native tree species by villagers, governments and monasteries has been traditional in many parts of the world, including East and Southeast Asia (FAO–UNESCO, 1992), but not in pioneer areas (Halenda, 1985). Timber tree species (e.g. Teak in Myanmar and Java) or multiple-use tree species may be planted in the fallow cycle of rotational slash-and-burn agriculture. The dual purpose is to boost productivity of the fallow and to restore the fertility of the soil. Michon (1993) describes the planting of *Shorea javania* Koord. et Vahl. for dammar (resin) and timber which has been practised for centuries in Sumatra. In Peninsular Malaysia, where fruit-tree gardens have an ancient tradition, organized planting of native tree species for site restoration began over a century ago. *Palaquium gutta* (Hook.) Burck, *Intsia palembanica* Miq., *Fragrea fragrans* Roxb. and other native species were planted on wasteland and tin tailings. An excellent

203

Fig. 7.1. Aerial view of the area of the experimental plantations and the old secondary forest which they replaced. Double rectangle = RP 76, inset is block 4A1; small rectangle = 65 years old TP 4B; circle in top-left corner indicates plantation of *Durio zibethinus* and *S. macrophylla*; centre-right circle indicates a 55-year-old plantation TP 7C of *S. macrophylla* over *Eusideroxylon zwageri*, Belian; centre-left circle indicates the area of approx. 90 years old secondary forest with a MDF-like canopy aspect. Semengoh F.R., 1991 (courtesy of Sarawak Government).

review of the subject, including the almost century-old controversy between the advocators of industrial tree monocultures of fast-growing exotics and the adherents of more naturalistic forest plantations with high-quality native timber trees, is given by Appanah and Weinland (1993).

The following five major degraded ecosystem types could be suggested to require restoration by forestry measures in the general context of economic and social development: (i) overlogged natural mixed forest on deep clayey udult to sandy–loamy humult ultisol soils (Section 7.2); (ii) secondary forest in various stages of development on deep clayey udult to sandy–loamy humult ultisol soils (Sections 7.3–7.6); (iii) secondary forest on deep sandy spodosols and medium deep aquic spodosols (Section 7.6); (iv) secondary woodland on medium to shallow sandy or stiff-clayey poor Kerangas forest (KF) soils; and (v) grassland and other types of arrested vegetation on deforested lands. The restoration of the third ecosystem type, good KF soil, may be feasible, but the fourth, degraded poor KF, is not. The last ecosystem type is outside the subject of this book and the reader is referred to the extensive literature on plantation silviculture (Evans, 1982; Bruenig, 1984c; Bruenig *et al.*, 1986; Evans and Wood, 1993) and to the reports by the IUBS/ICSU and MAB programmes on the ecology and management of savannah ecosystems.

7.2 Planting into Overlogged Upland Rainforests

Overlogged natural forests on good soils require and deserve restorative planting if they have been so severely depleted and degraded that natural recuperation takes more time than is desirable for economic or environmental reasons. This condition is indicated if the A- and B-layer canopy has been completely or almost completely (> 80%) removed and, more than 50% of the C- and D-layers have been smashed or badly damaged (crown and stem breakage, bark and root damage). Beyond this threshold it is unlikely that the remaining sound and healthy trees will be able to produce a satisfactory second harvest and sufficient regeneration and residuals for all subsequent felling cycles. The minimum safe standard is about 30–50 residuals at 30–80 cm diameter, 100 poles 10–30 cm diameter, and 200 saplings per hectare (see also Chin *et al.*, 1995). Planting may also be justified if logging has scraped away the D-layer on more than 30% of the area by tractor criss-crossing, turning of logs sideways, and on log landings and roadsides. Planting patches should be at least 0.5 ha (*c*. 70×70 m) and canopy gaps at least 25 m diameter wide to provide sufficient growing space for the planted trees (see Section 6.5). The rationale of these rather liberal threshold levels is that the rainforest ecosystem has a very high repair capacity, as many other forests, which is usually underestimated. The surviving residuals are usually more vigorous and competitive than the newly planted trees, whose growth they can successfully inhibit (Fig. 7.2). Foresters universally overestimate the need for planting and underestimate the dynamics of lateral growth of crowns in the canopy and of vertical growth of regeneration from the ground. As the species and genotypes of the residual growing stock and invaders are adapted to the local site and their rooting system well established, planted trees may find it difficult to compete. In contrast to this cautious approach, which uses as much as possible the natural dynamics of the forest, the Manual of Classification System for Logged-Over Forests in the Asia/Pacific Region sets higher and probably excessive standards for minimum residual stocking at 100 trees > 15 cm diameter, 400 saplings and 1000 seedlings per

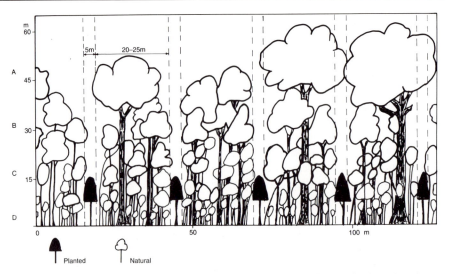

Fig. 7.2. Enrichment of modified tropical rainforest by planting rows of a valuable, shade-tolerant late-phase timber or multiple-use species. The purpose is upgrading the production functions after the forest had been creamed for highly valued or speciality timber species by traditional native creaming or commercial, very selective light logging. In selectively overlogged forest the canopy above 25 m height would have been almost completely removed and the subcanopy smashed (see Fig. 2.13) or, if repeated, would lead to a climber tangle (Fig. 2.14) in which planting is very expensive and risky.

hectare, of which less than 30% are species which are presently 'saleable in the export and domestic markets' (ITTO, 1993b, Table 1).

Very badly degraded forests, plundered and drained by repeated overlogging, or very heavily overlogged along forest roads, are often a tangle of aggressive climbers, such as *Merrimia* spp. (Figs 2.14 and 6.8). Experience with post-logging climber cutting shows that the cutting often sets the succession back and stimulates climber growth. In many cases the ecologically most natural and economically most feasible strategy is to wait. The same applied to the climber tangle which developed in semi-deciduous Dipterocarp-bearing forests after several aerial sprayings with dioxin containing weedicides in war zone D in Vietnam. I visited the area during a conference on biochemical warfare in 1983 (Bruenig, 1983). At that time, possibly about 20 years after the last spraying, and inspite of continued burning by the natives in the area, the climber tangle was already nearing the collapse phase, when the net primary productivity (NPP) turns negative and the mass of photosynthate-consuming stems pulls the photosynthesizing leaf layer down. In the tangle, seedlings, saplings and small trees were successfully initiating the first stage of natural succession back to a high forest (Bruenig, 1983). For political and monetary reasons, the delegation of the Soviet Union in 1983 advocated clearing, burning and planting, to be paid for by the USA. Similarly, but for other reasons, commercial interests in Sabah advocated clearing, burning and planting of overlogged and degraded forests, such as

Demarakot, rather than prudently assisting the already ongoing natural recovery process.

If the decision is for improvement and enrichment planting, only species which are adapted to the building-up and mature phases of primeval forest should be planted. Such trees are light demanding, but tolerate shade and root competition, grow fast with a narrow crown and produce high value. Strongly light-demanding, shade-intolerant, fast-growing pioneer species, such as *Paraserianthes falcataria*, *Aucoumea klaineana* or *Terminalia* spp., especially *T. superba*, are not suitable for patch- or strip-planting in logged or overlogged forest. They require clearing of such wide strips that with a between-line spacing of 15–20 m (final crop 70–50 trees ha^{-1}), eventually the whole area has to be tended as otherwise the crop trees become suppressed. This means repeated and expensive cleaning and thinning. Such species may be worth planting along excessively wide (80–100 m) logging roads and on log landings, provided the areas are large enough to make future commercial harvesting, extraction and utilization technically possible and economically feasible and provided equivalent local pioneer species do not invade naturally.

The practice of enrichment planting has a long tradition in the Asia-Pacific region. It ranges from the planting of fast-growing hardwoods in patches to complete line planting (Tang and Watley, 1976). In Peninsular Malaysia, tree planting trials began in 1876, and systematic enrichment planting experiments have been carried out since the early 1930s (Walton, 1932). In Sarawak, Nyatoh and Jelutong have been experimentally planted along cleared strips in depleted Peatswamp forest, PC1, since the 1920s, but survival rates of trees were poor, costs high and the project eventually faded out. The Forest Research Institute of Malaysia (FRIM) conducted line planting trials of several indigenous tree species in Tapah Hill Forest Reserve, Perak in 1972, including *Parashorea stellata*, *Shorea leprosula* and *Shorea parvifolia* (Hassan *et al.*, 1990), and with a number of native species during the 1980s in various areas of the Permanent Forest Estate (PFE) in Pahang and Selangor. On the whole, results here and elsewhere in the tropics have not been sufficiently encouraging to support the present faddish international call for planting in logged rainforests. Nor have the results of clear-felling and planting been convincing (Chapter 8). Natural recovery is in most cases still the safest course to take. The best and most effective, cheapest and appropriate course to secure sustainability is by replacing selective overlogging by orderly harvesting and management (reduced impact logging (RIL) and (SMS) solution silviculture management system).

7.3 Restoration in Secondary Forests on MDF Sites

Secondary forests are fundamentally different from logged and over-logged forests. The forests are less heterogeneous within and between sites, and generally less diverse. The trees of the initial colonizing phase are short-lived, fast-growing pioneers; the successional dynamics of growth and demographic changes are strong and fast. In a 35-year-old secondary forest after pepper cultivation on a Mixed Dipterocarp forest (MDF) site near Samarinda in East Kalimantan, Riswan and Kartaniwata (1988) found 121 tree species in 89 genera and 40 families on 0.8 ha; 30% of the species were secondary forest pioneers, mostly in the top canopy, and 70% were primary forest species, mostly in the subcanopy. The basal area density was about 50–60% of the average MDF value, the biovolume 38% and the biomass 25%. If the rate of invasion and build-up is maintained, the original forest structure would be restored within a century from now or 135 years from the abandonment of farming, except that overmature giant trees would of course be absent. Numerous other assessments confirm such rapid development rates on MDF and alluvial sites if the forest succession is not disturbed again. Secondary forest may contain commodity timbers, such as *Campnosperma* spp., *Canarium* spp., *Calophyllum* spp. and *Cratoxylum* spp., which may be as good for many

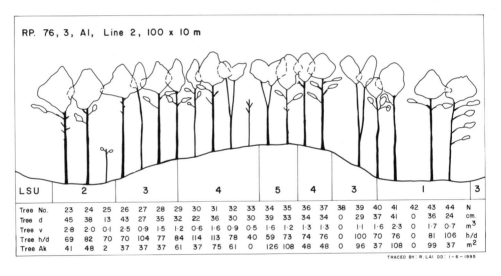

Fig. 7.3. Microheterogeneity of soil and terrain in relation to tree growth in Semengoh Forest Reserve, Sarawak. Example: profile of line 2, RP 76, block 3, series A, *Shorea macrophylla*, treatment 1 with starter fertilizing. State of the planted crop in August 1993, 20 years after planting. Understorey and undergrowth are shown in Figs 7.7 and 7.8. The soils are alluvial (1), transition alluvial-upland soil in small valley heads (2), deep (3) to moderately deep (5) soils of the Semengoh series red–yellow podzolic ultisol residual soil on Bau formation shale. The landform–soil units (LSU) are: 1, wet, gleyish clay, flat alluvium, often waterlogged; 2, well-drained gley-type clay, almost flat valley bottom, well drained; 3, well-drained clay–loam on gentle slope, erodibility low, deep; 4, -, on moderately steep slope, erodibility moderate, medium deep; 5, -, on summit of rounded ridge, prone to drying (soil profile FIL, Baillie, 1976, lies in this same LSU approximately 100 m to NW); Tree No., serial number of tree in the research plot; d, diameter over bark at breast height in cm; v, total above-ground stem volume over bark, estimated as v g.h.f (f = 0.55) in m^3; h/d, ratio of tree height to trunk diameter (mean = 75.5); Ak, crown projection area in m^2 (sum = 1143 m^2 on 0.1 ha = very dense).

purposes as any plantation-grown timber. Planting into secondary forest may be justified and feasible for high-value timber and non-timber production if costs and risks are low. It is ecologically and technically easier and less risky than planting in overlogged forest, provided intensive soil, site and vegetation survey and mapping support planning and the planted species (genotype, variety) match the usually heterogeneous soil and hydrological conditions (Figs 1.10 and 7.3, Table 1.5). The high-forest species usually can compete well with the secondary vegetation if a good start is given by wide line clearing and by fertilizing. Slash-and-burn clearing should be avoided because it sets succession back, causes nutrient losses and creates weed problems (grasses, sedges, bamboo, lianas). The existing secondary growth and invading newcomers should be used to restore biodiversity (Section 7.5). Halenda (1989) rightly suggests that integrative enrichment management of secondary fallow forest is an ecologically more feasible, and especially with respect to pests and diseases less risky strategy, than the usually practised slashing, burning and tree planting.

Site-adapted native species should be preferred, because they promise better resistance to site-specific stresses, strains and risks from abiotic and biotic factors, and the local people are used to them and their products. Among the A- and B-layer trees of the Malesian natural forest, the families *Anacardiaceae, Bombacaceae, Dipterocarpaceae, Euphorbiaceae, Meliaceae, Moraceae, Sapindaceae* and *Sapotaceae* contain multiple-purpose species that are particularly suitable for introduction into secondary degraded ecosystems (Fig. 7.4). According to Ashton (1988b) 'dipterocarps are exclusively components of the forest mature phase, only gradually invading successional forests after a closed canopy is fully established'. Many of the light-demanding A-layer dipterocarps, such as Engkabang and other *Shorea* spp., *Anisoptera* spp., *Dryobalanops* spp., *Dipterocarpus* spp. are also light-demanding in their youth. They do not require shade

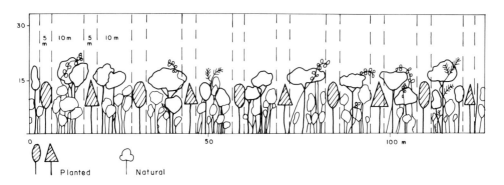

Fig. 7.4. First step of initiating the process of restoring degraded secondary vegetation to a mixed near-natural high forest by line-planting one or several fast-growing light-demanding, but shade-tolerant multiple-use species of the building or early-mature phase in natural high forest.

if they are gradually adapted to exposure. These dipterocarps agressively invade open secondary woodland, abandoned rubber and home gardens, landslides, road sides and landings, provided the seeds find a suitable, reasonably sheltered microsite to germinate safely, especially large seeds which quickly overheat in the open sunshine, and suitable soil in which they can root deep quickly. A-layer trees generally grow well, often better, in the open if the roots can quickly reach deep into the soil. In the open, relatively larger amounts of photosynthates are initially allocated to root growth which may reduce height growth until sufficient roots are developed to secure water supply. Sudden exposure, however, as in most plants, may be lethal. Since the 1920s, a real problem with native high-forest trees has been the procurement of suitable high-quality planting stock. Dick and Aminah (1994) reviewed the present state of vegetative propagation in Malaysia. The conclusion is that rooting of cuttings is a feasible method to propagate major commercial tree species, which would solve this problem.

7.4 Example: Multiple-purpose Plantations in Semengoh

Semengoh Forest Reserve lies 20 km south of Kuching in the forest growth district West-Sarawak Lowland. The mean annual rainfall at Semengoh (1960–1968) is 4064 mm; the A-pan evaporation at nearby Kuching airport is 1844 mm (45% of P). The monthly rainfall distribution shows a somewhat stronger peak during the northeast monsoon in December/January than Bintulu (Fig. 1.2). Episodic dry periods longer than 30 days are very rare and less common than in more coastal areas (Bruenig, 1971a). The terrain is undulating to low hilly with rounded ridge crests and intermittent alluvial flats. The substrate of the hillocks is argillaceous cretaceous sediments (shale and mudstone) (Baillie, 1970). The soils are very heterogeneous at small spatial scale (Fig. 7.3). In the flat flood plain the soil is mottled clay, and on low hills deeply weathered, moderately deep-rooted red–yellow clay–loam (Table 1.5, sandy clay–loam ultisol). The indigenous natural vegetation on the undulating low-hilly ground is typical Lowland Dipterocarp forest. Traditional shifting cultivation by the local Bidayuh communities created secondary forest which is now mostly 70–100 years old (Fig. 7.1). The secondary forest has a fairly open main canopy of larger trees (most common are *Tristania whitiana* Griff., *Calophyllum lowii* Planch. et Trian., *Eugenia* spp., *Palaquium* spp., *Cratoxylum arborescens* (Vahl.) Bl., *Artocarpus odoratissimus* Blanco, *Parartocarpus* spp., *Nephelium* spp., *Xanthophyllum* spp.) over a fairly dense undergrowth of small trees, shrubs, bamboo, sedges, gingers and ferns. Unfortunately, in the 1950s, a participatory *taungya* scheme (rice, Durian and Mahogany) of the Forest Department and the inmates of a nearby leper camp destroyed much of the surviving MDF. The scheme eventually failed because the camp inmates lost interest. Since 1926 and continuing into the 1970s several species of the genus *Shorea, Dipter-ocarpaceae,* have been planted experimentally on alluvial and upland sites in the northern part of the reserve to produce timber and illipe nuts. The planting lines were kept clean, but between lines regrowth and invasion of trees, palms and bamboo developed a species-rich diverse understorey. Planted trees and understorey in two plots, one planted in 1926 (RP 4B, *S. splendida* (de Vr.) Ashton), the other in 1973 (RP 76, *S. macrophylla* (de Vr.) Ashton), were enumerated in 1993 to obtain some information on the productivity of the planted trees and the state of the spontaneous understorey (Figs 7.5 and 7.6).

RP 76-3A1 is particularly interesting. An area of 6.5 ha secondary forest was more or less fully cleared in 1972–1973 and planted in February 1973 with Engkabang jantong (*Shorea macrophylla*) at spacing of 4.5–5 m along lines 10 m apart. In June 1977 each planted tree was fertilized with 250 g proprietary brand Nitrophoska blue, and twice again in the following 2 or 3 years. The trees were enumerated in 1978, 1985 and 1994.

Fig. 7.5. *Shorea macrophylla,* 20.4 years after 5×10 m line planting into secondary forest on undulating land and loamy–clayey udult ultisol. Semengoh F.R., RP 76, Block 4A1, 1993.

Fig. 7.6. Crown of the second *S. macrophylla* tree in the line in Fig. 7.5. The light-demanding but not crown-shy species develops straight boles and prunes well but tends to overstocking in plantations. Note the high diversity of the understorey (see Figs 7.5 and 7.6) which developed in spite of the overly dense canopy.

On 4–6 August 1993, three rows in 76-3A1 (0.3 ha) were studied in more detail. The 30 × 100 m rectangle straddles a broad ridge which rises 8 m above the alluvial flat (Fig. 7.3). The measured three rows (0.3 ha) contained 54 Engkabang trees (180 ha^{-1}), 82% survival of the planted 66. The basal area over bark was 5.3 m^2 (17.7 m^2 ha^{-1}). The arithmetic mean of breast-height diameter was 9.3 (1978), 23.8 (1985) and 35.5 cm (1993). The largest tree (no. 21, Fig. 7.3) was 7.9 (1978), 29.3 (1985) and 50.5 cm (1993). The tree density of the natural regrowth between the lines was 641 on 0.3 ha, or 2120 ha^{-1} (actual ground surface is 0.18 ha if the 4 m wide planting lines are deducted, but the canopy cover is 0.25 ha. See Figs 7.6 and 7.7). The basal area was 2.1 m^2 on 0.3 ha (6.9 m^2 ha^{-1}). The tree heights of the Engkabang range from 28–32 m on the foothill sites to 22–25 m on the broad ridge crests. The tree heights in the spontaneous C- and D-layer vegetation range from 2 to 22 m (Figs 7.5–7.7).

The periodic mean annual breast-height diameter increment of the

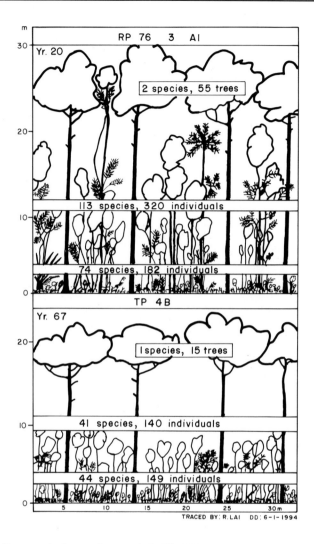

Fig. 7.7. Stand profile across planting rows in research plot RP 76-3A1, age 20 years (top) and in trial plot TP 4B, age 67 years (bottom, 0.08 ha) in 1993. The number of recorded species and measured trees is given in the layer in which they occur. The slender, tall tree causing crown shyness in the top canopy layer of RP 76 is *Mallotus* sp., *Euphorbiaceae*. Semengoh Forest Reserve, Sarawak, 1993.

Engkabang trees in the period 1973–1993 (19.5 years) was 1.8 cm, 1978–1985 (7 years) 2.1 cm and 1985–1993 (8 years) 1.5 cm. The largest periodic mean annual diameter increment of the five biggest trees was 3.0 cm in 1978–1985. The mean annual basal area increment 1973–1993 was 16% of the median basal area. The tree volume over bark was 266 m^3 ha^{-1} in 1993; of this 164 m^3 ha^{-1} was standing merchantable log volume, 148 m^3 ha^{-1} net of harvesting losses. The corresponding periodic mean annual increment (m.a.i.) values for Engkabang for the two periods were: 1973–1993 (19.5 years) 13.6 m³ ha⁻¹ tree volume, 7.6 m³ ha⁻¹ net log-timber volume; 1985–1993 (8 years) 27.8 m³ ha⁻¹ tree volume, 18.5 m³ ha⁻¹ net log-timber volume. The subcanopy during 1973–1993 contributed only 2 m³ tree volume over bark below merchantable size (mean diameter 6.4 cm) (Fig. 7.8). The high increments are explained by the deep, well-aerated, -watered and -drained soils well supplied with soils of types 1–3 (Fig. 7.3) and to some extent by the fertilizing at age 4 (1978), 5 and 6 or 7. The growth figures accord with the increments reported from plantations on

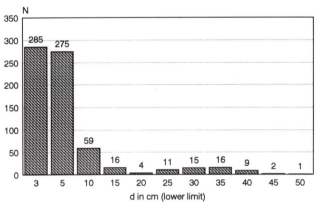

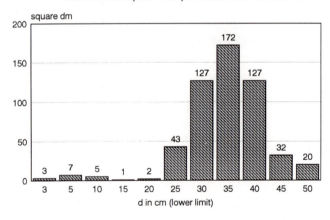

Fig. 7.8. Number of trees (N) and basal area distribution in 5 cm stem-diameter classes in a 20-year-old Engkabang plantation. The spontaneous undergrowth is rich in number of individuals and species (Fig. 7.7), but still low in basal area and biomass stocking. The vertical integration of the canopy is just beginning. RP 76-3A1, Semengoh Forest Reserve, Sarawak, 1993.

abandoned farmland on similarly favourable soils at FRIM, Kepong. Hahn-Schilling (1988, table 18) enumerated a 57-year-old plantation of *Shorea parvifolia* Dyer, compartment 17 E/F, on clayey–sandy ultisol over granite within the FRIM compound, Kepong. The crowns in the A- and upper B-canopies were densely closed, but the crowns transmitted sufficient light for a C-canopy to develop, which became so dense that it prevented natural regeneration (compare Fig. 1.18). The basal area density was 52.7 m^2 ha^{-1}; of this 41 m^2 was planted *S. parvifolia*. The volume over bark was 468 m^3 > 50 cm diameter, 651 m^3 > 10 cm diameter and 825 m^3 > 4 cm diameter. The m.a.i. of the standing crop was 8.2 m^3 merchantable timber over bark > 50 cm diameter and 14.5 m^3 > 10 cm (thickwood > 4 cm o.b.). I also recommended a growing schedule with much lower density, approximately 50% of the present *G*, to stimulate stem and crown diameter growth. The Semengoh Engkabang jantong plantation is, at age 20, similarly overstocked. Of the 220 trees planted per hectare in RP 76 in 1973, 180 had survived in 1993, producing a canopy density of >1.1. Crown stress must have developed about 1985 when thinning should have begun in order to reduce the density to 100 trees ha^{-1} in 1993 with a canopy projection of 0.8, which should have been maintained. This would have attained the same basal area (18 m^2 ha^{-1}), but a higher mean stem diameter (47 cm on the slopes and flats, Fig. 7.3). The merchantable log timber volume and value in 1993 would then have been considerably higher and the future value increment much higher than that achieved by the no-thinning regime. Fruit production would also have benefited from the improved crown illumination.

The silvicultural concept of growing high-value trees at wide spacing as fast as possible to large size has been criticized on the grounds that this produces wood of poor quality. This contention has been refuted by practical experience with the APP (Auermühle Production Programme) management system for pine, Douglas fir and oak in Germany (Bruenig, 1984b,c, 1986b). If the juvenile growth rate is fast and the juvenile core wide, the mantle of adult wood will consist of narrow-ringed, dense wood suitable for high-grade uses. Tests of the wood of plantation-grown light red meranti (*Shorea leprosula* Miq. and *S. parvifolia* Dyer) showed a wide variation of wood properties within a tree from pith to bark, but no consistent differences between 'wild' and plantation-grown trees. The latter appear to have a slightly less variable wood (Bosman *et al.*, 1993). This suggests that the concept of the ecologically and economically superior APP growth schedule, to grow élite trees fast at wide spacing in complex and diverse stands (Bruenig, 1984b,c, 1986b), is also suitable for rainforest trees and plantations. In all the Engkabang plantations in Semengoh, bamboo in the undergrowth was harvested by a nearby small papermill producing traditional Chinese papers and by local farmers for home use, but the amounts are unrecorded. Fruiting of Engkabang occurred, but the illipe nut yields were not recorded. Adoption of the APP schedule of stand structuring would have led to better crown development and faster diameter growth, and would have stimulated flowering and fruiting. The more open, large-crowned A-layer would also improve growth conditions for trees and bamboo and generally for biodiversity in the subcanopy.

7.5 Restoration of Biodiversity in RP 76 and TP 4B

In the same two research plots RP 76 and TP 4B in Semengoh Forest Reserve, the plant communities in the subcanopy and in the undergrowth were enumerated in quadrats, excluding mosses, epiphytic phanerogames and ferns. The 430 trees in the subcanopies of the two stands, 4B (0.08 ha) and 76-3A1 (0.18 ha), comprised 141 species in 28 families and at least 57 genera; 35 trees could only be determined to family level. Each of the 2 × 2 m subsample plots of the undergrowth flora and trees < 3 cm diameter contained on average 14 (4B) and 12 (76-3A1) species, and added a total of 80 species (Table 7.1). The grand total of recorded plant species, including the two planted Engkabang species, was 223 in 0.38 ha (Fig. 7.7). The subcanopy in 76-3A1 is vertically integrated and complex (Figs 7.5–7.8). It contained 113 species among 320 individuals on 0.18 ha. The total species number and composition by plant form (Table 7.1) in RP 76-3A1 was rather similar to that in TP 4B if the greater number of quadrats is considered. Natural Engkabang jantong regeneration occurred in five of the ten quadrats. It was absent in dense undergrowth and restricted to

Table 7.1. Number of plant species and individuals in the undergrowth sample quadrats of 4 m² each, five in TP 4B (20 m²) and ten in RP 76-3A1 (40 m²). The Engkabang plants are natural regeneration of the planted *Shorea splendida* (4B, in three quadrats) or *S. macrophylla* (76-3A1, in five quadrats). Differences between 4B and 76-3A1 are mainly due to former cleaning in 4B, soil differences and the different number of sampling quadrats and the small sample size. The species–area lines were closely parallel and still steeply rising at a rate of 6.0 per quadrat in TP 4B and 6.5 in RP 76-3A1.

	4B, 5 quadrats		76-3A1, 10 quadrats	
Category	Species	Individuals	Species	Individuals
Engkabang	1	3	1	7
Other tree species	17	26	24	55
Climbers	9	42	14	31
Rattan	1	1	2	14
Other palm species	2	21	3	9
Small-tree species shrubs	2	4	21	37
Herbs	7	18	7	21
Grasses, sedges	2	20	1	6
Bamboos	1	1	0	0
Ferns	2	13	1	2
Sum	44	149	74	182

more open and better illuminated patches. There is no evidence that Dipterocarps from the natural MDF in an arboretum area, about 2.5 km away, have yet reached either of the two plantation areas.

Hahn-Schilling (1988; see Section 7.4) enumerated two planted Dipterocarp stands behind the Forest Research Institute of Malaysia (FRIM). One plot of 0.8 ha *Dryobalanops aromatica* Gaertn.f. had been established in 1947 as single-species plantation on abandoned vegetable garden land. Forty years later the tree crop > 10 cm diameter contained 30 tree species that had spontaneously invaded, representing 18% of the 396 stems and 12% of the 33.8 m² ha⁻¹ basal area of the stand. The small-tree and seedling population < 10 cm diameter was dominated by naturally regenerated *D. aromatica* (48% of the individuals), associated with 26 other species, partly pioneers, but mostly invading high-forest species. Only four herbaceous species were found and a small number of climber species. The other plot, a 1.3 ha plantation of *S. parvifolia* Dyer, had a basal area of 51.5 m² ha⁻¹ and 564 living trees > 10 cm diameter. There were 67 tree species > 10 cm diameter, 22 tree species < 10 cm diameter, and seven herbaceous species, five climbers and three palm species. Both forest ecosystems are restoring by natural succession the spatial structure (first) and the species structure (subsequently) of the site-specific natural rainforest type. In this case, a nearby MDF, which had been logged early in the century but recovered to a near-primeval condition, served effectively as a source of seeds, spores, animals and microbes. This contrasts with the case in Semengoh, but the general trend of restoration of biodiversity in the Kalimantan (Section 7.3), Semengoh and FRIM examples is remarkably similar. Lugo *et al.* (1993), also noticing that species-rich understoreys developed spontaneously in tree crops planted on degraded land in Costa Rica, suggest 'to use plantations to accelerate successional processes and increase biodiversity [which] requires a broadening of conventional management objectives'. For ecological and economic reasons, silvicultural practices must be consistent with the goal of developing a species-rich understorey as a first step towards a mixed, complex near-natural forest. This approach is in accord with naturalistic silviculture in traditional European forestry, which has proved to be ecologically superior, and, from a business and national (socio-) economic point of view, at least equal to monocultures (Bruenig, 1970b, 1986b).

7.6 Conclusions on Restoration on MDF Sites

Plant biodiversity has increased rapidly in spite of the excessively dense monolayer of overlapping Engkabang crowns in both plots, RP 76-3A1 and TP 4B. The excessive density has suppressed development of stem diameters, basal area and stocking of merchantable volume below the site potential, causing loss of value yield. Also, the small crowns and dense canopy are not ideal for illipe nut production and for providing growing space and light for other useful products such as rattan and bamboo.

Overly dense stocking also makes trees whippy and weak, and consequently more susceptible to wind damage, pest and diseases. Policy-consistent and system-orientated management should aim at optimizing not only one single function but the overall performance and health of the forest ecosystem. This would mean in this case, and generally, to create an improved crop structure by appropriate initial spacing and subsequent canopy manipulation. The aim is to achieve faster growth and bigger final diameters of the stems and crowns of the commercial-timber producing species, thereby simultaneously improving vigour, vitality, resistance, multiple commercial value and versatility. The structure of the main canopy (big and healthy crowns, vertical integration) creates more favourable and diverse light and moisture conditions for a subcanopy of a variety of trees, rattan, other palms, bamboo and herbs, and a greater diversity of habitat for animals and microbes. The soil ecology and biology benefit from greater diversity of litter production, which improves water and nutrient cycling. The biodiverse forest has a higher amenity value for recreational use and higher levels of self-sustainability, utility, vigour and health, which reduces costs and risks of management for greater multiple-use value.

The tentative results of the preliminary pilot studies in Semengoh and at FRIM, Kepong, confirm that forest ecology in the rainforest biome is not a unique exception, different from all other forests in the world. The succession on deforested land and in degraded secondary forest restores the ecosystem surprisingly rapidly if the degrading factor or factors are removed. Restoration can be economically profitable if assisted by planting. The target should be the establishment of a site-specific, self-sustainable, complex, species-rich and dynamic natural forest. It remains unclear to what extent richness of tree species, functional species groups or guilds affect total diversity of plants, animals and microbes and to what extent these affect the self-regulation and the self-sustainability of the forest ecosystem. Experimental manipulation of ecosystems has indicated that a successive increase of species has, at first, a marked effect on ecosystem function, but that the response curve flattens quite early. In quoted examples, this happened at a plant species richness of > 10 (Schulze and Mooney, 1993, p. 505), which would correspond to the conditions in a fairly rich temperate mixed broadleaf forest. The thresholds in a tropical rainforest are not yet known. Even in temperate forests, it is at present not possible to demonstrate differences of function and properties, such as self-regulation, stability and viability, between forests of the same forest formation or sub-formation in different biogeographic regions which, for historical reasons, possess a different indigenous species richness. Examples are the temperate deciduous broadleaf–coniferous forests and the temperate and subtropical evergreen rainforests in Europe, America, Asia and Africa. The natural forests in these four regions differ noticeably in biodiversity, but not in functionality and sustainability. They appear equally viable and dynamically stable.

Short-rotation Tree Plantations

8

Short-rotation high-yield plantations differ fundamentally in structure, functions and objectives from naturalistic man-made forest (Fig. 8.1). They are intensively managed single-species or mixed 'artificial' forests that produce mostly wood, sometimes also additional non-timber forest products (NTFP), in rotations or production periods of a few years to 15–25 years. The primary objective of such plantations in the rainforest zone is to create a wood source and ensure supplies to a forest industry (e.g. Ministry of Forestry Indonesia, 1993). The desired high yields in terms of wood volume (over $15 \, m^3 \, ha^{-1}$) can only be achieved by planting fast-growing pioneer species and improved genotypes on prime sites of agricultural quality. High profitability is essential in a purely forestry enterprise, but less important in an integrated forestry–forest industry complex. The internal rate of interest is high, often above 10% per annum, due to short production periods and low volumes and values of growing stocks, if costs of holding stock are calculated (Bruenig, 1967a). The emphasis is on quantity and profitability, rather than on quality and value. Short-rotation plantations are usually integrated with the industrial processing of the wood to pulp, boards or low-grade sawnwood. In temperate and seasonal-tropical areas, short-rotation plantations are also being used as pioneer or *Vorwald* (Bruenig, 1971c) to restore mixed forests on barren lands which do not recuperate naturally. An example is the recultivation of degraded coastal land in the CERP project at Xiaoliang, Guandong, China. Pioneer forests of planted Eucalypts, pines and acacia were used to initiate successional restoration of biodiversity, first in the soils, then in the vegetation (He and Yu, 1984; Bruenig *et al.*, 1986b; Insam, 1990; Yang and Insam, 1991; Fig. 11.5). More recently new, some rather faddish, objectives have been claimed for short-rotation plantations of

219

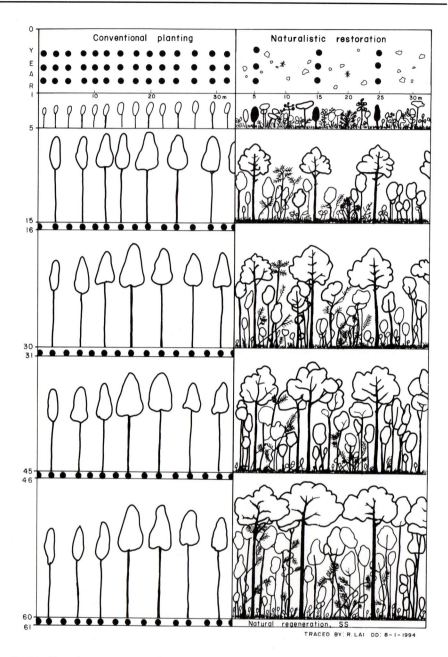

Fig. 8.1. Alternative approaches to afforestation in equatorial rainforest climate. On the left single-species planting on barren or cleared land. Four generations of short-rotation monoculture aim at maximizing volume and money yield. On the right, wide-spaced planting of autochthonous native fast-growing long-lived tree species in degraded secondary vegetation. The existing spontaneously invasive or underplanted mixed understorey initiates the process of restoration towards a near-natural multifunctional complex mixed forest. The newly planted trees are shown as solid black circles in the top rows.

fast-growing trees. The most fashionable and fanciful claim is that carbon pressure on the atmosphere is relieved significantly by photochemical fixation of atmospheric carbon (Section 2.11). In countries with declining growth and yields in overused or abused natural forest, fast-growing tree plantations are claimed to be a compensation for the decline. Further objectives are contribution to β-diversity and habitat quality in the landscape, including provision of migration routes between natural forests and totally protected areas (TPA) in agricultural landscapes and compensation for deforestation in the course of urban growth.

Afforestation (see Appendix 1) of abandoned agricultural (fields) and other types of deforested land with high-yielding, fast-growing trees faces three major problems: the yield potential is difficult to assess, the habitat for trees is strange and often hostile, and the most favourable soils carry the most obnoxious weed growth, such as *Imperata cylindrica*. Trial-and-error experience from the recultivation of degraded heath lands and abandoned fields in Germany by afforestation, the results of the Auer-mühle Production Programmes (APP) and the CERP coastal recultivation and biodiversity project at Xiaoliang, Guandong, China (Bruenig, 1984b,c, 1986b; Bruenig *et al.*, 1986b; Fig. 11.5), highlight the importance of careful matching of site, species, forest stand structure, β-diversity and socio-economic and political framework (ITTO, 1992a; Bruenig *et al.*, 1986, 1994). The initial failure of the Jari project beautifully demonstrates the importance of these points (Palmer, 1991). Land speculation and political opportunism often favour spectacular and easy afforestation rather than the difficult, tedious and time-demanding integrated management of natural succession which requires skills and understanding. Spectacular planting schemes are preferred to the controversial preservation and sustainable management of natural forests. The Philippine Masterplan for Forestry Development is one of many examples (Uebelhör and Abalus, 1991b).

8.2 Rationale and Risks

Fast-growing short-rotation tree plantations yield products that are fundamentally different from the products of natural forests managed under a selection silviculture management system (SMS), a tropical shelterwood system (TSS) or the Malaysian Uniform System (MUS) and that are, and will remain, worldwide in oversupply for a long time to come (Bruenig *et al.*, 1986a). The sustained-yield potential of the productive forest area of the world (primary economic productivity, PEP) lies between 9 and 12 billion m^3 (Bruenig, 1971c), with more than 90% in commodity grades. The recorded and unrecorded estimated removals in 1994 were at least 4.5 billion, possibly over 5 billion m^3, which is less than half the potential. The rationale of establishing short-rotation in forest regions that have the natural potential to produce high-quality timbers can only be to supply a demand in the home industry and market that cannot be supplied from the natural forests. The strong political preference for afforestation and reforestation programmes with fast-growing short-lived species may be explained by the overoptimistic vision of quick and visible successes, a naive belief in simplistic cost–benefit calculations, miscalculation of the risks being small because the rotation is short, and the reluctance to abandon the simple, almost habitual practice of simple tree planting for the sake of less well comprehended natural forest management. Realizing these unreflected preferences, the working group 'Role of Plantations' at the 14th Commonwealth Forestry Conference, Kuala Lumpur, 1993, was right when it stated that plantations cannot and should not be seen simply as substitutes for natural forests, but that they play a role in their own right. Communal plantations should be encouraged where there is demand, provided mixed forests are not the better alternative. Industrial plantations are economically viable if they are vertically integrated with the product-using industry. Plantations should only be established on deforested land, not replacing logged forest. Generally, monocultures with high-yield production regimes must be planned with caution and care to meet the usually exacting site requirements and avoid adverse effects on the hydrological site conditions, especially in the case of fast-growing species with high transpiration rates (Calder *et al.*, 1992), and must consider pest problems, narrowness of the genetic base, amenity considerations and problems of social benefits and social integration.

Only site-adapted and well-domesticated trees (Leaky and Newton, 1994) must be used. High-yield, short-rotation plantation forestry in the rainforest biome carries high ecological and economical risks of failing to achieve the production goal and of failing to find markets that pay attractive prices for the product. The expectations cited in the literature and summarized by Appanah *et al.* (1993) and Weinland *et al.* (1993; see also Evans, 1982; Evans and Wood, 1993) of achieving sustainable yields of 20–50 $m^3\,ha^{-1}$ with such species as *Acacia mangium, Anthocephalus chinensis, Eucalyptus deglupta, Gmelina arborea, Leucena* spp., *Octomeles sumatrana* and *Paraserianthes falcataria* refer to prime sites and ideal production conditions. In practice, increments and harvested yields amount to rarely more than 50% of the ideal, especially in the rainforest biome. Even in highly developed plantation forestry, such as in Germany, Britain or the USA, rarely more than 70–75% of the nominal yield of the site quality class is realized due to the unpredictable effects of climatic events, unpredictable mortalities and dieback, market failures and management deficiencies. Pre-investment calculations of cash flow, profitability and financial rotations are highly hazardous guesswork, particularly in the tropical rainforest biome. The tortuous course and eventual failure of the Jari project was, among many other reasons (Palmer, 1991) and according to my impressions on the spot, largely due to miscalculation of species performance, neglect of soil protection and management, and ignorance of the social and political framework. The nutrient situation after 15–20 years indicated that in the pine forest, magnesium and phosphorus may become limited after less than two rotations, calcium between two and three, and potassium between three and four. The conclusion was that '[pine] plantation forestry at Jari and presumably at other locations [with] similar soil conditions, is not sustainable without extra fertilizer input' (Proctor, 1992). The initial yield expectation of the Jari managers was to obtain 35 m^3 mean annual increment from *Gmelina arborea* and about 20–25 m^3 from pines. This equals the natural site potential (Bruenig, 1971c) and the planners did not allow for unavoidable losses (see biological net productivity (BNP) and PEP calculations in Section 1.13). The BNP and PEP of the Jari planners was unrealistic and naturally was not even fulfilled during the first crop generation. The essential preservation and management of soil biology, fertility and ecology (Swift, 1986; Insam, 1990; Sombroek, 1990; Yang and Insam, 1991) were grossly neglected. Constant fertilizing would pose problems of soil ecology, but particularly of the economics of producing low-value commodity timbers. The decisive mistake at Jari, however, was the failure to integrate the project socially and politically into the Brazilian environment. Useful overviews of the basic principles of plantation forestry, the need for holistic approaches and the available technologies to improve the chances of success and reduce risk of failure are given in Evans (1982), Ministry of Forestry Indonesia (1993), ITTO (1992a) and Lamprecht (1989, chapters 7 and 8).

8.3 Alternatives: SMS or Selective Logging and Conversion

The 14th Commonwealth Forestry Conference spoke against conversion of logged and overlogged rainforest to fast-growing exotic tree plantations. However, the topic of conversion is politically still fashionable. The need for a holistic and comprehensive analysis of costs and benefits before decisions are taken is often recognized, but action rarely implemented. Nykvist *et al.* (1994) described an example from Sipitang, Sabah. Mixed Dipterocarp forest (MDF) on very favourable orthic acrisol and some less favourable gleyic podzol on interbedded sandstone, siltstone and shale had been selectively logged in 1981. After only 6 years, in 1987–1988, the area was re-logged prior to conversion. Different methods of harvesting and conversion to plantation were compared in an experiment with different catchments. In series W4, the timber extraction was manual, followed by planting *Acacia mangium* into the slash; in W5, extraction was by crawler tractor, followed by burning before planting. The major result was that extraction by crawler tractor was about 30% cheaper than manual extraction (see Section 3.4; Fig. 2.11). Extraction by crawler tractor and burning increased nutrient losses by between 10 and > 100% for the different nutrients (Nykvist *et al.*, 1994, table 2). Burning reduced planting costs, but increased weed growth and subsequent costs of cleaning the plantation. Total conversion cost per hectare was RM1455 in W4 and RM1676 in W5. Soil disturbance, compaction and nutrient loss in W5 caused loss of productivity in the plantation, and costs were incurred for fertilizing and restoring the quality of water bodies. The authors presumed that a comprehensive analysis of all internal and external costs, risks and benefits would very likely have shown that it would have been more sensible and socially beneficial to retain the overlogged natural MDF and allow it to regenerate to produce high-quality timber. Leslie (1987) and Weinland *et al.* (1993) came to the conclusion that, depending on the chosen rate of interest, natural forest management can financially compete well with plantation management. This corroborates the results of a linear programming exercise to optimize the allocation of land, financial and human resources under constraints from nature conservation (Bruenig, 1967a). Natural rainforest management requires large areas, but less investment of financial and human resources. High-yield plantations require much less land, but much higher investments into establishment and management (Bruenig, 1967a, 1971c, 1973a). Natural forest management is the alternative which has the highest probability to be the least wrong: low-cost, low-risk, self-sustainable and adaptable crop and versatile production. For the profiteer, clearing of selectively logged forest for conversion to plantations is attractive if, as is usually the case, the returns from timber sales exceed the cost of establishing the plantation. In the case of forests overlogged by conventional selective logging, natural regeneration can be assisted by planting if selective logging has depleted and opened the forest beyond the thresholds discussed in Chapter 7.

The fragile Kerangas, Kerapah and Peatswamp forests are utterly unsuitable for any conversion, a fact which is well known to science, foresters and local farmers (Browne, 1952) but not always heeded. For restoration of deforested Kerangas land, Butt (1984) and Butt *et al.* (1983) recommend planting of trees, but Butt and Sia (1982) exclude specifically the five most fashionable fast-growing pioneer tree species and 12 timber and multiple-use species (Butt and Sia, 1982, table 3) from podzols, grey–white podzolic soils and gleys. Bruenig (1966, 1969c, 1974) restricts any tree planting for recultivation in Kerangas to deep soils with clay in the subsoil and with favourable, well-balanced hydrology, which excludes conversion of Kerangas on most sites (Fig. 11.6).

Forest Management Guidelines 9

9.1 Why are New Guidelines Needed?

Since the eighteenth century traditional forest practices have created integrated comprehensive multiple-use management systems for social forestry but during the second half of the twentieth century especially the development of democracy in society and of information technology in science have provided additional impetus for change. The supporting technical manuals and scientific professional textbooks provide the guidance needed to conduct social forestry in a sustainable manner. This extensive body of information is largely unknown and physically and intellectually inaccessible to non-professional outsiders. Foresters in many silviculturally less developed countries have the information, but not the opportunity to use it. Public concern about sustainability has provoked the emerging concept of certification and trademarking of timber to assist foresters in promoting the implementation of sustainability (Chapter 10). This requires generally accepted standards, which in turn require a framework of accepted norms of conduct. Concise and lucid guidelines provide terms of reference for the development of operational programmes in forestry practice in the silviculturally less developed countries.

Comprehensive guiding manuals defining principles and norms, and prescribing procedures of silviculture and management in tropical rainforests, have been available at least since the 1950s. Examples in Malesia are the Manual of Malayan Silviculture (Wyatt-Smith, 1963), the Sarawak Forest Manual, Forest Management Code and Forest Inventory Code (Forest Department Sarawak, 1959, 1961, 1982 and complementary Food and Agriculture Organization (FAO) reports), the Manual of Silviculture in Sabah (Forest Department Sabah, 1972, and later revisions by the Malaysian–German Sustainable Forest Management Project Sabah in the

1990s), the Handbook of Selective Logging in the Philippines (Bureau of Forestry, 1970), and the Pedoman Tebang Pileh Indonesia (Manual of the Indonesia Selection Felling System, Direktor AT Jenderal Kehutanan, 1980). In the rainforests outside Asia, the management procedures of the Uganda Forest Department (Dawkins, 1958), management rules and guidelines of the Queensland Department of Forestry (undated) and the recent provisional manual of the so-called CELOS management system (Graaf, 1986, 1991; Bodegom and Graaf, 1991, 1992; Tropenbos, 1986, 1995) are examples and important landmarks. In spite of this well documented and solid base of practical experience and guidance, dissemination of knowledge and enforcement of norms and codes of conduct fell short almost everywhere, except in Queensland and in individual concession areas of more committed companies. Forest misuse and abuse remained rampant, overlogging and underutilization persisted in tropical forests and elsewhere. Political *laissez faire* and poor policies led to non-sustainable practices of exploitation in natural tropical and boreal forests. The consequences eventually provoked international and national public concern, media reactions and campaigns by outsiders. Finally, the need for guidelines as terms of reference became evident. Comparable developments also happened outside forestry proper. The UNESCO programme 'Man and the Biosphere' (MAB) initiated the International Biosphere Reserves Action Plan, which used pilot models to show how to convert misuse, abuse and overuse of natural, cultural and human resources into sustainable conservation and management in practice, and at the same time to preserve outstanding ecosystems. The draft statutes of the World Network of Biosphere Reserves and the vision from Seville define criteria, principles, goals and activities (UNESCO Conference on Biosphere Reserves, Seville, 20–25 March 1995) that in basic philosophy very much resemble the gist of the ITTO guidelines (ITTO 1990b, 1992a, 1993a), especially in their comprehensive and holistic approach to integrated conservation and management, supported by socially constructive policies. The failures of forestries, as well as civilizations, are evidence that this approach has been lacking. To fill the gap, new guidelines are needed.

9.2 Example: the International Tropical Timber Organization (ITTO) Guidelines

ITTO decided in 1988 to prepare a set of global forestry guidelines. Indicative guidelines for the management of natural forests were drafted in 1989, for planted forests in 1990 and for the measurement of sustainability and for the conservation of biological diversity in production forests in 1991 (ITTO, 1990b,c, 1992a,b, 1993a). The guidelines are a structured check-list of basic principles and possible actions, and not operational manuals prescribing techniques and procedures. They provide essential orientation for the preparation of concrete, operation-orientated national and local manuals for planning and monitoring. Forest management, traditionally bound to the principle of sustainability, is in most countries specifically committed by national acts, such as the constitution and the forest law, to comply with defined criteria of sustainability. The guidelines are codes of conduct that give general guidance and basic rules to design strategies at national and management levels in order to achieve this compliance. The guidelines for the sustainable management of natural forest (ITTO, 1990b) are the core of the ITTO set of guidelines. They define 41 basic and global principles and recommend a corresponding set of 36 possible actions to comply with them. The subject cover is wide and includes forest policy, law and legislation, role of the forest service, enforcement, forest management performance, harvesting technology, security of the forest estate, incentives, monitoring and control of annual cut levels, protection of habitats and environment, research, public relations and information links to societal groups and political decision makers. Six appendices give mainly terminological clarifications and some additional detailed explanations.

9.3 The ITTO Guidelines for Planted Tropical Forests

The guidelines for the establishment and sustainable management of planted tropical forests (ITTO, 1992a) are similarly structured, but more complex and complicated than those for natural forest. 'Natural forest' is a well-defined and conceptually uniform entity, despite its extreme natural heterogeneity and complexity in reality. Planted forests, by comparison, cover a much wider range of conditions and objectives. The legal situation of land tenure and customary rights is diverse, often complicated and confusingly controversial. The crops are relatively simple, but the silvicultural and management systems are more complicated and diverse than in the natural forest, because the crops are less self-regulating and self-sustainable. The guidelines are correspondingly more complex and detailed than those for natural forest, and contain 64 principles and 72 recommended actions. At the core of the guidelines for planted forests is the principle that ecological, economic and social costs, benefits and risks are interdependent and interactive (see Chapters 7 and 8). Comparative evaluation among short-rotation fast-growing tree plantations, long-rotation high-quality timber tree plantations and natural forest management is only meaningful if the evaluation is based on holistic ecosystem aspects with respect to silvicultural management (Kollert *et al.*, 1993) and to the social and national economic effects. The ITTO guidelines realize the advantage of integrating the natural dynamics of any existing or potential natural vegetation with the planted forest to reduce risks, costs and uncertainties (Sections 7.4–7.6). The structure of the potential natural vegetation on the site is the standard by which to measure the feasibility of design of plantations and to assess the risks of damage from natural sources and failure due to existing political, legal and social weaknesses and unpredictable economic and social changes. The guidelines, therefore, stress the need for a system analysis of the policy, legal and social conditions before deciding on plantation establishment. Six appendices give concise definitions of terms and details on inventory, land capability survey, geographic information system, harvesting, plantation design and minimum prerequisites for sustainability and references to sources of information.

Timber Certification, Trademarking and Monitoring 10

10.1 Background and Purpose

Since ancient times, in all cultures and societies, excessive environmental degradation and overuse, abuse and misuse of natural and human resources have been connected with political failures and the misdemeanours of leaders. The development of true parliamentary democracy finally gave conservationism and environmentalism political weight when the public became aware that belching smokestacks, intensive agriculture and cash flow-based gross national product calculations do not symbolize cultural advancement, and when the media began to expose the ugly sides of economic development. Consumers have become more sophisticated, better educated and informed, and prepared to cooperate in the protection of nature, environment, human dignity and the sustainability of such natural resources as forests. Production processes and their products that do not meet the ancient, newly revived standards will lose out. Some of the major exporters of timber and timber products in the world, Canada, Finland, Sweden and Germany, and some of the major timber market countries, particularly the traditional forestry countries such as Germany, have recognized the trend and realized the advantages of early and effective action. The forests in Canada, Finland, Sweden and Germany are increasing in area and productivity. The prevailing practice in Canada, Finland, Sweden, Russia and the USA of large clear-fellings may be ecologically less harmful than environmentalists assume, but it cannot be reconciled with the ecological sensitivity of the public. The wasteful utilization and the lack of effective silviculture in boreal and tropical forests definitely offend the principles and criteria of sustainability. For a number of reasons, public attention was drawn by environmental action groups to the tropical forests first, rather than simultaneously to tropical and boreal forests, as Heske had in mind

(Heske, 1931a,b). A campaign against tropical timber and forestry began about 1980 in Switzerland, which became rather acrimonious and confrontational. Eventually the campaign succeeded in substantially reducing the volume of tropical timber traded in the European market. At the same time the malpractices in the tropical logging industry continued unabated. Finally, the issue became excessively politicized and positions hardened. The opinion of the German timber trade, trade unions and industry was that the issue was economic, social and technical, and not suitable for politicizing. Their answer was to establish the Initiative Tropenwald (Initiative on Tropical Forests, ITW) in 1992 (Section 10.3). The general task given to ITW was to study the situation and arguments, identify the real problems and propose practicable, comprehensive and coordinated solutions. Other groups formed about the same time are focusing more directly on schemes of certifying that timber has been sustainably produced in tropical, temperate and boreal forests. The most advanced of these are the Soil Association's Woodmark Programme, the scheme of the Societé General de Surveillance (SGS), Smart Wood Program, and the Green Label (ITTO, 1990–1995, vols 3 and 4). Lembaga Ecolable Indonesia (LEI) was founded to organize and supervise the certification of production in Indonesia. The Forest Stewardship Council (FSC) established itself in 1994 to be a global accreditor of timber certifying agencies. In addition, numerous individuals, organizations and companies are promoting their own concepts of certification to take advantage of the developing certification market. These latter activities may be short-lived, but cause much confusion and shake credibility while they are alive.

Professional foresters, timber traders and industrialists in Europe and Canada have understood that timber certification is not an issue of politics, but of efficient and sustained timber production and marketing. However, politicizing produced confrontation and stalemate. It is symptomatic that no other government of a tropical rainforest country followed the example of Sarawak and invited ITTO to review the state of their forestries, except Bolivia in 1994 (ITTO, 1990–1995, vol. 4, no. 5, 17).

Early moves towards timber certification by the producers and trademarking by the buyers could have prevented the decline of tropical timber in the high-quality, high-price markets. In the market, certification is an effective instrument to secure the sustainability of attractive market conditions. In the producing country, certification is an effective instrument to replace wasteful and costly selective logging by profitable and sustainable Selection Silviculture Management systems (SMS) and similarly adapted systems of naturalistic management. The further strength of eco-labelling or timber certification

'resides in its capacity to discriminate (through market signals) in favor of timber produced under sound environmental practices in contrast to the indiscriminatory nature of bans or boycotts. If introduced in a unilateral manner by a subset of importing countries, however, its effectiveness will be, to say the least, doubtful. It would lead to trade diversion and potentially perverse environmental results, not to mention an increase in trade disputes at GATT level. Addressing the issue at a multilateral level, countries would be able to minimize the problem of trade diversion. At the same time, standards would be determined in a cooperative manner, diminishing the risk of future trade conflicts.'

(Varangis *et al.*, 1993)

Voluntary, rather than governmentally decreed, national or regional schemes of timber certification or eco-labelling are unlikely to be impeded by the World Trade Organization (WTO) (Appleton, 1995). This should have been exactly the maxim and strategy of a vigorously forward-looking policy in the producer countries, but in reality governments imposed the opposite course, and at the same time prevented progress by feet-dragging in endless meetings and conferences. While sovereignty rights, discrimination and costs were used to justify delaying tactics, the real reason was fear of the transparency of timber logging and trade among politicians and profit-focused concession holders and loggers. By mid-1995 the market obviously was in no mood to wait any longer. German forestry companies in African countries and the Indio community-owned company Quintana Roo in Mexico, accepting that the timber market is a buyers' market and cannot be dictated, decided not to wait for the tortuous international and national dialogues to produce results and took the initiative in mid-1995 to develop a market-orientated certification scheme for the timber and timber products from their well-managed forest areas. A trademark association is being founded under public German and EU law, but independent of government, to become operational in 1996. The first properly and credibly certified timber will therefore be on the European market before mid-1996.

Johnson and Cabarle (1993) rightly list access to profitable markets as one of the six conditions that must be met if certification of tropical timber is going to contribute to achieving sustainability of forestry in the producing country. Only certified timber and timber products will in future have access to discriminating high-price and high-quality markets and only the assurance of sustained supplies to trade and industry will secure this access. This simple connection seems often not to be understood or to be purposely ignored. Powerful political and economic interests in some producer and consumer countries do not favour timber certification. In the producer country, timber certification may be seen as an undesirable hindrance to free individual access to the forest resource, in the consumer country to the national raw material supply policy. As a result of this and due to general political bungling of the issue, progress towards certification has been internationally sluggish during the period 1990–1995, when it would have been in the best long-term interests of the

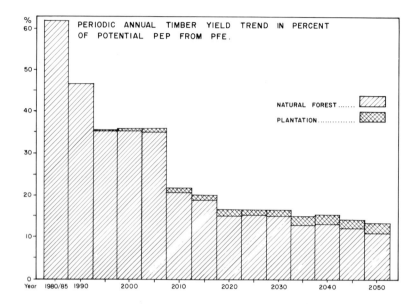

Fig. 10.1. Long-term effects of ignoring principles of sustainability and of adhering to selective overlogging on forest resource in Southeast Asia. The example is drawn from real life: the national rainforest area of a country in the Asia-Pacific region has been heavily overlogged but wastefully underutilized up to 1990. Only two-thirds or less of PEP has been recovered in wasteful selective logging (Fig. 2.15). Logging progress by area has been two to three times more rapid than necessary. Overlogging, illegal logging, volume adjustment and transfer pricing have reduced revenue and the long-term potential for timber yields. A level of 90% utilization efficiency in RIL/SMS since 1960 would have prevented the decline of growing stock. Plantations on about 3.6% of the PFE area cannot compensate the loss of volume, much less of value production, but can soften the fall.

forest countries and the trade to push the issue. This is particularly true for the Asia-Pacific region, where growing stocks and productivity of the forests sharply decline (Fig. 10.1), so that the rapid change to sustainable harvesting is more vital and urgent than in Africa and America (Grainger, 1985, 1986, 1989).

10.2 Principles, Criteria and Indicators of Sustainability

Principles are fundamental laws and codes of conduct. The fundamental principles of sustainability of forestry are the components of the body of professional attitudes and ethical conduct which have evolved in forestry over centuries as scientific knowledge and practical experience grew (Sections 3.1, 4.2 and 4.3). The criteria of sustainability are selected examples that express the principles and can be used in judgements and to define standards and norms for specific aspects of sustainability in forestry.

Examples of practicable criteria are the maintenance of the forest resource base, continuity of flows (increment, yield, income), environmental protection, social integration (employment, sharing), protection of species and nature, and heritage conservation. Some commonly proposed criteria, e.g. integrity of the forest ecosystem, are esoteric, cannot be quanitatively defined and are therefore not practical, but may also not be acceptable socially. Indicators are quantitative or qualitative parameters which can be scaled to assess the distance from 'critical load' levels. Criteria are usually well defined in statements of forest policy, forest laws, forest ordinances and in management codes and plans. Indicators are usually defined in inventory and management codes. Standards and norms are traditionally defined, scaled and prescribed in national management regulations and rules, and specifically prescribed in regional and local management and felling plans. Criteria and indicators have changed in response to growth of knowledge, technical and socio-economic development and social preferences (Hasel, 1985) and will continue to do so. Operational reality in management practice requires that check-lists are practical and limited to essential key criteria and key indicators 'which have the greatest effectiveness and number of linkages to others' (Udarbe *et al.*, 1993, 1994). ITW and the Institute for World Forestry, Hamburg, developed a project concept to identify such key criteria and indicators in a field test series, collating the lists produced by ITW, Woodmark, Smart Wood, LEI and the Dutch group DBB. The first field test was done in a German state forest management unit (Forstamt Bovenden in hill broadleaf–conifer forest) in late 1994. The second test was in Mixed Dipterocarp forest (MDF) in Kalimantan in February 1995 and the next tests will be in African and American rainforests later in 1995. The results are expected in 1996 (see also Prabhu in ITTO, 1990–1995, vol. 4, no. 5).

In accord with traditional forestry planning, the tested lists distinguish national, regional and management unit levels, and encompass the whole gamut of traditional social multipurpose forestry. The quantification of indicator standards at any level is not as simple and straightforward as the definition of safe minimum standards and critical loads in water and air pollution monitoring and control. The value scaling of the various functions of forests and forestry is affected by the constantly changing perceptions of professional, civil and political power groups. Even in the case of very simple stand structure assessment, accepted indicator standards may prove wrong. A private forest owner and I decided, after a catastrophic windthrow in 1972 in Lower Saxony, Germany, to replant at unconventionally wide spacings of $3 \times 1\,m$ $(3300\,ha^{-1})$ for Scot's pine and $4 \times 1.6\,m$ $(1600\,ha^{-1})$ for Douglas fir. The idea was to save expenses and later costs, to establish a more natural and self-sustainable forest and to create a more sustainable forest management system. The sustainability principle 'orderly silvicultural management' includes the criterion 'volume increment' which among others had the indicator 'density of stocking'. The safe minimum standard, officially decreed by the state, at the time was 14,000 ha^{-1} for Scot's pine and 6000 ha^{-1} for Douglas fir. Our concept therefore did not meet the requirements of 'orderly forestry' and sustainability. We were consequently refused the state subsidy for reforestation after calamities. Twenty years later, these stands are now considered a demonstrative example of holistic sustainability and of management for biodiversity, ecological stability and economic efficiency (Bruenig, 1995). Another problem of standards in forestry is statistical sampling errors and systematic measuring errors. Goenner (1992), enumerating the number of bird species, found a difference of 30% between line and spot sampling of the same area of MDF in the Sepilok virgin forest reserve, Sabah, with line sampling being more reliable. Spiders have been useful indicators of recovery of biodiversity in the CERP recultivation area at Xiaoliang (Yu and Pi, 1984) and birds and insects for the comparison of MDF and plantations in Sabah (Khen *et al.*, 1992), but how to derive hard and fast scales and thresholds for these indicators and how to verify them is a vexing puzzle. Estimates with large measuring and sampling errors of variable parameters are not enough; high precision and accuracy are needed when market chances and survival of a business are at stake and liabilities may have to be decided in court.

Sustainability is not a principle of nature, where only change is constant, but a human concept which changes according to social consensus. Consequently, definitions of criteria and scales of indicators cannot be persistent and universal. Different natural site and forest conditions and different technical, economic and social environments require different sets and scales of indicators and different safe-standard levels. Therefore, key indicators and standards must, as much as possible, relate directly to conditions that are the expression of the basic keystone processes of viability, productivity, self-regulation, participation and other essential properties of the natural and social ecosystems. Biomass, volume stocking, tree densities, growth and timber increment are not keystone indicators but only supplementary indicators of sustainability. Keystone indicators are features of the physiognomy and structure of the A, B, C and D storeys, tree-stand curve, soil organic matter (SOM), soil biology and hydrology. Criteria, indicators and standards must be adaptable to the specific natural, technical, economic and social conditions of each case, but fundamental principles apply universally.

Sustainability of biodiversity is particularly intractable. The stability of populations of plant and animal species is naturally very dynamic, the more so the smaller the area of assessment. Biodiversity in relation to forest management cannot be assessed at the scale of a forest management unit alone, but must be seen in a landscape and regional geographic dimension. However, the analysis of temporal dynamics of geographic populations of plant and animal species is severely restricted by the short period of time available for observation, usually at the scale of a few years to, at best, decades, while the dynamic oscillations or unidirectional changes of species populations may run over scales of hundreds or thousands of years (Maurer, 1994). This is very much longer than any census, trend analysis and monitoring of size and diversity of species populations could be conducted to detect accurately and reliably the local and regional effects of forest management on biodiversity. A way out of this dilemma is to assess habitat quality, rather than biodiversity directly. Sustainability must not be assessed by a perfectionistic list of non-assessable indicators of non-quantifiable properties, judged by non-verifiable arbitrary standards. Forestry practice requires a few operational and practicable key indicators which are judged in context with the state of the ecosystem. The key indicators are linked with properties of the ecosystem which are difficult to measure and for which useful indicators are not available. Non-assessable properties, such as 'resilience', 'tolerance', 'stability' and 'viability' may be linked by professional best judgement to key indicators, while 'integrity' is esoteric, unrealistic and vague as a concept, which is not useful as a criterion and has no practical indicator. In essence, after many expensive and costly years of circular and esoteric arguments in meetings and media about sustainability, criteria, indicators and standards, I suspect that the outcome will be a reconfirmation of those simple parameters of forestry and forest physiognomy and structure, ground vegetation and soil and humus conditions which have indicated health and vigour to qualified and experienced foresters for hundreds of years. What is new will be the application of advanced technology for the monitoring of compliance, such as remote sensing of canopy structure and condition, soil disturbance, road and skidtrail locations and densities, and water quality (Böhm *et al.*, 1995; Kuntz *et al.*, 1995). A new dimension is being opened by the use of sophisticated data and information systems in forest management and monitoring for sustainability. Forestry datasets include much environmental, biological and social information of general relevance to conservation of nature and of culture. The data and information systems (for examples see Ashdown and Schaller, 1990) for securing sustainability in forestry should therefore be, as far as possible, harmonized and linked with the UNESCO–MAB World Network of Biosphere Reserves (Biosphere Reserves: the Vision from Seville for the 21st Century, UNESCO conference, Seville, 20–25 March 1995).

10.3 The Dualistic Certification–Trademarking Concept

The German Bundestag in 1987 initiated a permanent commission to report regularly on the global climate. The commission conducted an inquiry on the tropical forests that is the most thorough and substantial study of its kind (Deutscher Bundestag, 1988, 1990, 1994a,b; Deutscher Bundestag, Enquete Commission, 1992; German Federal Government, 1993). The reports raised public concern about the wisdom of using tropical timber. In the course of the public discussions, the Initiative Tropenwald (ITW) was founded in 1992 by the timber trade, timber industry and trade unions (Section 10.1). ITW was mandated to promote a consensus on sustainability between all sectors, to prepare a list of sustainability criteria and indicators, to develop a concept and proposal for the infrastructure and procedure to monitor the flow of tropical timber in the EU timber market, and to promote a pilot field test of lists of criteria and indicators to assess their practical suitability for use in a three-tiered forest management assessment procedure. In 1992, a working group was formed by ITW that included representatives of the tropical producer countries, conservation non-governmental organizations, including the World Wide Fund for Nature (WWF) and International Union of Conservation of Nature (IUCN), consumer associations, institutions from other member states of the EU, representatives of the German tropical timber trade, timber industry, producer countries, international science and the German government. By mid-1994, consensus was achieved and the list and field test procedure were prepared and handed over to the International Centre of Forestry Research (CIFOR) for implementation (Section 10.1). The development during 1995 of a coordinated certification–trademarking scheme of timber-flow monitoring was the most difficult of the four tasks.

Timber certification has two main goals: to promote the ecological, economic and social sustainability of forestry in the world, and to promote the sustainability of markets that pay the price which timber deserves. The process of timber certification has two main interlinked but autonomous components: assessment and certification of sustainability of forest resource management by a surveillance scheme in the producer country, and assuring accountability and identification of origin of the product in the market place by a trademarking scheme in the consuming country or trade region. Basic principles of the dualistic ITW scheme are that both certification and trademarking must be voluntary, uniform in each producing forest and consuming market region, transparent, accountable, creditable, trust-building, and reliably accurate. The principle of voluntary certification and trademarking has advantages over mandatory eco-labelling. It is simpler to administer and it does not collide with GATT 1947 and WTO. The principle of uniformity has the essential advantage that the trademark becomes more easily established in the market as a recognized and familiar symbol of sustainability which signals acceptability. The proliferation of certification and trademarking schemes would confuse the market and would not inspire confidence and trust in the consumer (ITW, 1994c,d). ITW sees certification and trademarking as a coordinated production- and market-orientated programme without any element of discrimination, which creates justifiable confidence in the consumer that buying and using timber contributes to environmental and nature protection generally and to economic and social development in the producing region. Producers are assured of a sustainable position in high-price markets if they guarantee sustainable supply of timber as an environmentally and socially friendly product. This assurance can only be given if it is reliably based on the reality of sustainable timber production. The prospect of a sustainable and attractive market should be sufficient incentive to encourage producers to adopt the principles of sustainable forestry in their own interests (ITW, 1994a,b). Four large German forestry enterprises in equatorial Africa and a community-based company in Mexico have decided in 1995 to have their management units credibly assessed by certifying agencies which are accredited by the Forest Stewardship Council (FSC). This shows that the private sector which knows reality on both sides, the producing forest and the consuming market, has realized the long-term advantages of certification and is determined to make good use of them (Section 10.1).

The field surveillance of sustainability of forest management applies the norms or codes of conduct of traditional practices of sustainable forestry. Technically this poses no fundamentally new challenges or prohibitive difficulties. The difficulties are rather more related to indolence and to attempts from outside the profession to impose unrealistic and impractical standards and linkages that are irrelevant to the issue. Otherwise, the assessment is operationally simple and easy to handle if the honest will, mutual trust and operational readiness exist among all parties to apply common sense and make good use of existing knowledge and experience (Sections 2.8, 3.4, 3.5, 4.3, 4.4, 6.3–6.7 and 10.2). Some timber companies, especially concessionnaires with large areas and relatively secure tenure in Central Africa (Gabon, Congo, Zaire), already operate technically mature sustainable management systems inspite of the problems with regeneration and stimulation of growth which are connected with the low yield per hectare (Sections 3.4, 3.5, 6.3 and 6.5). In the Asia-Pacific region the problem is more intractable, partly as a result of the particular pressures in the East Asian market, partly due to the pecularities of the concession systems and the socio-political environment. The high density of merchantable timber stocking are an almost irresistible temptation to overlogging and wasteful underutilization. The rugged terrain and fragile soils pose additional technical difficulties. Many companies already operate sophisticated data and information systems and apply electronic aids to timber marking and operational monitoring. It poses no great technical difficulties to link the data of the tree in the harvesting block (position, species, size, quality) with the data on the logs from that tree as they progress through the various stations of processing to the final product at the point of shipment or local sale. All that is required at the procedural level is a hierarchical, integrated, multisectoral spatial information system to link the various specific spatial or geographic information systems (GIS) and data systems. These companies have the facilities needed for certification by surveillance agencies and would qualify for certification provided they adopt proper and orderly timber harvesting procedures and implement an appropriate management system.

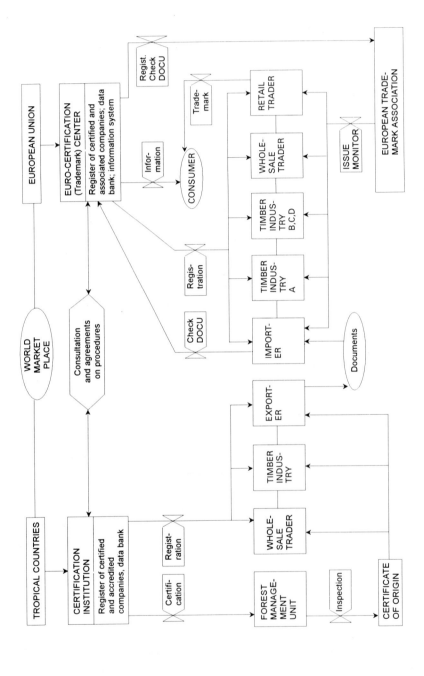

Fig. 10.2. Structure and flows of the procedure in the dualistic certification–trademarking scheme proposed for the EU by Initiative Tropenwald (ITW, 1994d). The centres for certification and trademarking are autonomous, established under civic law and controlled by general government supervision. The centres monitor the performance of logging, trading and processing companies that are committed as registered members of the certification and trademarking associations. Product flows are checked and documented under national surveillance of compliance with internationally accepted standards both in the producing and consuming areas. (From ITW, 1994a,c.)

More complicated and not yet technically developed is the procedure of tracing timber and timber products in such a complex and diverse market place as the EU. Efficient monitoring requires insight and familiarity with the most complex structure and flows of the market. The multilevel trade network and the complicated agglomerate of processing factories and end-product manufacturers from industrial complexes to small family businesses and individual artisans is so complex that it is difficult to elucidate and comprehend for non-professional outsiders. The ITW concept visualizes the creation of an infrastructure that includes a trademark surveillance centre and a trademark association of the timber buying and processing companies who have the essential experience and insight in the problems (Fig. 10.2). The dualistic concept clearly divides responsibilities between producer region and consumer region (Fig. 10.2). An efficient, practical, unified and reliable procedure for data and information flow from the producer's certification agency to the corresponding trademarking agency in Europe, or to any other consuming region, is crucial for success. The procedures, which have still to be worked out, are data transmission to Europe and data processing, storage and data and information flow in Europe, information transfer to the consumers, the public and the media, issue and surveillance of a legally protected trademark. The respective roles and responsibilities of the centre and the trademark association must still be defined and harmonized. Finally, the procedure must be tested in practice before certification, trade monitoring in practice and timber trademarking can be introduced as a routine procedure at large scale. The high-price markets give no chance or time to wait until the year 2000. If the high-price markets for tropical timber products are to be sustained, there is no time to lose; the sophisticated market needs, and the buyers definitely want, certification and trademarking in 1996.

10.4 Trade Policies and Tree-species Conservation

It is well known that trade policies are inefficient instruments for correcting domestic distortions. In other words, they are poor instruments with which to pursue environmental goals. Yet, if one were constrained to choosing among trade policies to advance these goals, it seems that eco-labelling holds more promise for promoting sustainable forestry than either bans imposed by exporters or boycotts applied by importers.

(Varangis *et al.*, 1993)

We have shown that eco-labelling and timber certification are not a trick to use trade to impose codes of conduct on others to one's own satisfaction, but there are cases where this has been attempted. Very surprisingly Germany, and also The Netherlands, recently attempted to have the timber of tree species included in the Convention on International Trade with Endangered Species (CITES) (ITTO, 1990–1995; IUCN, 1988–1995). The agreement was never intended for controlling timber trade of certain species (I was involved in the formulation of the text for the agreement, and remember discussions on this point in Bonn) and it is technically an utterly unsuitable instrument for this purpose, even if it were possible to prove the reality of trade threatening extinction or depletion below viable stock levels, and if it were possible to identify origin and species of timber reliably along the trading routes.

However, the following Malesian timber species were proposed for inclusion in Appendix 2 of CITES: Merbau (*Intsia palembanica* Miq.), Tapang (*Koompassia excelsa* (Becc.) Taubert), Ramin (*Gonystylus bancanus* (Miq.) Kurz) and also Amazonian, Caribbean and African Mahogany species. Species qualify for inclusion if they are potentially or actually endangered as species or populations by international trade. Excess of market demands over supply, or in the case of Ramin adjustments of annual felling quotas, have been interpreted as indicating a developing scarcity caused by dwindling stocks due to commercial overuse that is 'considered to have a decimating effect on large, commercially exploitable populations'. None of these three tree species is endangered as a species or as a viable population, neither by trade nor otherwise. Ramin is very site tolerant and occurs as a rare to frequent, locally common species on an extremely wide range of sites in the Peatswamp forests (PSF) PC1 and 2, and in Kerangas and Kerapah forests (KF and KrF) in Borneo and Peninsular Malaysia (Anderson, 1961a, 1964, 1983; Bruenig, 1969b,c, 1974, 1989e,f; Corner, 1978; Bruenig and Sander, 1983; Newbery, 1991). In Sarawak, PSF PC1 in the permanent forest estate is sustainably managed (Section 6.7). Harvesting has reduced the hump in the bimodal stand curve (see Fig. 6.4 for an example of a hump > 100 cm diameter in MDF) but does not threaten Ramin as a species or as a member of the PC1 community. Some areas of PC1 are included in totally protected areas (Forest Department Annual Reports 1991, 1992) and most of the Ramin-bearing KF and KrF will never be logged. Similarly, Merbau occurs widely spread and locally common in MDF. In contrast to Ramin, it regenerates very easily, is easy to propagate and is commonly planted inside and outside the forest. Tapang and its generic relatives are possibly the least endangered species of all the Bornean primeval forest species. Tapang would survive even if an unbelievable complete slash-and-burn holocaust of all the forests in Borneo eliminated all other species. The species regenerates easily and the natives do not fell the big trees because the wood is extremely hard and the crowns harbour honey-bees. Loggers also usually avoid the species, which is a common sight as a giant relic overtowering even very heavily overlogged forest. Even more ludicrous and scurrilous than the official German and Netherland initiatives are current campaigns by some action groups to propose species of Ebony (*Diospyros* spp.) and some common African commercial timber species, including mahoganies, sapele, sepo, utile and in Malesia even common Dipterocarps, for inclusion in CITES. The uncertainties of ascertaining the status of a species as threatened, endangered or extinct, and the difficulties of timber identification are well known. Even in such thoroughly researched and monitored areas as the Harz National Park in Germany, extinct species reappear (e.g. *Pseudorchis albida* (L.) A.&D. Löve in Dierschke, 1994). In Sri Lanka, several tree species have been recently re-discovered which had been officially declared extinct (Jayasuriya, 1995).

Effective and practicable protection of very rare, endemic and possibly potentially threatened tree species can only be accomplished by preserving their habitats in sizes and conditions that permit the species to survive. In addition, it would be sensible in the Asia-Pacific region generally to refrain from excessive promotion of rare and lesser used or lesser known timber species. The high stocking of true merchantable timber of well-known commercial species already causes overlogging. The crucial issue is how to restrain felling intensities (volume cut and frequency of re-entry) to preserve habitat structure and biodiversity (Section 6.4). It is naive to try and protect arbitrarily selected species by monitoring the international trade of their timber. In Africa and America threats are more real for some individual species which produce special quality timbers for export and local use and which grow on special sites in isolated habitats, such as true mahogany, Rio palisander, rosewood and true ebony. CITES may help in these cases, but would be ineffective unless the species and habitats are at the same time protected effectively as totally protected areas.

11 Failures, Major Obstacles, Trends and Needs

11.1 Potential for Success and Failure

The zonal tropical rainforest ecosystem is a rich storehouse of information and is equipped with a large capacity to be resistant and tolerant to exogenous perturbations, disturbances and long-lasting changes. It has a great richness of tree species, tree-species associations and tree guilds. These form mutually dependent, symbiotically or competitively interacting and potentially vicarious units within the ecosystem in a functional structure which is not unlike that in the human brain. This may explain the well-known fact that the rainforest ecosystem is resistant to disturbance, capable of buffering impacts without change, can resiliently bounce back after change but also can tolerantly pass through adversities successfully without change (Sections 1.4, 1.9 and 1.11). The zonal tropical rainforest is more easy to manage sustainably than any other forest formation if interferences are kept within the range to which the soil and the vegetation systems are adapted. The principles and criteria for such a strategy and the key indictors for success are adequately understood, supported by practical experience and scientific knowledge (Chapters 2 and 3). Viability and biodiversity can be restored in damaged ecosystems (Chapters 6 and 7). The sustainable management of zonal natural rainforest is sufficiently profitable to be financially attractive at business and management unit levels to private business as an alternative to exploitative selective logging. At a national socio-economic level, non-sustainable selective logging causes high social and business costs that exceed revenues, and reduces the balance of the natural resources account. Sustainable forestry is rare in reality. Even the urgent and immediately profitable transition from selective logging to sustainable orderly and proper harvesting makes only slow, if any, progress. Clear indications of change can be seen only in a few countries and in individual management units, notably in areas where the management is firmly in professionally capable hands and where the interests of the state and the private sector are not in conspiratorial collusion in the interests of easy and quick cash flows.

The chances for replacing selective logging by sustainable harvesting and management largely depends on the motivation of all people involved. The necessary motivation will only evolve if the incentives are fairly strong and the benefits equally enjoyed with responsibilities shared by the government, its civil servants, private forestry and forest industry sectors, their labourers and the local population. The crucial key criterion in this respect is that the political and economic leaderships set examples of high ethical and moral standards themselves. An example of forest management improvement by equitable sharing of rights and responsibilities is described from Ghana by Sargent *et al.* (1994). An essential component of such a strategy is to unify the responsibility for utilization (e.g. RIL), silviculture and management in one professionally qualified hand. The private sector needs to be motivated by security of long tenure, conditional on compliance and good behaviour. The practicability of such a strategy is indicated by the successful advance of natural forest management towards sustainability by German-owned forestry companies in Gabon, Congo and Zaire, and by some companies in the Malesian region. These companies own large concession areas in relatively secure tenure; their code of conduct accords with the ethics of traditional multiple-use social forestry. This also includes the creation of a company-integrated, appropriately participating, trained and skilled labour force, instead of the exclusively money-orientated casual and unskilled, underpaid and socially insecure sub-subcontractor only too common elsewhere. Another example of success is the Indio-owned communal forestry enterprise Quintana Roo, Mexico (Janka *et al.* in Bruenig and Poker, 1991, pp. 41–49). The rash and brash order by the Australian government to close rainforest SMS practice and research in Queensland has eliminated this most successful example of approaching sustainability. In all examples of success, the concessionnaires are professionally motivated, well-qualified and work large areas under long-term tenure and effective control by government. In contrast, other companies and whole countries in Africa and in the Asia-Pacific region still have the many-tiered concession-cum-contractor system which dilutes responsibilities and creates inequalities which demoralize. The concessionnaire acts as multiple fee-collecting licence holder who contracts the exploitation out to contractors, who in turn hire logging subcontractors and these in turn sub-subcontractors for the felling and extraction. The latter two are financially squeezed, but have to bear the ultimate responsibility and risk. This system offers no incentives towards sustainability. Cash flow and profit, and economic and physical survival are the only relevant motives. Under these unfortunately prevailing conditions, the chances to meet the ITTO sustainability target 2000 are slim. The cash-flow orientation also hinders the progress of establishing an adequately sized and distributed multipurpose permanent forest estate (PFE) with sufficient and adequate totally protected areas (TPA). Powerful pressure groups want to log the stateland areas before they are protected as PFE and TPA.

Urban people are the most active participants in the predatory acquisition (Fig. 11.1) and the depletion of natural resources, but the urge to acquire is not confined to them. Local indigenous people also jump on the bandwagon, if their leaders realize that this is profitable and possible. It is common that local people object to sustainable management and conservation because they fear interference with their habitual slash-and-burn shifting agriculture (shag) (e.g. from the Congo by Setzer in Bruenig and Poker, 1991, pp. 21–32). There are also problems if the management of the forest resource is entrusted to the local indigenous people without adequate safeguards. For example, the Environment Protection and Indigenous Land Reclamation Association (APARAI), a non-governmental organization in the state of Rondonia, Brazil, states in a project description:

> Indiscriminate timber logging by Indians and local communities has already been well documented. Approximately 905,000 m^3 of high-quality timber is reported to have been logged during 1987–90 in only eight areas of the State of Rondonia, where there are twenty indigenous communities. Another clear example is the Mequéns indigenous area in Southern Rondonia, where 90% of hardwood species have already been logged. These results are alarming for they reveal that 50% of the communities are currently selling hardwood timber species, while the remaining 50% are not doing it simply because they no longer have any. This problem is not only limited to the indigenous areas; it also extends to the colonists, who often ignore timber resources, burning them and deforesting areas indiscriminately. Therefore, there is a need to develop appropriate forest settlement policies to regulate colonist settlements in the forest, while avoiding massive urban invasions and the resulting consequences of this migration. Studies have shown that since 1985 the income of colonists has gradually been decreasing, partly because of the timber shortage.

(ITTO PD15/1992 (F))

In spite of the pleas of successive world congresses of Food and Agriculture Organization (FAO) and International Union of Forest Research Organization (IUFRO) and of the results of United Nations Conference on Environment and Development (UNCED), the global situation has not improved, but rather worsened. The initiator and founder of world forestry, Professor F. Heske (1931a), branded greed-propelled profiteering as the cause for the unsustainable use of forest resources in the boreal and some temperate countries. Even then he saw

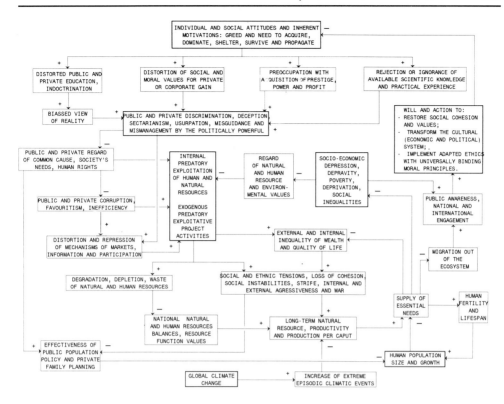

Fig. 11.1. Some of the crucial feedback loops between major external and internal key conditions (elements) and causal factors which cause the current global cultural decline by spiritual and material resource degradation in the form of plundering, squandering, degrading and destroying natural and human resources. The continued degradation of the environment, depletion of natural resources and the destabilizing abuse of societal systems and human resources obstruct the attainment of any kind of sustainability. The positive feedbacks among and between politics for power, prestige and money, political misdemeanour and mismanagement, technological incompetence, bureaucratic bungling, greediness of entrepreneurs, general corruption, demotivation and inefficiency in the public civil and military services, and in some parts of the world also the unbridled population increase especially at the impoverished and underpriviledged social base, destabilize each and the entirety of the various natural and cultural ecosystems. Natural resources are depleted, national assets wasted, wealth inequitably distributed, suffering and poverty of the deprived ordinary people perpetuated. Free or forced emigration out of the ecosystem (forest, landscape, country) may bring temporary relief to the source ecosystems (e.g. Bosnia, Africa), but new problems are created in the refuge areas and asylum-offering ecosystems (e.g. Germany). Public awareness of deprivation will eventually force action towards change but motivation must finally come from a revival of ethical values and the practice of moral behaviour by the political and economic power-holders. + means that increase or decrease of the emitting source element causes the receiving sink element to change in the same direction; – means that the receiving element changes in the opposite direction. (Adapted from Bruenig *et al.*, 1986; Bruenig, 1989d; Bossel and Bruenig, 1992.)

indications of predatory forest use spilling over into the tropics. By mid-century, the practice of destructive and wasteful logging began on a large scale. The abuse has remained unchecked to the end of the century (Baaden, 1994). A recent report of the Inter-government Panel on Climate Change (IPCC, 1992) states, on the forest decline from all sources:

> The process of reduction of tropical forest on the African continent, first noted in the tropical rainforest areas of Zaire and Uganda 9000 BC (Livingstone and van der Hammen, 1978) with the arrival of humans, will continue unabated and even be reinforced by global warming, resulting in an accelerated reduction of tropical forest.

Worldwide in the rainforests, agricultural expansion reduces the forest area, selective logging and clear-felling reduce the biomass, the viability and vigour of the growing stock, and abuse by both destroys the productivity of the soil (Figs 2.4, 2.13, 2.14 and 10.1) and the viability of the biosphere.

11.2 Key Obstacles: Information Deficiencies

In the past three decades, rainforest forestry as a science and art was marked by two mostly extraneous events. One was the negative and destructive campaign against tropical forestry which made the lay public believe that foresters and ecologists know next to nothing about the tropical rainforest. The wealth of available knowledge and experience was purposely ignored and the transferability of scientific knowledge from other regions denied. The other, positive and constructive event was the undeterred continuance of international scientific cooperation and individual efforts in research on rainforest ecosystems. Unimpressed by the campaign, most scientists and foresters continued to test existing ideas and to create new knowledge. Some, but very few, scientists joined the anti-forestry campaign bandwagon. Their pretensions 'that we know next to nothing and that the tropical rainforest cannot be sustainably managed' caused considerable confusion among the public and in fact supported the 'wasted asset' theorem and consequently exploitative practices. Such band-wagon phenomenon is not unknown in forestry science (Prause and Randow, 1985), and common in international cooperation. A key argument of the anti-tropical forestry campaign was that we had absolutely no knowledge about the effects of modification and manipulation of forest structure on the functioning of trees and the ecosystem, and on their economic and social functions. Sustainability of rainforest management was declared a utopia. In reality, available knowledge and experiences would have been sufficient, if judiciously applied, to stop forest decline and restore sustainability. Private and public forestry were denied the political and financial support which is essential to implement the necessary changes. Instead, governments and non-government organizations discussed in endless cycles of mostly futile and unqualified conferences the fact that they did not understand the rainforest and sustainable forestry. While debates flourished, serious research and practical implementation fell further behind needs (Queensland Department of Forestry, undated, 1983; de Graaf, 1982, 1991; Tropenbos, 1986–1987; Figueroa *et al.*, 1987; Hadley, 1988; Wyatt-Smith, 1988; Lamprecht, 1989; Goudberg *et al.*, 1991; Abdul Rahman *et al.*, 1993; Udarbe *et al.*, 1993).

The major obstacle to improvements, and one of the driving forces behind the unabated decline of the tropical rainforest resource, is not so much lack of knowledge but the failure to utilize effectively existing knowledge (Bruenig, 1971c; Wiebecke and Bruenig, 1974; Lubchenco *et al.*, 1991). This failure very often is intentional. If forest rent and cash flow from logging is simply added to gross national product, a politically welcome economic growth is demonstrable even if the natural forest resource account balance is negative. Forest conversion is hailed as economic progress, while comprehensive economic–social cost–benefit analyses (Section 8.3) would in many cases show the opposite (Johansson, 1993). Such analyses are naturally shunned by power groups whose vested privileges depend on the *status quo ante*. Consequently, the information deficit is upheld. Inside forestry and at the scientific–technical level, information deficits are unintentionally caused by poor information infrastructure. The research and practice of management and forest policy work in mutual isolation. Directions in research and practice are determined by personal preferences and the common reluctance to risk conflict with power groups. In addition comes the human preference for customary procedures, cherished habits and fixed opinions. Bilateral cooperative research projects are strongly influenced by the short-term interests of foreign funding agencies and their governments. Foreign research institutes or individual foreign researchers pursue their own goals at the expense of forestry progress in the host country. Forestry planning and evaluation, especially in research, are usually closed-shop affairs that are convenient but inefficient (Fig. 11.2). Blueprints for efficient, transparent, comprehensive and integrated planning and reporting are available (Fig. 11.3) but rarely applied in research or in practice. This closed-shop protective exclusiveness has contributed to the deterioration of the esteem and support of forestry in government and its image with the public. Information to the national political and administrative decision-making level are well filtered *ad hoc* reports in relation to acute events. Realistic in-depth problem analyses and transparent information networking (Fig. 11.4) are rare exeptions. The overall consequence is that rainforest research in forestry lacks relevance, and operations of forest services and of the private forestry sector lack adequate goal orientation, professional quality, efficiency and effectiveness. Effective remedies are known and could be applied but factors outside forestry counteract these, usually very effectively.

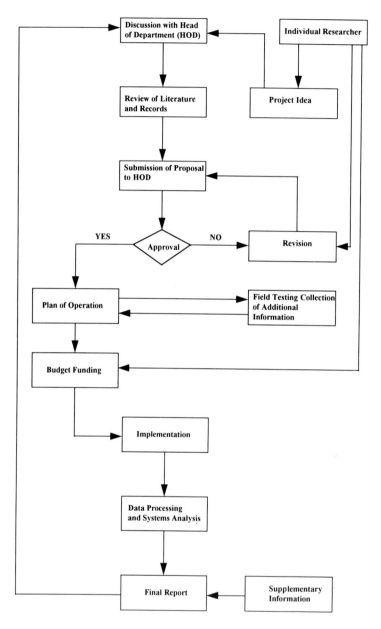

Fig. 11.2. The customary, ineffective basic structure of the process of planning and implementing normal budget-funded research projects. The example is taken from a manual of the research operating procedure of a forest department in a tropical rainforest country, chiefly designed for conventional silvicultural research. The superficially appealing simplicity of structure, however, guarantees inefficiency of analysis during planning and synthesizing during evaluation. Consequently, research does not address relevant problems, applies inappropriate methods inefficiently, and is unable to progress in a coordinated manner by a holistic concept. The consequence is the failure of scientific and application-oriented research, and of technical development projects, to have their results appreciated, adopted and applied in practice. Similar conditions apply to technical departments in the public forest services.

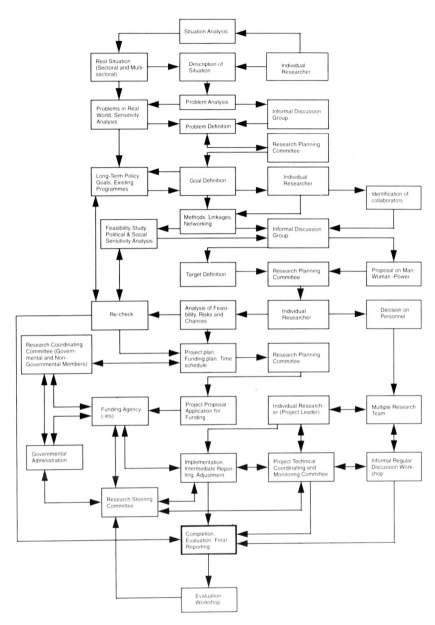

Fig. 11.3. Efficiency, relevance and acceptance of research and technology development in practice can be improved by optimizing the infrastructure and interactive process of information creation and transfer. Processes of planning, implementing, monitoring, evaluating and, finally, disseminating and accounting must be transparent and multisectorally integrated. The diagram shows an example of infrastructure and one-way or two-way information flows by which deficiencies (Fig. 11.2) can be overcome and the quality of research or of development of technology raised without excessive and stifling administrative red-tape and at little, if any, extra cost. The need for infrastructural and operational improvement, greater efficiency and relevance, better information exchange and dissemination, collaboration and participation, are universal in forestry research and practice.

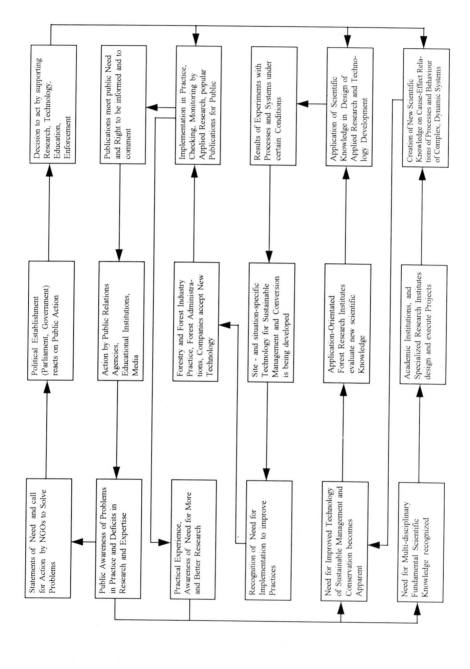

Fig. 11.4. The flow of information from scientific research through application-orientated research and technology development into the practice of forestry and forest industry, and finally into the social and political environment where preferences for actions are evolved and fed back into practice, technology and science.

11.3 Key Obstacles: Information Biases

The reality and seriousness of the threat to the tropical rainforest and its sustainability has been undisputed since the 1930s (Heske, 1931b, 1937/38). In the 1960s, mass media introduced the principle of marketability of news. Horror scenarios and negative news became fashionable: stories were circulated that the rainforests might disappear within the lifetime of those living at that time (Richards, 1952; Dilmy, 1965). In the 1980s it was added that logging was to blame. Loggers and involved politicians reacted by circulating their story that nothing was wrong with selective logging. Both stories, for different purposes, were biased and harmful to progress towards sustainability. This tendency to bias was further exaggerated by the ideological activists of the green and culture-phobic movement of suburban North America, to whom wilderness was sacred and civilization hateful. The forestry world was, in a simplistic dialectic, divided into 'villains' who were the loggers and foresters, and 'victims', the forests and the indigenous forest dwellers (North, 1995). During the same period, idealistic scientists declared shifting agriculture, for example in Kalimantan, sustainable; slashing and burning was likened to natural rainforest gap formation. Social forestry was marketed, especially in the neotropics, as a new concept incompatible with traditional forestry. These North American cliches, with many others, eventually spilled over to Europe and Asia. The campaigns against forestry in the tropical rainforest and the tropical timber trade deployed emotionally appealing horror visions of rainforest holocaust and extinction of species, including the indigenous rainforest dwellers. Political and professional bandwagon jumpers saw their chance and joined the action. The arguments were that commercial timber felling irretrievably destroys the 'integrity' and upsets the 'homoeostatic equilibrium', stability and biodiversity of the fragile rainforest ecosystem. If the forest were only left alone, it would be stable and harmonious. Counter-arguments, based on scientific research results, that timber harvesting can be sustainable and compatible with biodiversity conservation, and may even increase tree biodiversity (Bruenig, 1989c,d; Sections 1.8, 2.8 and 4.5; Table 1.6), were denounced as criminal (Gardner, 1992). The insistence was that we did not possess the scientific knowledge and understanding needed to interfere with the rainforest ecosystem, except in the traditional ways of forest dwellers. The arguments were that 'We do not yet understand these ecological systems, we even do not know why there are so many species', and that knowledge from temperate forests is not applicable in the rainforest (Barthlott and Fittkau cited by Barbara Veit in *Südd. Ztg.,* 1992, 60, 53). Therefore, the rainforest must be left untouched.

It is a truism that we shall never have a 'complete' or anywhere near complete knowledge of such a complex and dynamic ecosystem as the rainforest. There will also never be perfect, complete and final 'sustainability', which does not even exist in nature. We know enough already about the basic features of the chemical, physical and biological processes and structural states, and of the reactions of the rainforest to disturbance and change, to design a cautious approach towards sustainability (Appendix 1). We also have a great deal of practical experience with biodiversity management in temperate and tropical forestry practice, and the possibilities and limitations of transfer of information both ways are fairly clear. Knowledge can with prudent discretion be pooled. MacArthur (1972) in a review of patterns, diversity and dynamics of insects and birds in temperate and tropical environs found that underlying causal relationships in both biomes are similar, but their manifestations differ in accord with the differences in the configurations of the systems and the relative strengths of factors. Applying this assumption to management practice, we successfully used rainforest experience to design more stable, healthy and self-sustaining forests in Germany (Bruenig, 1969c, 1973b, 1984c, 1989a). The basic human urges and motivations are universal and the attempt to separate eastern, western, northern and southern values as incompatible is political scare-mongering and anachronistic. The value goals and the principles of social forestry and personnel management are relevant and transferable, if the differences in the phasic development of different cultures are considered. The denial of existing knowledge and the insistence on the fundamental non-transferability between climatic and cultural zones have been serious obstacles to promoting sustainability in tropical forests. The popular biases of the anti-forestry campaigns have damaged the sustainability of those, particularly European, markets that pay high prices for high quality, which are indispensable as a support for the effective implementation of sustainable conservation and management in tropical forests. The intolerant arguments of sectarian 'ecologism' have by now worn thin (North, 1995). In Germany, Greenpeace in early 1995 abandoned its tropical forest campaign. Other campaining groups had lost faith in their biased arguments during 1994. But the damage has been done. Good markets for processed timber and timber products have been lost, poor exploitative markets for round timber benefited and flourished, progress towards sustainability was bogged down by futile arguments, and wasteful selective logging had an undeserved extended lease of life.

11.4 *Probable General Trends in and for Rainforestry*

The tropical member countries of ITTO pledged in 1991 to achieve sustainability according to the ITTO definition by the year 2000, the so-called ITTO target 2000. The pledge was essentially in their own national interests, but the time horizon is overoptimistic and technically the target is naive. Sustainability is a dynamic and complex concept which concerns many facets of forestry and society. Reinstating the principle of sustainability will require revisions and upgrading of concession systems, legal clarification and material accommodation of native customary rights (NCR), upgrading of standards of education and professional training, and institution of norms of qualification as requirements for employment. Forestry and forest industries must be reintegrated. This process will be a continuously on-going approach towards sustainability. It is laudable to wish, but utterly unrealistic to expect, 'sustainability' to be achieved in the rainforest countries by the year 2000. Until 1995 not much had changed except for the commencement of a great number of projects which will produce results after the target date. Governments keep feet-dragging and effective change towards sustainability before the year 2000 can obviously only come from private companies which take the lead. The few lead companies with professional expertise and a sense of responsibility, and the few places with political and social stability and favourable infrastructural and economic conditions, such as Queensland, Indonesia and Malaysia, bear a heavy responsibility. They have to pave the way for those less fortunate countries which suffer from tribal and sectarian internal strife, social inequality, profuse corruption and autocratic mismanagement, economic regression and social misery and poverty. These need guidance and encouragement from the leading countries. The majority of timber concessionnaires, loggers and labourers in the tropical rainforest lack competence and any motivation except for money. They will fall further behind the pacemakers; sustainability, as a dynamic and holistic concept and as an operating principle in forestry, will continue to progress very unevenly in the tropical rainforest biome. Correspondingly uneven will be the market opportunities for the leading companies and countries, and the stragglers.

Recently published statements on the probable future international timber supply and demand situation (Grainger, 1985, 1986, 1989; Deutscher Bundestag, 1990, 1994a,b; Arnold, 1991; Amelung and Diehl, 1992; Barbier *et al.*, 1993) consistently indicate the following trends that are relevant to forest policy and the formulation of integrated strategies in tropical rainforest conservation and management. Only 10 of the 33 tropical forest countries that are still net exporters of timber and timber products will remain net exporters beyond the turn of the century. The regional timber demand will continue to rise, while the forest area, natural and economic forest productivity, and actual economic yield of the forest resource will continue to decline, especially in the Asia-Pacific Region (Fig. 10.1). The removals of high-quality and commodity timbers will increase in Central Africa and America, but will remain within the potentials. The very large temperate hardwood growing stocks and increments, especially in the USA, will increasingly be utilized and compete with tropical hardwoods in the Asian and European markets. Industrial commodity timbers, especially softwoods, will continue to be oversupplied from natural forests in the boreal zone and from planted forests. Domestic timber and fuelwood consumption in the tropical countries will rise as populations and incomes grow. In countries with rising standards of living and quality of life, there will be a shift from basic needs for fuelwood and simple utility timbers to higher quality grades and finished products. High-quality timber in the form of semi-finished or finished products from tropical rainforests will retain a preferential niche in the interlaced and interdependent national and international supply and demand networks (Gane, 1992). Tropical timber products in the long run will strengthen their position especially in the European market, if productivity and quality in tropical forestry and forest industry are raised to competitive levels and sustainability is achieved. The home markets will absorb industrial timber, fibre-wood and xylochemical wood as a by-product of harvesting and silvicultural tending or produced in planted forests outside the natural rainforest. The global demand for industrial commodity-grade wood in the international market can increasingly and more cheaply and efficiently be met by timber from sustainable forestry in the overproductive forests in temperate and boreal areas (World Bank, 1991b).

Even if selective logging is completely and effectively replaced by orderly harvesting and management before the year 2000, sustainability will not be ensured. The effects of current resource plundering will be felt through several felling cycles. The second felling cycle will have to be much longer and will yield much less value than anticipated by most national policy makers. Every effort must be made in the second and subsequent felling cycles to increase log diameters and to boost the production of high-grade timber. Some form of SMS will most likely remain the most suitable and feasible concept of integrated rainforest conservation, restoration and management. In addition to high-quality timber, SMS can also supply commodity timbers and non-timber forest products. Communal forestry, agroforestry, mixed timber and multiple-use plantations and pure industrial plantations will supplement, but not replace or compensate, natural forest management. These systems will probably be less self-sustainable, be more expensive, suffer more from competition and the effects of climate and social changes than natural forest management. Amenity and recreational values of forests in themselves and as components of the landscape will without doubt obtain a higher rating when the standards of living of the general population improve and when the present culture-phobic and incoherent era of 'aesthetics of ugliness', utopian 'ecologism' and unsustainable materialism has flip-flopped back to a revival of more humane and sustainable cultural values, attitudes and lifestyles.

11.5 Need for Change and Action

Two diametrically opposed, forceful external disturbance factors, radical profiteering and environmental radicalism, impacted on the tropical rainforests and tropical rainforestry during the pantropical exploitative phase 4 of the history of tropical forestry (Section 3.4). The more destructive, apart from agricultural expansion, of the two was profiteering by social groups who knew nothing about forestry but a lot about power and money. The opposing force knew little about rainforest ecology and nothing about forestry, but a lot about dialectics. Biocybernetically, this should have caused rainforestry to react and adjust rapidly; however, foresters were not empowered to act. The approach towards a solution to the problem came only gradually at the beginning of phase 5 (Section 3.4). The traditional forestry principles of sustainability began to gain political recognition again. The concepts revived of forests as common heritage, forest ownership as social obligation, and forest utilization as integrated component of general land-use and economic and social development (Bruenig *et al.*, 1975, 1986a, 1994; Bruenig, 1976, 1993b; Smith, 1978; Bossel *et al.*, 1992). The causal forces of rainforest degradation were identified as the vested interests of the rich and powerful, the legal and material helplessness of the poor (Remigo, 1992; Fig. 11.1) and *laissez-faire* policies of governments. Empowerment of forestry is needed to initiate the cooperative and effective restoration of the basic principles of sustainability in all categories of forest. This process is on the way in some countries and some companies, as described above for some cases in Mexico, Brazil and Sarawak in Malaysia, but most countries and companies have still to progress from declaration of intent to implementation of action (Bruenig *et al.*, 1989, 1991; Grammel, 1995; Prabhu *et al.*, 1993).

The essential global political framework for the change is a legally binding forest convention as a universal code of conduct to replace the 'non-legally binding authorative statement of global principles for a global consensus on the management, conservation and sustainable development of all types of forests' (for a review of post-UNCED actions see Grayson, 1995). Forest certification and forest product trademarking schemes (Chapter 10) can be powerful incentives for change and against market distortions. Harmonization and institutionalized global linking of forestry research and technology advancement will improve the prospects for success (Sombroek, 1986; Swift, 1986; Estes and Cosentino, 1989; ITTO, 1992a; Nichol, 1993; CIFOR, 1994). At national level the most urgent need is for rational and fully integrated planning of forest, land and socioeconomic development. Regional forest management plans must be revived, based on comprehensive land capability survey (soils, site conditions, natural resources, environmental conditions, present land-uses, claims, rights, ownerships; Fig. 11.6), and socio-economic and social system analyses (Bruenig *et al.*, 1986a, 1994; Bossel, *et al.*, 1992; Johansson,

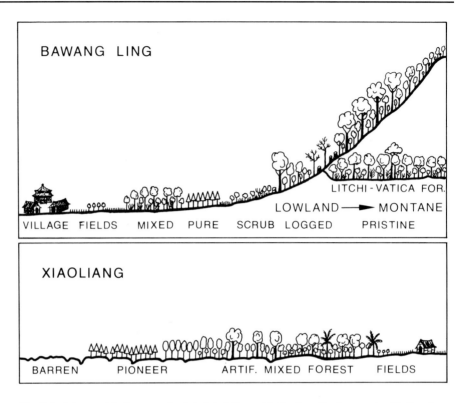

Fig. 11.5. Schemes of (top) conservation to halt depletion and (bottom) recultivation to restore biodiversity and productivity. Top: Bawangling CERP-Project area on Hainan, China; tropical landscape under heavy pressure of overuse by indigenous ethnic minorities (especially from the Li minority) and from immigrating Han Chinese. Integrated land-use development aims at ecological, economic and social sustainability by preserving the few more or less primeval forest areas which have survived and by introducing sustainable management in the overlogged modified forests and by restoring biodiversity and cultural diversity in the deforested areas. Bottom: Xiaoliang, Guangdong, mainland China; the absent biodiversity, productivity and functionality of the barren coastal landscape are being restored by establishing a sequence of pioneer forest, mixed semi-natural forest and finally integrated forestry and agriculture in a diversified landscape. In both cases the aim is an integrated land-use pattern with site- and use-specific biodiversity as a tool of management towards self-sustainability and as a source of diverse products. (From Bruenig *et al.*, 1986; Bossel and Bruenig, 1992.)

1993). Management for biodiversity in forestry at α- and β-levels is an essential tool to achieve self-sustainability which includes protection of common and rare, endemic (localized) or widespread species (or other taxons) of animals and plants (Ellenberg *et al.*, 1992). Sustainability will remain utopian without the immediate interest of the people involved in the processes of forest conservation and management and protection and the participation of an educated public (Fig. 11.7). The change from unconcerned profiteer or hireling in forest resource plundering to responsible participant in sustainable forestry requires clear signals of intent and honestly implemented action at the top of the social and political pyramid. The revival and restoration of traditional East–West ethical and moral standards in business and politics will be a slow process and hardly be effective by the target year 2000. The elimination of collusion of interests between government and the private sector is an essential part of restructuring the concession system and indispensable for the revival of motivation among all concerned with forestry. Concessionnaires must be competent and not simply beneficiaries. They must be made, and feel, personally responsible and committed to the principles of sustainability and social fairness as a non-negotiable mandatory obligation, even if both, sustainability and fairness, elude precise quantification. The same applies to owners of NCR of forest land and of usufruct in the forests.

The realities of global climate, environment and social change (Bruenig, 1987b, 1989c, 1991a; IPCC, 1992) and of rainforest decline (Bruenig, 1977a, 1981, 1985, 1989e; Amelung and Diehl, 1992; Baaden, 1994) are beyond reasonable doubt. There is also no doubt that time is running out. Decisive and rapid action is needed to change the tide of forest decline and to prepare forests and forestry to face changes whose specifics cannot be predicted with certainty (Sections 2.10, 3.3 and 6.8). This is easier to achieve technically than socially. The struggle for survival during biological evolution has deeply ingrained in the human genetic heritage the urge to acquire, preferably by capture rather than by culture, life-supporting territory and resources (Markl, 1986). This basic urge is supplemented by four important complexes of unpredictably flip-flopping primeval emotions and basic motivations which have survived unchanged during the evolution of civilization (Doerner, 1992). These four flip-flopping basic motivations concern the wish for security (or denial) of material and intellectual existence; the preservation and propagation (or ascetic denial) of genetic and other heritage; the acquisition of power and recognized excellence within a solid social order (or its denial by anarchy and permanent cultural revolution); and finally the craving for social acceptance and integration (or provoking social rejection by abject attitudes and behaviour). Success in biological and cultural evolution has favoured those with acquistive motivation and aggressive behaviour, often fuelled by substantial criminal energy. Culturally less sophisticated societies still glorify aggressiveness as a virtue in everyday life, especially in politics and business. These primeval attitudes and their flip-flopping make the road towards sustainability in rainforest particularly thorny and tortuous, and progress tedious. The primordial human urge is to acquire wealth and power by plundering resources, such as forests, lands, labour or warriors. Culture is basically the attempt of humanity to contain and bridle these dangerous urges which arise from our genetic heritage (Section 2.1).

Woodwell (in Rama Krishna and Woodwell, 1993) argues that to stop the misuse and abuse of the biosphere requires

Site quality	Natural Forest Functions			Self-sustain-ability	Priority option	
	Protection	Social	Economic		Land use	Management system
+	+	+	+	+	EP, PP	SM, CR
				−	EP, PP	SM, CR
			−	+	EP, PP	SM
				−	PP, TP	SM
		−	+	+	EP	SM
				−	EP, PP	SM, CR
			−	+	TP	CR
				−	TP	CR
	−	+	+	+	EP	NP, AF
				−	EP, OU	IP, AC
			−	+	PP	SM
				−	TP	CR
		−	+	+	EP, OU	SM, AF
				−	OU	AC
			−	+	TP, PP	CR, SM
				−	TP	CR
−	+	+	+	+	PP, TP	CR
				−	TP	CR
			−	+	TP	CR
				−	TP	CR
		−	+	+	PP	AF
				−	TP	CP
			−	+	PP	SM
				−	TP	CP
	−	+	+	+	EP	SM
				−	PP	AF, CR
			−	+	PP, TP	SM
				−	TP	CR
		−	+	+	PP, TP	SM, CP
				−	TP, OU	CP
			−	+	TP	CP
				−	TP, OU	CP

Fig. 11.6. Simplified binary decision diagram of the complex course of decision on priority options for forestry or conversion to other land uses. Determinants are natural conditions of the site, features and functions of the forest ecosystem and the physical and economic accessibility (adapted from Bruenig, 1974). The diagram illustrates the general approach towards allocation of sites to different systems of forest management and conservation or to non-forest uses in rainforest landscapes. Site quality corresponds to forest productivity as follows: +, NPP > 25 t ha^{-1} a^{-1}, NBP stemwood > 8 m^3 ha^{-1} a^{-1}, PEP logwood > 3 m^3 ha^{-1} a^{-1}; −, lower

Fig. 11.6. (*continued*) productivity than 25, 8 and 3 m^3 respectively. Site quality: compound effect of physiography, exposure, soil and climate. Protection: species, nature, soil, water, climate, environment. Social: recreation, amenity, heritage, science. Economic: accessibility, manifold productivity, profitability, income. Self-sustainability: capability of the ecosystem to maintain itself and its functions without silvicultural assistance. TP, totally protected non-production forest; EF, economic production forest; PP, protection and production forest; SM, selection-silviculture management system; CR, conservation, preservation, research and recreation; CP, conservation and preservation, other uses severely restricted; IP, industrial tree plantation; NP, naturalistic tree plantation; AF, agroforestry systems (diversified); AC, agriculture crops; OU, other land uses (buildings, infrastructure).

> mutual [social] coercion which is mutually agreed upon: governmental regulation to protect the common interest in a viable landscape from the private desire for immediate gain at public expense. Such a transition, however unpopular as political philosophy in a neo-conservative world, offers the only alternative to chaos.

I consider this an overoptimistic assessment of the role and prospects of governmental regulation which underestimates the inherent dangers such a course involves even in democracies. It would be more realistic and less risky if 'governmental regulation' is replaced by 'high ethic and moral standards, binding but democratically evolved, agreed and controlled, and legally adaptable social norms regulating the conduct of a culturally sophisticated society of free and well educated, articulate people who recognize and live up to their social obligations'. Revival of the basic civilized values of human culture, which have practically identical roots in the East, West, North and South, may bridle and domesticate the biological heritage and transform acquisition urges and power and prestige megalomanias to positive forces of cultural advance. Otherwise, the persistence of the current social and political evils will continue to threaten not only the sustainability of forestry and natural resource management (Figs 11.1 and 11.8) but also the very basis of human existence and the chances for a dignified survival of a viable biosphere and all its inhabitants. Obviously, a forest department could not hope to cope with its part of this task by simple enforcement of law and order.

Fig. 11.7. Correct and sufficient information and public education are indispensable requisites of substainable, integrated forestry: a television team is briefed on SMS in MDF. Sarawak, 1994.

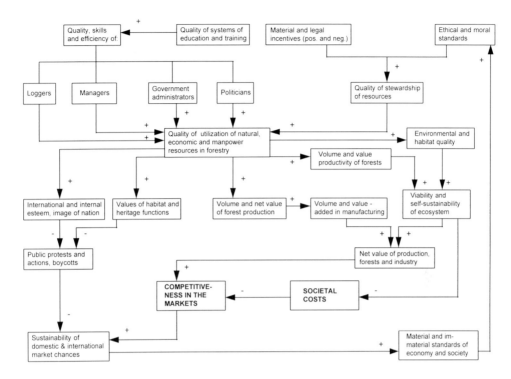

Fig. 11.8. Some of the critical key elements which determine the success of maintaining effective and sustainable standards of education, training, incentives, morals and ethics. The quality of these standards determines the degree to which the national forest resource is enabled to fulfil its functions for the benefit of the people. Environmental quality affects self-sustainability; this in turn affects social costs and competitiveness in the markets, these again affecting market chances, which in turn determine standards of living and quality of life on which morals and ethics depend. The standards of morals and ethics, based on the very similar traditional spiritual and socio-cultural values of East (Asia) and West (Europe, not USA), determine the manner in which natural and human resources are treated. + means that both elements change in the same direction – if the issuing element increases, the receiving element increases (and vice versa); – means a change in the opposite direction.

Appendix 1: Glossary

The following terms are commonly used, but frequently misunderstood or misused. The terms are arranged broadly by subject and linkages, and not alphabetically.

1 Ecological Terms

Ecology. The science of the habits of living things or organisms (aut-ecology) and of ecosystems (syn-ecology) at all levels of scale or functions, including the sociopolitical ecosystem level (political ecology as the art and practice of eco-politics).

Ecosystem. A coherent unit of interactive natural (physical, chemical, biotic) and/or cultural (anthropogenic or man-made) elements. The boundaries of a concrete ecosystem for research, description and practice are intentionally defined and accordingly delimited with a purpose in view. Any ecosystem is embedded in a complex network of hierarchically ordered natural and cultural ecosystems which differ by scale or function (natural, economic and sociopolitical ecosystem function levels from local to global scale levels). Ecosystems interact within and between levels. At each level ecosystems are composed of compartments and consist of biotic and abiotic, natural and man-made elements. Ecosystems are open systems and characterized by the kind and intensity of physical, chemical and biological processes of input, cycling and output, and by structure and organization. Primeval forest ecosystems possess singular site-specific dynamic and structural features which determine the capacity for change, adaptation, resistance, resilience, elastic response and repair which together determine stability, self-sustainability and viability of the forest. Forest ecosystems are either primary-natural (primeval, pristine, virgin forest), modified-natural forest (e.g. logged forest), secondary-natural (e.g. pioneer forest after deforestation) or anthropogenic (e.g. planted forest established by afforestation or reforestation) (see Sections 2 and 3).

System analysis. Ordered and logical arrangement of data about a real system in a word, graphic or mathematical model describing this system with a specified purpose in view, and the testing and validation of the model; in a more general sense the analytical process of data collecting, processing and representation, and finally data synthesis in a systemic context.

System dynamics. The feature of a real system to change its structure, organization, processes and functions in the course of time. The system can be a natural or a cultural ecosystem.

Diversity. Heterogeneity and variety, e.g. of the various biotic and abiotic, natural or anthropogenic features of the composition, morphological structure, processes and functions of natural or cultural ecosystems. Often used to denote the patterns of mixtures of species in ecosystems, meaning the evenness, i.e. the uniformity or equality, of the shares of the component species in the total number of individuals or, in tree and other plant communities, in the basal area or biomass of the ecosystem.

Species richness. Number of species (or taxa) per unit area or geographic region, or per number of individuals, also termed 'speciosity' or somewhat misleadingly 'species diversity'.

Biodiversity. Diversity (see above) of the biotic components of ecosystems at the levels of organization, such as genes, species, populations, communities (e.g. tree community or forest ecosystem) and regions (landscape ecosystems, biogeographic units) (for details see Solbrig, 1991a,b) (compare 'ecosystem'). Biodiversity is more than, but includes, species richness. It denotes the entirety of the life-forms in a system at any level. At α-level in forestry it denotes, as dominance diversity, the pattern of mixture of species within a community (evenness or unevenness of mixture) in terms of their contribution to the number of individuals or biomass of the community; at β-level it denotes the patterns of between-community diversity within a geographic unit at the scale of landscapes; at γ-level it denotes differences at larger regional scale. Biodiversity at α- and β-levels are crucially important elements of sustainable forest conservation and management. Biodiversity determines structural diversity and organizational complexity, and is the key to ecological and economic self-sustainability and sustainability of forest management (see Section 3).

Keystone. 1 The central principle of a system on which all the rest depends (*The Concise Oxford Dictionary*). 2 The central stone at the top of an arch which bears the lateral and vertical stresses and binds the structure of the arch together. A single arch becomes unstable and will eventually collapse if the keystone is lost. A complex vault composed of multiple arches will hardly lose stability if one of the arches loses its keystone.

Keystone species. A plant, animal or microbial species which binds together an interactive feedback loop in the trophic and functional networks of an ecosystem. Alan (*Shorea albida*) is the keystone species in the natural monoculture of Sarawak Peatswamp forest, PC3 and 4. Its removal causes collapse and restoration is uncertain and tedious. Keystone tree species in species-rich, complex forests may locally disappear without affecting the capability of the forest ecosystem to survive, regenerate and function (e.g. the different effects of the ice ages on tree-species diversity in Europe and North America, also the effect of recent chestnut blight and spruce budworm in northeast America). So far no extinction of keystone species in tropical rainforests has been recorded, but there has been elimination at local scale by deforestation or clear felling. Elimination of seed-dispersing animals by severe and prolonged hunting pressure has occurred and apparently affects distribution patterns and, thereby, diversity of tree species. It could be speculated that this may have some effect on ecosystem functions in the long run.

Aerodynamic canopy roughness. In forest climatology, the dimensionless measure (z_0 in cm) of the resistance caused by the configuration of the crowns and the canopy which is encountered by air currents as they pass over the canopy of the forest. Aerodynamic roughness affects air eddies and free and forced convective air flows in a forest canopy and influences the airflow in the atmosphere to about 500 m above the surface in flat terrain. It is an important climatological parameter and indicator which integrates many climatic and ecological processes and conditions.

2 Vegetation

Kerangas (syn. Heath forest). A word of the Iban language in Borneo for land which is unsuitable for growing padi because the soil is too infertile. 'Kerapah' is a variety of Kerangas distinguished by impeded drainage and consequent peat bog formation. Winkler (1914) of Hamburg University coined the term 'heath forest' for secondary white-sand savannah and *Agathis* Mixed Kerangas forest vegetation on raised beaches in South Kalimantan because the 'sub-xerophyllous' nature of the vegetation reminded him of his native Atlantic and sub-Atlantic heaths. The term 'heath forest' has subsequently been adopted to denote forests on kerangas lands. This is unfortunate because 'heath' suggests ecological conditions which do not exist. Kerangas forests are a primary tropical formation of very diverse site-vegetation subformations and forest associations from tall (up to 50 m) to low (~ 15 m) closed forest with a xeromorph and sclerophyll physiognomy on kerangas soils. Distinct boundaries may be the result of sudden change of soil, but also of historic events, especially forest destruction by fires. Definition and delimitation of Kerangas forests are easy because of its distinct physiognomy and structure.

Caatinga. The evergreen caatinga in the American tropical rainforest zone is equivalent to the Bornean kerangas.

Padang. Open scrub or woodland vegetation on infertile kerangas and peatswamp soils, often degraded by human activities. The equivalent vegetation in tropical America is termed evergreen bana.

3 Terms of Forest Technology and Land Use

Biocybernetics. The science, art and practice of regulating individual processes which operate in complex dynamic, natural, technical and social ecosystems at all levels from cell to biosphere. In complex dynamic ecosystems, biocybernetic mechanisms do not produce a static equilibrium (homeostasis) but a dynamic oscillation about means with occasional extremes and collapses. Biocybernetic principles contribute to self-sustainability in forestry (see Appendix 2).

Biotechnology. The intentional manipulation by human beings of the processes, structures and components of biological systems for certain purposes. Figure A1.1 illustrates the nature and relations between modern-traditional and modern-new biotechnology in forestry. The various disciplines can be combined in interdisciplinary and multi-disciplinary research projects and in integrated management systems. Natural (biological) biotechnology has developed unintentionally in the course of biological evolution.

Forestry (in late Latin *forestis silva*, meaning wood outside the walls of a park, derived from classic Latin *foris*, outside). The science, art and practice of conserving, creating and sustainably managing forest lands by generating or regenerating, tending and harvesting forest tree crops and managing non-forested components of the forest land. Since its inception, forestry traditionally aims at maintaining diverse forest functions to serve equally diverse purposes by means of multiple-use management designed to meet the manifold and continuously changing needs of the owner and of the society (e.g. city forest management in medieval Europe, Brandis teak taungya technique and teak management system in nineteenth century Myanmar). Forestry operates at forest management unit, landscape, regional, national, international and global levels.

Forest. An area of land covered by trees which create a characteristic stand climate (forest microclimate), including clear-felled areas in the

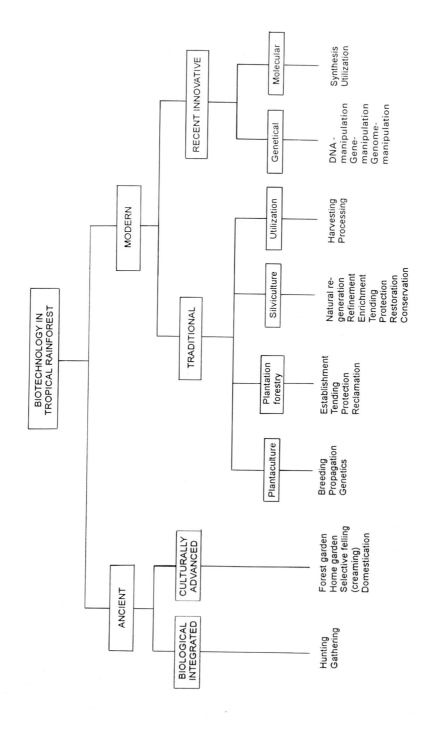

Fig. A1.1. The various types of biotechnology in forestry.

permanent forest estate which, after harvest of the previous forest stand, are in the early stages of natural regeneration or reforestation.

Closed forest. An area of land covered by trees which form an accumulative canopy covering more than 10% of the ground (crown projection/land surface ratio). The 10% threshold level is very low from an ecological and silvicultural point of view, especially in humid and wetter climates.

Open forest. An area of land covered by trees which form a forest (see forest above) with an accumulative canopy covering less than 10% of the ground (crown projection/land surface ratio).

Primary forest (syn. primeval, pristine, virgin, old growth forest). Forest which has never been subject to human disturbance, or has been so little subjected to hunting, gathering and single-tree felling (e.g. for collecting fruit or incense) that the ecosystem could buffer the effects elastically and thereby retain or regain its primeval natural structure, functions and dynamics. Any changes will be ephemeral variations within the range of the dynamics of natural perturbation responses and primary succession.

Secondary forest (partly syn. pioneer forest on land deforested by man). Forest which has developed by natural secondary succession on deforested land, such as land abandoned after shifting or settled agriculture, pasture or mining.

Natural forest. A forest composed of native trees and having a structure resembling, or identical to, the native primary forest (see Primary forest, Modified natural forest, Manipulated natural forest).

Planted forest (partly syn. artificially established forest). Forest which has been established by planting or sowing on barren land, grassland, land cleared of secondary forest or scrub, or on land cleared of primary, modified or manipulated forest.

Afforestation. Establishing a forest on treeless land, except in regeneration of clear-felled areas in the permanent forest estate naturally or by planting, which is reforestation.

Reforestation. Establishing a forest by planting or sowing trees on land where the forest has been removed by regular clear-felling within the rotational framework of ordered sustained-yield management; also on such land after a natural catastrophe (e.g. storm throw, fire, pest).

Clear-felling (syn. clear-cutting). Strictly, the felling and extraction or burning of all trees in an area large enough to develop an open-field microclimate. Selective logging and selection silvicultural management by definition essentially exclude clear-felling, but careless and blanket girth-limit selective logging may amount to clear-felling in large patches. Natural catastrophes such as fire, storm throw or lethal defoliation over large tracts may ecologically be equivalent to clear-felling. These events and clear-felling may be ecologically neutral, beneficial or even essential for self-sustainability, e.g. natural fire or burning in mor-accumulating forests or lightning and storm causing gaps and initiating regeneration in stagnating final stages of succession.

Deforestation. The clearing of forest from large tracts of land which consequently remains unforested either as barren land or as agricultural crop.

Logging. As a forestry term it denotes the harvesting of timber in log form, including felling, cross-cutting, extraction from the felling area and transportation from the forest.

Creaming. In timber logging, the harvesting of single trees for special products, such as ornamental cabinet woods, durable heavy hardwoods for boat and house building, wood used in musical instruments or for incense or dye production, usually from easily accessible locations such as river-banks ('river-scratching').

Selective logging. Indiscriminate harvesting of all merchantable timber (defined by size = diameter and utilizable length of trunk, and/or quality) of commercial tree species. In essence, simply the liquidation of the merchantable forest growing stock irrespective of the consequences to regeneration, structure, functionality and productivity of the forest, and therefore not in accord with the principles and criteria of sustainability (see Sustainability and Selection felling).

Selection felling. Selection, marking and directional felling of trees under strict silvicultural rules in a harvesting system which is ecologically and environmentally compatible and is in strict accordance with the traditional rules of good forest management and the principles of sustainability (see Sustainability and Selection-silviculture management).

Modified natural forest (partly syn. logged forest, manipulated forest). The natural forest cover has been retained but has been affected by uncontrolled timber exploitation (creaming, selective logging) or planned and controlled timber harvesting (selection felling). Harvesting of non-timber products (tapping of latex, collecting of cane, fruits, medicines, incense, etc.) may also modify structure, function and dynamics.

Manipulated natural forest. Forest, which may have been original primary or modified natural forest, which has been manipulated by silvicultural operations.

Tree-species status (adapted and amended from Binggeli, 1994):

- Native (indigenous): species naturally occurring in an area or having invaded an area in times before human cultural history.
- Introduced (alien, exotic): species deliberately or accidentally released by humans into an area in which it did not occur before.

- Reintroduced: species that have occurred in an area but have become extinct.
- Invasive (naturalized, neophyte, adventive): introduced species expanding by self-regenerating populations in a free-living state in the wild.
- Weed (pest): any plant, either native or introduced, interfering undesirably with the objectives or requirements of people.

Silviculture. The science and art of establishing, tending, protecting, preserving and regenerating forests of any of the kinds defined above. In the system of forest science, silviculture is placed between the basic natural sciences and the applied sciences of management, engineering, economics, politics and sociology. In practice, silvicultural systems are traditionally distinguished and classified by the criteria: (i) soil exposure (full, partial, none), (ii) nature of regenerating (natural, artificial, mixed), (iii) structure (uniform – even-aged; uneven – uneven-aged), (iv) kind of establishment (kind of manipulation of canopy and method of planting or natural regeneration, (v) seedling, saplings, size and age of residual trees for the next generation, kind of mixture). Silviculture can be 'artificial' by applying chemicals intensively, genetically improving trees (see Biotechnology) and using mechanical treatments to maximize one or a few product functions; or 'naturalistic' by mimicking nature in the design of the geometric and taxonomic structure of the forest and of the exchange and cycling processes and feedback mechanisms, but without doctrinally copying the potential natural floristic or geometric structure, utilizing the natural dynamics of the ecosystem structure in order to minimize the amounts of artificial inputs, to increase self-sustainability and the capability to serve manifold functions and multiple uses.

Selection-silviculture management. Set of rules and their implementation in practice which imitate or mimic nature by means of:

- single-tree harvesting according to biological and economic maturity and availability of regeneration causing minimum damage and disturbance;
- natural regeneration and, as a rare exception, assisted natural regeneration;
- maintaining a site-specific natural mixed community of tree and non-tree species and of the site-specific forest stand and canopy structure;
- applying basic biocybernetic principles of ecosystem design in order to utilize optimally the natural site potential and dynamics of the ecosystem to secure low levels of risk and achieve high levels of self-sustainability.

Agroforestry. The science, art and practice of combined forestry and agriculture. Forestry is integrated with agriculture, husbandry or aquaculture at crop level (agrosilviculture) and/or business (management-unit or village) levels.

Shifting agriculture (shag) (syn. slash-and-burn agriculture, traditional native cultivation, swidden, taungya). An early form of agriculture before invention of irrigation and manuring of permanent fields. Shag uses natural fallow regrowth to restore soil organic matter and nutrients, to suppress weeds, and to control pests and diseases by slash burning before planting. Shag consists of two components: (i) the fallow cycle and (ii) migration into new, primeval lands when soils have become exhausted, eroded and weed covered in the course of repeated cropping cycles. It comprises a multitude of diverse site- and culture-specific cropping systems. Shag requires high rates of inputs of land and labour but low inputs of money. Productivity is extremely variable, but generally low. A common sequence of events is: felling and burning of primeval or secondary forest; cropping for one to several years; several years fallow; decline of productivity; migration to new, pristine territory, often usurping territories under native customary rights of other communities or tribes by militant conquest.

Sustainability. The capacity of a system in its entirety to endure, last, persist and survive. Within the time dimension of human perception, it relates to permanence and steadiness of states and processes in contradiction to the fact that there is no steady-state in ecosystems. In forestry, sustainability is a comprehensive ethical principle, substantiated according to agreed criteria which are monitored by agreed indicators of certain forest properties and judged by recognized standards of ecological, technical, economic and social state or flow variables (e.g. critical load, threshold value). Sustainability encompasses the totality of the natural and anthropogenic states and processes by which quantities and qualities of energy, material or immaterial goods or functions of the system are passed on to the unpredictable future, while at the same time the constantly changing present needs of the system and its environs are served appropriately, adequately and satisfactorily. This applies to the natural ecosystem (natural or planted forest, natural or cultivated landscape) and to the cultural (technical, economic or sociopolitical) ecosystems. The principles of sustainable development (see below) require that the liquidation of forest growing stock outside the permanent forest estate (PFE) is done prudently, which means full utilization of the timber and minimum damage to soil and water bodies, and reinvestment of the net returns in the national economy. The relation between sustainability and biodiversity is not simple, but complex and diverse. Sustainability is not equivalent to equilibrium.

Sustained yield. In forestry, the term 'sustained yield' refers to a state of the forest in which the same or an increasing, steady yield of timber or non-timber products can be enjoyed in perpetuity without impairing, possibly improving, climate, soil, water bodies and growing stock. This sustained-yield concept is equivalent to the concept of *Nachhaltigkeit* (meaning in this context to keep something back and refrain from immediate consumption or use, in favour of future use) that has developed in German forestry since the thirteenth century. It was first scientifically formulated and published by v. Carlowitz (1713), elaborated and extended by Hartig (1795, 1820), Hundeshagen (1828) and Speidel (1984) (see the following terms).

Sustainable forest management. The management of a forest as a diverse and dynamic self-sustainable renewable natural resource in such a manner that its continued and lasting (permanent) persistence, viability, vitality, flexibility, resilience and adaptability as well as its natural–ecological, environmental, economic and social (cultural) values and multiple social utilities are ensured and, if possible, enhanced for the benefit of the present and future generations of humankind. During the eighteenth and nineteenth centuries, sustainability of forest management was linked to certain contemporary, produce-orientated objectives of management which are in contrast to the principles of sustainability (see above and below). The working definition of sustainable forest management adopted by the German Initiative Tropical Forests (ITW) for drafting a scheme of certification for sustainably produced tropical timber reads:

> Sustainable management of forest is the totality of those direct and indirect measures of utilization, cultivation and protection in a forest ecosystem which secure the lasting existence and natural development of the forest, the adequacy of its functions and the preservation of its species richness and diversity of life forms on which the fulfilment of its economic, ecological, social and spiritual functions depends.

(translation by the author)

Sustainable forestry. The condition of sustainable forest management extended to the whole gamut of the art and practice of scientific forestry within the framework of the wider biospheric and cultural systems of a region or nation, of which forestry in an integrated part.

Sustainable development. The planned or spontaneous changes of matter, quality and organization of any entity (e.g. ecosystems at the various levels of the hierarchy) which can be sustained for a time which is foreseeable but hardly predictable, without risk of degradation. In connection with economics, a change which implies that the aggregate real value of the economic assets of natural resources, industrial investment and educated workforce should at least be sustained at present levels but preferably enhanced to benefit adequately future generations. In forestry, s.d. relates to the real value of compounded cost, benefit and substitution of all material and immaterial goods and services from the forest over long time horizons. These conditions imply that social fairness and equality are realized and basic human rights guaranteed.

Self-sustainability. The capability of a forest ecosystem to maintain itself without aid by human interference. Analogously, the capability of a cultural ecosystem to sustain itself without exogenous assistance, such as unilateral food, technical, financial or intellectual inputs from external subsidizing sources.

4 Forest Categories Distinguished by Forest Functions

Protective forest. Forests which are exclusively (totally protected area) or partly designated to fulfil protective functions for the preservation of soil, water, climate, fauna and flora, nature, environment, heritage, scientific value or landscape amenity.

Productive forest. Forests which are designated for the sustained production of timber or other forest products or any combination of them, also including patches of non-productive forest or deforested land.

Multiple-purpose forest. Forests which are designated to fulfil a combination of productive and protective functions. In practice, often classified either as productive or as protective forest, depending on the priority ranking accorded to its functions. Multiple-purpose forests and their management are traditionally designed to supply multiple and changing demands of society.

5 Forest Categories Distinguished by Legal Status

Permanent forest estate (PFE). Lands which are legally constituted under statutory law and gazetted to be under forest forever. They include the functional categories protective forest, productive forest and multiple-purpose forest. The forest may be specifically designated to the following categories.

Production forest. PFE areas with the primary function of production.

Totally protected area (TPA). PFE areas legally established with the exclusive function of protection of species, nature, ecosystems, heritage and amenity in the form of national park, nature reserve or wildlife sanctuary; also areas established under management plan prescriptions as primeval forest or virgin forest reserve to protect ecosystem types, habitat and refuge areas, seed sources and gene pools.

Protection forest. PFE areas with the primary function of protection of soil, climate or water.

Community reservation. An area set aside usually for the primary functions of protection and preservation of the habitat of forest-dwelling ethnic groups.

Communal forests. A forest owned under statutory law by a village or town usually for the primary functions of protection, production, employment, recreation and heritage.

Private forest. Forest owned under statutory law by private person(s) or corporation, usually with the primary functions to contribute positively to the livelihood of the owner and to social stability and economic development with respect to any or all forest functions (Figs 4.1 and 4.2).

State forest. Forest owned by the state; functions as for the private forest but with different priorities and preferences.

Stateland or crownland forest. Forest areas that are not part of the PFE; stateland forests may be designated for conversion to other land uses or may eventually be transferred to PFE as land-use and resources develop and allocations change.

6 *People, Population and Forest Use*

People. All persons comprising a nation, community or race; a body of persons belonging to a place, a class, a profession or other groupings of society.

Population. In terms of genetics, a community of plants or animals in a habitat which share a common gene-pool. Ecologically, a community of individuals of the same species of plants or animals in a habitat which interact. Anthropologically, the inhabitants of a forest area, village, town, region or nation.

Indigenous people. People who, because of historic affiliation or for political reasons, are considered to belong to a site. The term is often abused by interest groups to gain economic or political advantages.

Aboriginal people. People who are assumed to have inhabited a tract of land from the earliest time of human evolution or since time immemorial, or have more recently occupied pristine, uninhabited land first. The aboriginal status is difficult to prove and is subject to political whims and conveniently defined criteria. The status is often at the core of disputes of customary usufruct and territorial rights between indigenous and native people, aboriginals and recent immigrants.

Native people (partly syn. indigenous, partly aboriginal). Communities of human beings who have been born in a habitat. The term is often twisted and abused by interest groups to gain economic or political advantages.

Forest-dwelling people. Human beings who permanently live within the forest precincts and are fully dependent on the forest for their livelihood. Forest-dwellers are an ecologically completely integrated biological component of the natural ecosystem living on gathering and hunting, and frequently being hunted themselves. The term is frequently misused for rural people living near, but outside, the forest.

Appendix 2: Biocybernetic Principles of System Design

The eight basic biocybernetic rules for system design as defined by Vester (1980) can be adapted and applied to planning and management in forestry as follows.

Rule 1

In a complex of feedback loops the number of linkages with *negative sign* must exceed those with *positive sign* (Figs 2.3 and 11.1).

Rule 2

Function must be independent of quantitative growth: the forest stand and the estate can continue to be ecologically and economically robust, viable and serviceable even if, for example, the market does not expand or the increment of timber does not increase.

Rule 3

Design for broad *functionality* by means of production and product diversity: production schedules should produce a variety of timbers and other products in a mixed forest instead of maximum rates of a single product in a monoculture; the maximizing of biomass or timber according to yield table models in densely grown, uniform, single-species stands is a classic example of a non-biocybernetic concept.

Rule 4

Use *inherent dynamics* of the forest ecosystem: successional and regenerational dynamics can be used in forest management to achieve the target of sustainability with reduced levels of cost, uncertainty and risk, e.g. succession-forest management systems (Bruenig, 1984c, 1986b) mimicking, but not pedantically copying, nature. The inherent dynamic of a pioneer secondary forest is to accumulate leafage and biomass rapidly, to scavenge and cycle nutrients and water intensely, and collapse early, a 'strategy' which is mimicked by uniform plantations with fast-growing trees.

Rule 5

Multiple utilization: products, structures, operations and organizational units of the forest and the management unit should be designed to serve several ecological, economic and social functions.

Rule 6

Recycling principle: the harvested timber or biomass must be in a sustainable relation to the net primary productivity and leave enough litter in the forest to maintain health, self-sustainability and ecosystem functions.

Rule 7

Symbiosis between components: high biodiversity and suitable species mixture (e.g. N-fixing and mull-forming trees) in the forest favours symbiotic interaction and mutualism between plants, animals and microorganisms more than it creates competition, in contrast to conditions in single-species forests.

Rule 8

Compatibility between biological, technological and economic structures and mechanisms: creation of a forest structure by silviculture to suit both nature and technology, simultaneously improving health and vigour of the trees, stability of the forest, working conditions for tending and harvesting and the value of timber (e.g. Auermuehler Production Programmes, Bruenig, 1984b).

References and Further Reading

This list includes sources quoted in the text and sources quoted in the 1989 report on the causes and effects of deforestation, which are now adequately clarified and only treated briefly in the present text.

Aaron Ago Dagang (1993) *Wood Residues and Harvesting Damage in Harvested Forest of Sarawak, Malaysia*. Forest Department Sarawak, Kuching, ITTO Project PD 74/90.

Abdul Rahman, Sharulzaman, I. and Weinland, G. (1993) Stand development of naturally regenerated stand of Kapur (*Dryobalanops aromatica* Gaertn.f.) after heavy thinning. In: *International Conference on Tropical Rainforest Research: Current Issues, 9–17 April 1993*. Universiti Brunei Darussalam (in press).

AIFM (1993) Growth and yield function applicable to the ASEAN region. AIFM Growth and Yield Section, Kuala Lumpur.

Alder, D.C. (1992) Simple method for calculating minimum diameter and sustainable yield in mixed tropical forest. In: Miller, F.R. and Adam, K.L. (eds) *Wise Management of Tropical Forests 1992*. OFI, Oxford, pp. 189–200.

Allen, J.C. (1985) Soil response to forest clearing in the United States and the tropics: ecological and biological factors. *Biotropica* 17, 15–27; *Forestry Abstracts* 46, 772, no. 6314.

Amelung, T. and Diehl, M. (1992) *Deforestation of Tropical Rainforests: Economic Causes and Impact on Development*. Kieler Studien no. 241, commissioned by Greenpeace. J.C.B. Mohr, Tübingen.

Anderson, A.B. (ed.) (1990) *Alternatives to Deforestation: Steps Toward The Sustainable Use of Amazon Rain Forest*. Columbia University Press, New York.

Anderson, D. (1987) *The Economics of Afforestation, a Case Study in Africa*. World Bank Occasional Papers, no. 1 new series. Johns Hopkins University Press, Baltimore.

Anderson, J.A.R. (1961a) The ecology and forests types of the peatswamp forests of Sarawak and Brunei in relation to their silviculture. PhD thesis, University of Edinburgh.

Anderson, J.A.R. (1961b) The destruction of *Shorea albida* forest by an unidentified insect. *Empire Forestry Review* 40, 19–28.

Anderson, J.A.R. (1964) The structure and development of the peatswamps of Sarawak and Brunei. *Journal of Tropical Geography* 18, 7–16.

Anderson, J.A.R. (1983) The tropical peatswamp of eastern Malesia. In: Gore A.J.P. (ed.) *Mires: Swamp, Bog, Fen and Moor. Part B, Ecosystems of the World 4B*. Elsevier, Amsterdam, pp. 181–199.

Anderson, J.A.R. and Muller, J. (1975) Palynological study of a holocene peat and a miocene coal deposit from NW Borneo. *Reviews of Palaeobotany and Palynology* 19, 291–351.

Anderson, J.A.R., Jermy, A.C. and Cranbrook, G., Earl of (1982) *Gunung Mulu National Park, a Management and Development Plan*. Royal Geographical Society, London.

Anderson, J.M. and Spencer, T. (1991) *Carbon, Nutrient and Water Balances of Tropical Rainforest Ecosystems Subject to Disturbance. Management Implications and Research Proposals*. MAB Digest 7, UNESCO, Paris.

Anderson, J.M., Proctor, J. and Vallack, H.W. (1983) Ecological studies in four contrasting lowland rainforests in Gunung Mulu National Park, Sarawak. III. Decomposition processes and nutrient losses from leaf litter. *Journal of Ecology* 71, 503–527.

Anderson, M.H. (1990) Extraction and forest management by rural inhabitants in the Amazon estuary. In: Anderson, A.B. (ed.) *Alternatives to Deforestation: Steps Toward the Sustainable Use of Amazon Rain Forest*. Columbia University Press, New York.

Andriesse, J.P. (1962) *Field Classification of Sarawak Soils*. Technical Paper no. 1, Soils Division, Department of Agriculture, Sarawak.

Andriesse, J.P. (1974) *Tropical Lowland Peats in South-East Asia*. Communication 63, Department of Agricultural Research, Royal Tropical Institute Koninklijk Instituut voor de Tropen, Amsterdam.

Anonymous (1994a) Researching the economic value of non-timber products from tropical forests. *INFOMAB* 21, 14–16.

Anonymous (1994b) Research hypotheses. Adapted from Godoy, R.A. and Bawa, K.S. *INFOMAB* 21, 16.

Anonymous (1995) IMAZON logging improvement noted. *ISTF News*, 16, 1: 2 and 10.

Appanah, S. (1987) Insect pollination and the diversity of Dipterocarps. In: Kostermans, A.J.G.H. (ed.) *Third Round Table Conference on Dipterocarps*. UNESCO Regulatory Office Science and Technology South-East Asia, Jakarta, pp. 277–391.

Appanah, S. and Weinland, G. (1992) Will the management systems for hill Dipterocarp forests stand up? *Journal of Tropical Forest Science* 3, 140–158.

Appanah, S. and Weinland, G. (1993) *Planting Quality Timber Trees in Peninsular Malaysia. A Review*. Malayan Forest Record no. 38. FRIM, Kuala Lumpur.

Appanah, S., Weinland, G., Bossel, H. and Krieger, H. (1990) Are tropical rainforests non-renewable? An enquiry through modelling. *Journal of Tropical Forest Science* 2, 331–348.

Appleton, A.E. (1995) Tropical timber and the WTO agreement: a legal perspective. *IUCN Forest Conservation Programme Newsletter* 21, 4–6.

Arnold, M. (1991) *Forestry expansion: A Study of Technical, Economic and Ecological Factors*. Oxford Institute Paper no. 3. OFI, Oxford.

Arruda Pinto, A. de (ed.) (1977) *Tropicos Humidos*, vols 1–3. EMBRAPA-CNP, Belem.

Ashdown, M. and Schaller, J. (1990) *Geographic Information Systems and their Application in MAB-projects, Ecosystem Research and Monitoring.* German National Committee for the UNESCO programme MAB, Bonn.

Ashton, P.S. (1964) *Ecological Studies in the Mixed Dipterocarp Forests of Brunei State.* Clarendon Press, Oxford.

Ashton, P.S. (1972) The quarternary geomorphological history of Western Malaysia and lowland forest phytogeography. In: Ashton, P. and Ashton, M. (eds) *The Quarternery Era in Malesia. Transactions 2nd Aberdeen–Hull Symposium on Malesian Ecology.* University of Hull Department of Geography Misc. Ser. no. 13, pp. 35–62.

Ashton, P.S. (1988a) Conservation of genetic resources. In: *The Case for Multiple-use Management of Tropical Hardwood Forests.* International Tropical Timber Organisation (ITTO), Yokohama.

Ashton, P.S. (1988b) Dipterocarp biology as a window to the understanding of tropical forest structure. *Annual Review of Ecology and Systematics* 19, 347–370.

Ashton, P.S. (1989) Dipterocarp reproduction biology. In: Lieth, H. and Werger, M.J.A. (eds) *Tropical Rainforest Ecosystems. Ecosystems of the World 14B.* Elsevier, Amsterdam, pp. 219–240.

Ashton, P.S. (1995) Towards a regional forest classification for the humid tropics of Asia. In: Box, E.O. *et al.* (eds) *Vegetation Science in Forestry.* Kluwer Academic Publ., Dordrecht, pp. 453–464.

Ashton, P.S. and Bruenig, E.F. (1975) The variation of tropical moist forest in relation to environmental factors and its relevance to land-use planning. *Mitt. Bundesforsch. anstalt Forst-Holzwirtsch.* no. 109, 59–86.

Ashton, P.S. and Hall, P. (1992) Comparison of structure among mixed dipterocarp forests of north-western Borneo. *Journal of Ecology* 80, 459–481.

Baaden, J. (1994) Zur Deforestation tropischer Wälder und ihre ökosystemaren Klima-, Boden- und Bioeffekte, dargestellt an Beispielen in Süd- und Südostasien. Diplomarbeit, Geogr. Inst., J. Gutenberg Universität, Mainz.

Baillie, I.C. (1970) *Report on the Detailed Soil Survey of the Experimental Afforestation Sites.* Report no. F3, Forest Department Sarawak, Kuching.

Baillie, I.C. (1972) *Further Studies on the Occurrence of Drought in Sarawak.* Report no. F7, Forest Department Sarawak, Kuching.

Baillie, I.C. (1976) Further studies on drought in Sarawak, East Malaysia. *Journal of Tropical Geography* 43, 20–29.

Baillie, I.C., Ashton, P.S., Court, M.N., Anderson, J.A.R., Fitzpatrick, E.A. and Tinsley, J. (1987) Site characteristics and the distribution of tree species in mixed dipterocarp forest on tertiary sediments in central Sarawak, Malaysia. *Journal of Tropical Ecology* 3, 201–220.

Barbier, E., Burgess, J., Bishop, J., Aylward, B. and Bann, C. (1993) *The Economic Linkages between the International Trade in Tropical Timber and the Sustainable Management of Tropical Forests.* Final Report: ITTO Activity PCM (XI)/4, LEEC, IIED, London.

Barnett, T. (1989) Report on the state of the logging industry in Papua New Guinea. Port Moresby, Government of Papua New Guinea. Partly published report of a Royal Commission.

Basnet, K., Likens, G.E., Scatena, F.N. and Lugo, A.E. (1992) Hurricane Hugo damage to a tropical rainforest in Puerto Rico. *Journal of Tropical Ecology* 8, 47–55.

Baumgartner, A. and Bruenig, E.F. (1978) Tropical forests and the biosphere. In: *Tropical Forest Ecosystems, a State-of-Knowledge Report Prepared by UNESCO/ UNEP/FAO.* UNESCO, Natural Resources Research, Paris, pp. 33–60.

Baur, G.N. (1962) *The Ecological Basis of Rainforest Management.* FAO, Rome.

Bawa, K.S. and Krugman, S.L. (1991) Reproductive biology and genetics of tropical trees in relation to conservation and management. In: Gomez-Pompa, A., Whitmore, T.C. and Hadley, M. (eds) *Rainforest Regeneration and Management.* Parthenon, Carnforth and UNESCO, Paris, pp. 119–136.

Beccari, O. (1904) *Wanderings in the Great Forests of Borneo.* Facsimile reprint in 1986. Oxford University Press, Singapore.

Becker, P. and Wong, M. (1993) Drought-induced mortality in tropical heath forest. *Journal of Tropical Science* 5, 416–419.

Beer, J.H. de and McDernoff, M.J. (1989) *The Economic Value of Non-Timber Forest Products in Southeast Asia.* IUCN/WWF, Gland.

Bier, A. (1933) *Der Wald in Sauen [The Forest in Sauen].* Der Deutsche Forstwirt, no. 86–89. Erde and Kosmos Publ., Schoenau.

Binggeli, P. (1994) Misuse of terminology and anthropomorphic concepts in the description of introduced species. *British Ecology Society, The Bulletin* XXV, 10–13.

Blaikie, P. (1989) Environment and access to resources in Africa. *Africa* 59, 18–40 (cit. in Bryant, R.L., 1992).

Blakeney, K.J. (1993) Global Environment Facility, Initial Executive Project Summary. World Bank, Washington.

Blanford, H.R. (1948) Forest Practice, (a) Silviculture, 3. Tropical Evergreens. Fifth Empire Forestry Conference 1947. *Empire Forestry Review* 27, 106–109.

Blaser, J. (1993) Forest management of natural forests in Malaysia. *ITTO Tropical Forest Update* 3, 4–5.

Blockhus, J.M., Dillenbeck, M., Sayer, J.A. and Wegge, P. (eds) (1992) *Conserving Biological Diversity in Managed Tropical Forests.* IUCN Forest Conservation Programme Workshop. IUCN, Gland and Cambridge.

Bock, C. (1881) *The Headhunters of Borneo.* Oxford University Press, Oxford.

Bodegom, A.J. van and Graaf, N.R. de (eds) (1991) *The CELOS Management System: a Provisional Manual.* Foundation Bos, Department of Forestry of Agricultural University and Scientific Information Division, Wageningen.

Boettcher, J.H. (1987) [Relation between vegetation structure and species richness of birds]. MSc thesis, Faculty Biology Hamburg University.

Böhm, H.D.V., Haisch, S. and Friauf, E. (1995) Environmental helicopters with modular sensor concept, demonstrated on the example of forestry monitoring. *21st European Rotorcraft Forum,* St Petersburg, Russia, 30 August–1 September 1995 (in press).

Bormann, F.H. and Berlyn, G. (eds) (1981) *Age and Growth Rate of Tropical Trees: New Directions of Research.* Yale University, School of Forestry and Environmental Studies, Bulletin no. 94, New Haven.

Bosman, M., Kort, I. de, Genderen, M. van and Baas, P. (1993) *Radial Variation in Wood Properties of Naturally and Plantation Grown Light Red Meranti* (Shorea, Dipterocarpaceae). Rijksherbarium, Hortus Botanicus, NL 2300 RA Leiden.

Bossell, H. (1989) Modelling forest dynamics: moving from description to explanation. In: *Workshop on Modelling Forest Dynamics in Europe, Wageningen, 17–19 October 1988,* p. 13. Department of Forestry, Agricultural University, Science Information Division, Wageningen.

Bossel, H. (1994) *TREEDYN3 Forest Simulation Model*. Forsch. zentr. Waldökosysteme, Univ. Göttingen, ser. B, vol. 35.

Bossell, H. and Bruenig, E.F. (1992) *Natural Resource Systems Analysis*. Feldafing, Schriftenreihe der Deutschen Stiftung für Internationale Entwicklung (DSE).

Bourdeau, P. *et al.* (eds) (1989) *Ecotoxicology and Climate*. SCOPE Series no. 38. John Wiley, Chichester.

Boyle, T.J.B. and Sayer, J.A. (1995) Measuring, monitoring and conserving biodiversity in managed tropical forests. *Commonwealth Forestry Review* 74 (1), 20–25.

Brosius, J.P. (1988) A separate reality: comments on Hoffmann's 'The Penan: hunters and gatherers of Borneo'. *Borneo Research Bulletin* 20, 81–106.

Brosius, J.P. (1991) Foraging in tropical rainforests: the case of the Penan of Sarawak, East Malaysia. *Human Ecology* 19, 123–150.

Brown, K. and Adger, W.N. (1994) Economic and political feasibility of international carbon offsets. *Forest Ecology and Management* 68, 217–229.

Brown, L.R. Brough, H., Durning, A. *et al.* (1992) *State of the World 1992*. Earth Scan Publications, London.

Browne, F.G. (1949) Stormforest in Kelantan. *Malay Forestry* XII, 28–33.

Browne, F.G. (1952) The kerangas lands of Sarawak. *Malay Forestry* XV, 61–73.

Browne, F.G. (1955) *Forest Trees of Sarawak and Brunei*. Government Printing Office, Kuching.

Bruenig, E.F. (1957) Waldbau in Sarawak. *Allgemeine Forst-Jagdztgeitung* 128, 156–165.

Bruenig, E.F. (1958) Voraussetzungen und Ziele der Forstpolitik in Sarawak. *Allgemeine Forst-Jagdztgeitung* 129, 53–62.

Bruenig, E.F. (1961a) *An Introduction to the Vegetation of the Bako National Park*. Government Printer, Kuching.

Bruenig, E.F. (1961b) *(1) Forest Survey and Inventory Code. (2) Forest Working Plans Code*. Sarawak Forest Department, Government Printer, Kuching.

Bruenig, E.F. (1963) The history of forest inventories in Sarawak. *Malay Forestry* XXVI, 141–159.

Bruenig, E.F. (1965a) The management of forest estates through working plans. *Malay Forestry* XXVIII, 46–55.

Bruenig, E.F. (1965b) A guide and introduction to the vegetation of the Kerangas Forests and the Padangs of the Bako National Park. In: *Proc Symp. Ecol. Res. Humid Trop. Vegetation. Kuching, Sarawak, July 1963*. Government of Sarawak, Kuching and UNESCO, Tokyo, pp. 289–318.

Bruenig, E.F. (1965c) Die Kuda-Kudamethode der Stammholzwerbung in Sarawak. *Forstarchiv* 36, 162–164.

Bruenig, E.F. (1966) [The Heath forests of Sarawak and Brunei]. Thesis, Hamburg University and published in 1968, *Der Heidewald von Sarawak und Brunei. I. Standort und Vegetation. II. Artenbeschreibung und Anhänge*. Mitteilung der Bundesforschungsanstalt Forst-Holzwirtschaft no. 68. vols I and II.

Bruenig, E.F. (1967a) *Financial Aspects for Planning the Conversion of Natural Mixed Dipterocarp Forest by Natural Regeneration or Plantation Establishment*. (1) Manila, GTZ-Seminar proceedings; (2) Rome, FAO, Proc. 1st Session of the Committee on Development of Tropical Forestry, 1967; (3) Forstliche Produktionslehre, Europ. Univ. Series, XXV, Bern, P. Lang, pp. 318.

Bruenig, E.F. (1967b) On the limits of vegetable productivity in the tropical rainforests and the boreal forests. *Journal of International Botany Society* 46, 314–322.

Bruenig, E.F. (1969a) On the seasonality of droughts in the lowlands of Sarawak (Borneo). *Erdkunde* 232, 127–133.

Bruenig, E.F. (1969b) The classification of forest types in Sarawak. *Malay Forestry* XXXIII, 143–179.

Bruenig, E.F. (1969c) Forestry on tropical podzols and related soils. *Tropical Ecology* 10, 45–58.

Bruenig, E.F. (1970a) Stand structure, physiognomy and environmental factors in some lowland forests in Sarawak. *Tropical Ecology* 11, 26–43.

Bruenig, E.F. (1970b) Multiple-use management in Germany's forests. *Journal of Forestry* 68, 718–722.

Bruenig, E.F. (1971a) On the ecological significance of drought in the equatorial wet evergreen (rain) forest of Sarawak (Borneo). In: *Transactions First Aberdeen–Hull Symposium on Malesian Ecology.* University of Hull, Department of Geography, Misc. Ser., no. 11, pp. 66–97.

Bruenig, E.F. (1971b) Die Sauerstofflieferung aus den Wäldern der Erde und ihre Bedeutung für die Reinerhaltung der Luft. [The oxygen output from the world forests and their role in keeping the atmosphere.] *Forstarchiv* 42, 21–23.

Bruenig, E.F. (1971c) *Forstliche Produktionslehre.* P. Lang Publ., Frankfurt.

Bruenig, E.F. (1973a) *Silvics and Silvicultural Management and Humid-Tropical Forests,* chapters 4, 22 and 65. Institute for World Forestry, Hamburg.

Bruenig, E.F. (1973b) Some further evidence on the amount of damage attributed to lightning and wind-throw in *Shorea albida* forest in Sarawak. *Commonwealth Forestry Review* 52, 260–265.

Bruenig, E.F. (1973c) Species richness and stand diversity in relation to site and succession in forest in Sarawak and Brunei, Borneo. *Amazoniana* 4, 293–320.

Bruenig, E.F. (1973d) Biomass diversity and biomass sampling in tropical rainforest. In: Young, H.E. (ed.) *IUFRO Biomass Studies.* IUFRO, University of Maine, pp. 270–293.

Bruenig, E.F. (1974) *Ecological Studies in Kerangas Forests of Sarawak and Brunei.* Borneo Literature Bureau for Sarawak Forest Department, Kuching.

Bruenig, E.F. (1976) Classifying for mapping of kerangas and peatswamp forest as examples of primary forest types in Sarawak. In: Ashton, P.S. (ed.) *The Classification and Mapping of Southeast Asian Ecosystems.* Department of Geography, University of Hull, Misc. Ser. no. 17, pp. 57–75.

Bruenig, E.F. (1977a) The tropical rainforest – a wasted asset or an essential biospheric resource? *Ambio* 6, 187–191.

Bruenig, E.F. (ed.) (1977b) *Transactions of the International MAB–IUFRO Workshop on Tropical Rainforest Ecosystems Research, Hamburg–Reinbek, 12–17 May 1977.* Chair of World Forestry, Hamburg–Reinbek.

Bruenig, E.F. (1981) The world's forests, a declining resource. *Universitas* 23 E, 9–16.

Bruenig, E.F. (1983) La guerre chimique et les dynamiques et la gestion des ecosystemes forestiers tropicaux. In: *Symposium Internationale sur les Herbicides et Defoliants Employes dans la Guerre: les Effects a Long Terme sur l'Homme et la Nature.* Com. Nat. d'Invest, Cons. Guerre Chim. Hanoi, pp. 53–63.

Bruenig, E.F. (1984a) Forest research and planning in South and Southeast Asia. *Applied Geography and Development* 23, 46–54.

Bruenig, E.F. (1984b) The means to excellence through control of growing stock. In:

First Weyerhaeuser Science Symposium, Tacoma, 30 April–3 May 1979. Weyerhaeuser Company.

Bruenig, E.F. (1984c) Designing ecologically stable plantations. In: Wiersum, K.F. (ed.) *Strategies and Designs for Afforestation, Reforestation and Tree Planting.* Pudoc, Wageningen, pp. 348–359.

Bruenig, E.F. (1985) Deforestation and its ecological implications for the rainforests in Southeast Asia. Proc. Symp. FRIM and IUCN, Kepong 1–2 September 1983. IUCN Commission on Ecology Paper no. 10, IUCN. *The Environmentalist 5,* Supplement, 17–35.

Bruenig, E.F. (1986a) Lowland-montane ecological relationships and interdependencies between natural forest ecosystems. *Tropical and Subtropical Ecosystems* 4, 1–21.

Bruenig, E.F. (1986b) Aspects of current forestry practice and silvicultural trends in West Germany affecting fresh waters. In: Solbe, J.F. de L.G. (ed.) *Effects of Land Use on Fresh Waters: Agriculture, Forestry, Mineral Exploitation, Urbanization.* Ellis Horwood, Chichester, pp. 378–397.

Bruenig, E.F. (1987a) Tropical forest areas as a source of biological diversity. In: *European Conference on Biological Diversity – a Challenge to Science, the Economy and Society,* Dublin, 4–6 March 1987.

Bruenig, E.F. (1987b) Die Entwaldung der Tropen und die Auswirkung auf das Klima. *Forstwissenschaftliches Centralblatt.* 106, 263–275.

Bruenig, E.F. (1987c) The forest ecosystem tropical and boreal. *Ambio* 16, 68–79.

Bruenig, E.F. (1989a) [Forest ecosystem research in the tropical rainforest as stimulator of silviculture in Germany]. In: Hartmann, G. (ed.) *Amazonien im Umbruch/Amazonas em Transformacao.* Reimer, Berlin.

Bruenig, E.F. (1989b) Tropical forest resources. In: *Resource Management and Optimization,* vol. 7. Harwood Academic Publishers, London, pp. 67–95.

Bruenig, E.F. (1989c) Ecosystems of the world. In: Bourdeau, J., Haines, A., Klein, W. and Murti, C.R.K. (eds) *Ecotoxicology and Climate.* SCOPE no. 38. John Wiley, Chichester, pp. 29–41.

Bruenig, E.F. (1989d) Use and misuse of tropical rainforest. In: Lieth, H. and Werger, M.J.A. (eds) *Tropical Rainforest. Ecosystems of the World, vol. 14B.* Elsevier, Amsterdam, pp. 611–636.

Bruenig, E.F. (1989e) *Die Erhaltung, nachhaltige Vielfachnutzung und langfristige Entwicklung der Tropischen Immergrünen Feuchwälder (Regenwälder).* Institute for World Forestry, Hamburg.

Bruenig, E.F. (1989f) Oligotrophic wetlands in Borneo. In: Lugo, A.E., Brinson, M. and Brown, S. (eds) *Forested Wetlands. Ecosystems of the World, vol. 15.* Elsevier, Amsterdam, pp. 299–334.

Bruenig, E.F. (1990) Nature and resource conservation in the kerangas and kerapah forests of Sarawak. In: Kiew, R. (ed.) *The State of Conservation in Malaysia.* Malayan Nature Society, Kuala Lumpur, pp. 29–41.

Bruenig, E.F. (1991a) Overview: the state of knowledge of global change with respect to the tropical zone; and the effects of the most probable trend of climate change on humid tropical forest ecosystems. In: Pongpan Kongton, Suree Bhumibhamon, Wood, H. and Kansri Boonpragob (eds) *Global Change: Effects on Tropical Forests, Agricultural, Urban and Industrial Ecosystems.* ITTO/Science Soc. Thailand, Bangkok, pp. 11–16 and 51–61.

Bruenig, E.F. (1991b) Pattern and structure along gradients in natural forests in Borneo and in Amazonia, their significance for the interpretation of stand dynamics and functioning. In: Gomez-Pompa, A., Whitmore, T.C. and Hadley, M. (eds) *Rainforest Regeneration and Management*. UNESCO, Paris and Parthenon, Carnforth, pp. 235–243.

Bruenig, E.F. (1992) Integrated and multi-sectoral approaches to achieve sustainability of ecosystem development: The Sarawak forestry case. In: Stott, P. (ed.). The political ecology of Southeast Asia's forests. *Global Ecology and Biogeography Letters*, Vol. 3, pp. 253–266, 1993.

Bruenig, E.F. (1993a) Impressions of Brunei Forestry in 1958. In: *Sixty Years of Forestry in Brunei Darussalam*. Forestry Department Brunei Darussalam, pp. 82–83.

Bruenig, E.F. (1993b) Research and development programme for forestry in Sarawak: a pilot model approach towards sustainable forest management and economic development. In: Lieth, H. and Lohmann, M. (eds) *Restoration of Tropical Forest Ecosystems*. Kluwer Academic, Amsterdam, pp. 173–183.

Bruenig, E.F. (1994) The role of NGOs in rainforest conservation. Friends of the Earth, Hong Kong FOE News, nos 21 and 22.

Bruenig, E.F. (1995) [APP-target orientated multiple use concept: a review after 20 years]. *Allgemeine Forstzeitschrift, AFZ* (in press).

Bruenig, E.F. and Csomos, S. (1995) Harmonized ecosystem research methodology for an integrated forest research programme in Southeast Asia. In Box E.O. *et al.* (eds). *Vegetation Science in Forestry*. Kluwer Academic Publishers, Dordrecht, pp. 613–620.

Bruenig, E.F. and Droste, H.J. (1995) Structure, dynamics and management of rainforests on nutrient deficient soils in Sarawak. In: Primack, R.B. and Lovejoy, T.E. (eds) *Ecology, Conservation and Management of Southeast Asian Rainforests*. Yale University Press, New Haven, 304 pp.

Bruenig, E.F. and Heuveldop, J. (1994) Social functions of forests: Some basic facts and figures on the effects on the carbon-balance. Research Report, University of Hamburg, Chair for World Forestry.

Bruenig, E.F. and Huang, Y.W. (1989) Patterns of tree species diversity and canopy structure and dynamics in humid tropical evergreen forests in Borneo and in China. In: Holm-Nielsen, L.B., Nielsen, I.C. and Balslev, H. (eds) *Tropical Forests*. Academic Press, London, pp. 75–88.

Bruenig, E.F. and Klinge, H. (1977) Comparison of the phytomass structure of tropical rainforest stands in Central Amazonas, Brazil, and in Sarawak, Borneo. *The Garden's Bulletin, Singapore*, pp. 81–101.

Bruenig, E.F. and Muellerstael, H. (1987) Water relations and gas exchange of tropical rainforest tree species under stress: *Manilkara* sp. from high caatinga. In: Sethuraj, M.R. and Raghavendra, A.S. (eds) *Tree Crop Physiology*. Elsevier, Amsterdam, pp. 121–138.

Bruenig, E.F. and Poker, J. (eds) (1989) *Management of Tropical Rainforests – Utopia or Chance of Survival?* Nomos, Baden-Baden.

Bruenig, E.F. and Poker, J. (1991) Is sustainable utilization of the tropical evergreen moist forest possible? In: Erdelen, W., Ishwaran, N. and Müller, P. (eds) *Tropical Ecosystems, System Characteristics, Utilization Patterns, Destruction, Conservation Concepts*. Verlag Josef Margraf, Saarbrücken–Weikersheim, pp. 91–106.

Bruenig, E.F. and Sander, N. (1983) Ecosystem structure and functioning: some interactions of relevance to agroforestry. In: Huxley, P.A. (ed.) *Plant Research and Agroforestry*. ICRAF, Nairobi, pp. 221–247.

Bruenig, E.F. and Schmidt-Lorenz, R. (1985) Some observations on the humic matter in kerangas and caatinga soils with respect to their role as sink and source of carbon in the face of sporadic episodic events. In: Degens, E. *et al.* (eds) *SCOPE/UNEP Special Issue.* Institute of Palaeo-Geology, University of Hamburg, no. 58, pp. 107–122.

Bruenig, E.F. and Schneider, T.W. (1992) Assessing and monitoring biological diversity in sustainable management of tropical rainforest. In: *2nd Princess Chulabhorn Science Conference, 1962, Bangkok.* Chulabhorn Research Institute, Office of Scientific Affairs, Mahidol University, Bangkok.

Bruenig, E.F., Buch, M.v., Heuveldop, J. and Panzer, K.F. (1975) Stratification of tropical moist forest for land-use planning. *Plant Research and Development* 2, 21–24.

Bruenig, E.F., Heuveldop, J., Smith, J. and Alder, D. (1978) Structure and functions of a rainforest in the International Amazon Ecosystem project: floristic stratification and variation of some features of stand structure and precipitation. In: Singh, J.S. and Gopal, B. (eds) *Commemoration Volume of Tropical Ecology on the Occasion of the 70th Birthday of F. Misra: Glimpses of Ecology.* International Scientific Publications, Jaipur, pp. 125–144.

Bruenig, E.F., Alder, D. and Smith, J. (1979) The international MAB Amazon Rainforest Ecosystem Pilot project at San Carlos de Rio Negro: vegetation classification and structure. In: Adisoemarto and Bruenig, E.F. (eds) *Transactions of the Second International MAB–IUFRO Workshop on Tropical Rainforest Ecosystems Research, Jakarta, 21–25 October 1978,* Chair of World Forestry, Hamburg–Reinbek, pp. 67–100.

Bruenig, E.F., Schneider, T.W. and Ollmann, H. (1986a) Wald und Holzproduktion: muss die Holzerzeugung gesteigert werden. *Allgemaine Forstdzeidschrift AFZ* 42, 501–503.

Bruenig, E.F., Bossel, H., Elpel, K.P., Grossmann, W.D., Schneider, T.W., Wan, Z.H. and Yu, Z.Y. (1986b) *Ecological–Socioeconomic System Analysis and Simulation: a Guide for Application of System Analysis to the Conservation, Utilization and Development of Tropical and Subtropical Land Resources in China.* National MAB Committee, Bonn and German Foundation for International Development (DSE), Feldafing.

Bruenig, E.F., Buhmann, S. and Poker, J. (1991) Structure and distribution patterns as basic parameters for the analysis and modelling of ecosystems in the tropical evergreen moist forests. In: Erdelen, W., Ishwaran, N. and Müller, P. (eds) *Tropical Ecosystems, System Characteristics, Utilization Patterns, Destruction, Conservation Concepts.* Verlag Josef Margraf, Weikersheim, pp. 1–10.

Bruenig, E.F., Grossmann, W.D., Kahn, H.D. and Kasperidus, M. (1994) *Ecological and Socio-economic Modelling for Agroforestry and Social Forestry Development in Southeast Asia.* DSE, Feldafing.

Bruenig, E.F., Lee, H.S., Chai, F.Y.C. (1995) Natural forest management in Sarawak: risks, uncertainties, options and strategies. *3rd Conference on Forestry and Forest Products Research 1995,* FRIM, Kepong, Kuala Lumpur, 2–4 October, 1995.

Bruijnzeel, L.A. (1987) A review of the hydrological aspects of tropical forests with special reference to the study of nutrient cycling. In: *Brit. Ecol. Soc. Symp. on Mineral Nutrients in Tropical Forests and Savanna Ecosystems, Stirling, 9–11 September 1987.* British Ecology Society.

Bruijnzeel, L.A. (1991) Nutrient input–output budgets of tropical forest ecosystems: a review. *Journal of Tropical Ecology* 7, 1–24.

Bruijnzeel, L.A. (1992) Managing tropical forest watershed, for production: where contradictory theory and practice co-exist. In: Miller, F.R. and Adam, K.L. (eds) *Wise Management of Tropical Forests 1992.* OFI, Oxford, pp. 37–76.

Bruijnzeel, L.A., Waterloo, M.J., Proctor, J., Kuiters, A.T. and Kotterin, B. (1993) Hydrological observations in montane rainforest on Gunung Silam, Sabah, Malaysia, with special reference to the 'Massenerhebungs Effekt'. *Journal of Ecology* 81, 145–167.

Brunei Forest Department (1993) *60 years Forestry in Brunei Darussalam.* Bandar Sri Begawan.

Bryan, M.B. (1974) *A Study of Stand Conditions and Production Potential in the Peatswamp Forests of Sarawak.* FO: DP/MAL/72/009, Working paper 27, Kuala Lumpur.

Bryant, R.L. (1992) Political ecology, an emerging research agenda in Third-World studies. *Political Geography* 11, 12–36.

Bureau of Forestry (1965) *Handbook of Selective Logging,* 2nd edn. 1970. Bureau of Forest Development, Manila–Quezon City.

Burkill, L.H. (1935) *A Dictionary of the Economic Products of the Malay Peninsula,* vols I and II. Oxford University Press, Oxford.

Burslem, D.F.R.P., Grubb, P.J. and Turner, I.M. (1995) Responses to nutrient addition among shade-tolerant tree seedlings of lowland tropical rainforest in Singapore. *Journal of Ecology* 83, 113–122.

Butt, G. (1984) *Forest Soils Research, Past, Present and Future, a Terminal Report.* Forest Research Report SS8, Forest Department Sarawak, Kuching.

Butt, G. and Petch, B. (1985) *Erosion Mapping in Logging Areas: Ulu Niah (Pilot Project)* Forest Research Report SS14, Forest Department Sarawak, Kuching.

Butt, G. and Sia, P.C. (1982) *Guide to Site–Species Matching in Sarawak, 1, Selected Exotic and Native Plantation Species.* Forest Research Report SS1, Forest Department Sarawak, Kuching.

Butt, G., Ting S.P. and Sia, P.C. (1983) *A Site Evaluation of Reforestation Areas in Sampadi Forest Reserve.* Forest Research Report SS6, Forest Department Sarawak, Kuching.

Calder, I.R., Hall, R.L. and Adlard, P.G. (eds) (1992) *Growth and Water Use of Forest Plantations.* John Wiley, Chichester.

Callister, D.J. (1992) *Illegal Tropical Timber Trade: Asia Pacific.* TRAFFIC International, Cambridge.

Carlowitz, H.C.v. (1713) *Silviculture oeconomica oder hauswirtliche Nachricht und naturmässige Anweisung zur wilden Baumzucht.* J. Fr. Braun, Leipzig.

Carpenter, R.A. (ed.) (1981) *Assessing Tropical Forest Lands: their Suitability for Sustainable Uses.* Tycooly International, Dublin.

Carter, S.E. and Murwira, H.K. (1995) Spatial variability in soil fertility management and crop response in Mutoko communal area, Zimbabwe. *Ambio* 24, 77–84.

Cassels, D.S., Gilmour, D.A. and Bonell, M. (1985) Catchment response and watershed management in the tropical rainforests in north-eastern Australia. *Forest Ecology and Management* 10, 155–175.

Cassels, D.S., Bonell, M., Gilmour, D.A. and Valentine, P.S. (1988) Conservation and management of Australia's tropical rainforests: local realities and global responsibilities. *Proceedings Ecological Society Australia* 15, 313–326.

Castri, F. di and Younés, T. (eds) (1990) *Ecosystem Function of Biological Diversity.* Biology International Special Issue no. 22, IUBS, Paris.

Catinot, R. (1965) Sylviculture en forêt dense africaine. *Bois et Forêts Tropiques*, 100, 5–18; 101, 3–16; 102, 3–16; 103, 3–16; 104, 17–29.

Chai, D. (1993) Tour report on helicopter logging at Bintulu, Sarawak. Forest Department Sabah, Forest Management Planning, Forest Research Centre Sepilok, Sandakan, Library no. R143.

Chai, E.O.K. (1984) The response of commercial *Shorea* spp. to silvicultural treatments in the hill mixed dipterocarp forest of Sarawak. Thesis, Forestry Faculty, Agricultural University Malaysia, Serdang.

Chai, E.O.K., Lee, H.S. and Yamakura, T. (1994) Preliminary results of the 52 hectares long term ecological research plot at Lambir National Park, Miri, Sarawak, Malaysia. In: *Abstracts, session 1 of the workshop on 'Long Term Ecological Research in Relation to Forest Management', Monbusho Japan*, 25–27 July 1994. HIID Harvard and Forest Department Sarawak, Kuching.

Chai, F.Y.C. (1986) Silvicultural treatment in mixed swamp forest. In: *9th Malaysian Forestry Conference, October 1986*. Forest Department Sarawak, Kuching.

Chai, F.Y.C. (1991) Diameter increment models for the mixed swamp forests of Sarawak. MF Thesis, Department of Forestry UBC, Vancouver.

Chai, F.Y.C. (1995) Above-ground biomass of a secondary forest in Sarawak. Kuching, Forest Department Sarawak, report in Forest Research Library. *Journal of Forestry Science, Kepong* (submitted).

Chai, F.Y.C., Kho, S.Y. and Chung, K.S. (1994) *Growth and Yield of a Logged-over Mixed Dipterocarp Forest in Sarawak*. Forest Department Sarawak, Kuching and ASEAN Institute of Forest Management, Kuala Lumpur.

Champion, H.G. (1936) *A Preliminary Survey of Forest Types of India and Burma*. Indian For. Rec. (N.S.), silvics, 1.

Champion, H.G. and Seth, S.K. (1968) *A Revised Survey of Forest Types of India*. Manager of Publications, New Delhi.

Chen, P.C.Y. (1990) *Penans, the Nomads of Sarawak*. Pelanduk Publ., Petaling Jaya.

Chiew, K.Y. and Garcia, A. (1989) Growth and yield studies in Yayasan Sabah forest concession area. In: Wan Razali *et al.* (eds) *Growth and Yield in Tropical Moist-mixed Forests*. FRIM, Kepong, pp. 192–204.

Chin, S.C. (1992) Curiouser and curiouser: forestry in Sarawak. *Walleceana* 68–69, 1–5.

Chin, T.Y., Gan, B.K. and Weinland, G. (1995) *Silviculture of Logged over Forests in Peninsular Malaysia: Current Practice and Future Challenges*. 12. Malaysian Forestry Conference. Forestry Department Sarawak, Kuching (in press).

Chua, D. (1986) *Tree Marking for Directional Felling and its Effects on Logging Efficiency and Damage to Residual Stand*. Research Report FE 1/85, Forest Department Sarawak, Kuching.

Chua, D.K.H. (1993) *A Case Study on Helicopter Harvesting in the Hill Mixed Dipterocarp Forests of Sarawak*. Research Report FE 2/93, Forest Operations Branch, Forest Department Sarawak, Kuching.

Chung, K.S. (1985) *Comparison of Land Value of Natural Forest, Forest Plantation, Oil Palm and Cocoa in Selected Localities*. Forest Department Sarawak, Kuching.

CIFOR (1993) *CIFOR News*, no. 1. September 1993, Center of International Forestry Research, Bogor.

CIFOR (1994) *CIFOR News*, Special Anniversary Edition. CIFOR, Jakarta.

Clark, D.A. and Clark, D.B. (1994) Climate-induced annual variation in canopy tree growth in a Costa Rican tropical rainforest. *Journal of Ecology* 82, 865–872.

Clarke, E.C. (1964) *A Report on Silvicultural Research and the Silvicultural Treatment* of Exploited Mixed Swamp-Forest in the Peatswamp Forests of Sarawak 1960–1963. Research pamphlet 45, Forest Research Institute, Forest Department Malaya.

Cleary, M. and Eaton, P. (1992) *Borneo, Change and Development.* Oxford University Press, Singapore.

Clüsener-Godt, M. and Sachs, I. (eds) (1994) *Extractivism in the Brazilian Amazon: Perspectives on Regional Development.* MAB Digest 18, UNESCO, Paris.

Collins, N.M., Sayer, J.A. and Whitmore, T.C. (1991) *The Conservation Atlas of Tropical Forests, Asia and the Pacific.* MacMillan, London.

Conway, S. (1986) *Logging Practices. Principles of Timber Harvesting Systems.* Miller Freeman Publ., San Francisco.

Cooper, J.I. and Tinsley, T.W. (1977) Background note concerning viruses liable to influence exploitation of tropical rainforests. In: Bruenig, E.F. (ed.) *Transactions International MAB–IUFRO Workshop on Tropical Rainforest Ecosystems Research.* Chair of World Forestry, Hamburg–Reinbek, pp. 249–262.

Corner, E.J.H. (1978) The freshwater swamp-forest of South Johore and Singapore. *Garden's Bulletin Singapore,* (Suppl. 1), 266 + ix pp.

Coto, Z. (1994) Certification/eco-labelling; Indonesian perspective. In: *Seminar on Trade of Timber from Sustainably Managed Forest, 5–6 April 1994, Kuala Lumpur, Malaysia,* p. 6.

Cousens, J.E. (1952) The tropical forest experimental station Puerto Rico. *Malay Forestry* XV, 49–50.

Cox, P.A. and Elmquist, T. (1991) Indigenous control of tropical rainforest reserves: an alternative strategy for conservation. *Ambio* 20, 317–321.

Dalling, J.W. and Tanner, E.V.J. (1995) An experimental study of regeneration on landslides in montane rainforest in Jamaica. *Journal of Ecology* 83, 55–64.

Dames, T.G.W. (1962) *Report to the Government of Sarawak on Soil Research in the Development of Sarawak.* FAO, Rome.

Darmstadter, J. and Toman, M.A. (eds) (1994) *Assessing Surprises and Nonlinearities in Greenhouse Warming.* Resource for the Future, Washington, DC, 158 pp.

Daryadi, L. and Sormin, B.H. (1994) Sustainably managed forest's certification and ecolabelling: Indonesia's perspective. In: *Seminar on Trade of Timber from Sustainably Managed Forest, 5–6 April 1994, Kuala Lumpur, Malaysia,* p. 12.

Davidson, J. (1983) Forestry in Papua New Guinea: a case study of the Gogol woodchip project near Medang. In: Hamilton L.S. (ed.) *Forest and Watershed Development and Conservation in Asia and the Pacific.* East–West Centre, Honolulu and West View Press, Boulder, pp. 19–138.

Dawkins, H.C. (1958) *The Management of Natural Tropical High Forest with Special Reference to Uganda.* Imperial Forestry Institute, Oxford University, Oxford.

Dawkins, H.C. (1959) The volume increment of natural tropical high-forest and limitations on its improvement. *Empire Forestry Review* 38, 175–180.

Dawkins, H.C. (1963) The productivity of tropical high-forest trees and their reaction to controllable environment. Thesis, Oxford University and Commonwealth Forestry Institute, Oxford.

Dawkins, H.C. (1988) The first century of tropical silviculture – successes forgotten and failures misunderstood. In: McDermott, M.J. (ed.) *The Future of the Tropical Rain Forest.* Oxford Forestry Institute, Oxford, p. 8.

Denmead, O.T. (1964) Evaporation sources and apparent diffusivities in a forest canopy. *Journal of Applied Meteorology* 3, 383–389.

Deutscher Bundestag (1988) *Schutz der Erdatmosphäre: Eine internationale Her-*

ausforderung. Zur Sache; 88, 5. Deutscher Bundestag, Bonn.

Deutscher Bundestag (1990) *Protecting the Tropical Forests. A High-priority International Task.* Bonner University Press, Bonn.

Deutscher Bundestag, Enquete Commission 'Protecting the Earth's Atmosphere' (1992) *Climate Change – a Threat to Global Development.* Economica Verlag, Bonn.

Deutscher Bundestag (1994a) *Protecting the Earth's Atmosphere.* Third report of the Enquete Commission, Bundestag. Deutscher Bundestag, Bonn.

Deutscher Bundestag (1994b) *Schutz der Grünen Erde.* 3. Report der Enquete Kommission 'Schutz der Erdatmosphäre'. Economica Verlag, Bonn.

Dick, J. McP. and Aminah, H. (1994) Vegetative propagation of tree species indigenous to Malaysia. *Commonwealth Forestry Review* 73, 164–171.

Dierschke, H. (1994) *Tuexenia* 14, 399–402.

Dilmy, A. (1965) The effect of fire used by early man on the vegetation of the humid tropics. In: *Proceedings Symposium on the impact of man on humid tropics vegetation, Goroka, 1960.* Admin. Terr. PNG and UNESCO, Tokyo, pp. 119–126.

Direktorat Jenderal Indonesia (1980) *Pedoman Tebang Pileh Indonesia.* Directorate General of Forests, Jakarta.

Doerner, D. (1992) Statement made during a meeting of the German National Committee for the UNESCO Programme 'Man and the Biosphere' (MAB), Bonn, February 1992.

Douglas, I., Spencer, T., Greer, T., Kawi Bidin, Waidi Sinun and Wong, W.M. (1992) The impact of selective commercial logging on stream hydrology, chemistry and sediment loads in the Ulu Segama rainforest, Sabah, Malaysia. *Philosophical Transactions Royal Society, B* 335, 397–408.

Driessen, P.M. and Rochimah, L. (1976) The physical properties of lowland peats from Kalimantan. In: *Peat and Podzolic Soils and their Potential for the Future.* Soil Research Institute, Bogor, Indonesia, pp. 56–73.

Droste, H.J. (1995) *Sabal F.R., Logging and Regeneration, Summary of Preliminary Findings.* Chair of World Forestry, Hamburg.

Droste, H.J. (1996) *Rainforest Ecosystems Sabal Mulu and Lambir, Feasibility LTER Report,* DFG Br 316/11–1 and 12–1, Hamburg, Chair for World Forestry, Hamburg University.

Droste, H.J. and Bruenig, E.F. (1992) *Structure and Dynamics of Natural Rainforest in Sabal Forest Reserve: Analyses of Dead Tree Biomass.* Chair of World Forestry, Hamburg.

Droste, H.J., Hahn-Schilling, B. and Hoch, O. (1995) Growth potential and quality of planted dipterocarps in Peninsular Malaysia. *Plant Research and Development* (in press).

Dubois, J.C.L. (1991) The present status of research into management of the rain forests of Amazonian Brazil. In: Gomez-Pompa, A., Whitmore, T.C. and Hadley, M. (eds). *Rain Forest Regeneration and Management.* UNESCO, Paris and Parthenon, Carnforth, pp. 431–436.

Dunn, F.L. (1982) *Rainforest Collectors and Traders, a Study of Resource Utilization in Modern and Ancient Malaya,* 2nd edn. Monograph 5, Malaysian Branch, Royal Asiatic Society, Kuala Lumpur.

Dykstra, D.P. and Heinrich, R. (1992) Sustaining tropical forests through environmentally sound harvesting practices. *Unasylva* 43, 2 no. 169, 9–15.

Ebisemiju, F.S. (1993) Environmental impact assessment: making it work in developing countries. *Journal of Environmental Management* 38, 247–273.

Eggeling, W.J. (1947) Observations on the ecology of Budongo rainforest Uganda. *Journal of Ecology* 34, 20–87.

Eggers, H. (1890) *Die Mahagoni-Schlägerungen auf Santo Domingo.* Globus, Braunschweig, pp. 193–195.

Eidmann, H. (1942) Grundprobleme der Kolonialen Forstzoologie. *Mitteilungen Hermann Goering Akademie der Deutschen Forstwissenschaft* I, 115–142.

Eisemann, K. (1991) Entwicklungsmöglichkeiten der Holzwirtschaft (Industrie und Handwerk) in Sarawak/Malaysia. Thesis, Department of Biology, University of Hamburg.

Ellenberg, H. (1973) Die Ökosysteme der Erde. Versuch einer Klassifikation der Ökosysteme nach funktionalen Gesichtspunkten. In: Ellenberg, H. (ed.) *Ökosystemforschung.* Springer Verlag, Berlin, pp. 235–265.

Ellenberg, H. Jr (1993) Bird species richness and aerodynamic roughness of forest canopies in the Sachsenwald. Unpublished research record, Institute for World Forestry, BFH, D21031 Hamburg.

Ellenberg, L., Esser, J. and Niekisch, M. (1992) [*Thoughts and Recommendations on the Technical Cooperation in Nature Conservation*]. GTZ, Eschborn.

Embrapa-Cpatu (1986) *Symposium on the Humid Tropics,* vols 1 and 2. Belem.

Estes, J.E. and Cosentino, J. (1989) Remote sensing of vegetation. In: *Global Ecology.* Academic Press, New York, pp. 75–111.

Evans, J. (1982) *Plantation Forestry in the Tropics.* Oxford University Press, Oxford.

Evans, J. and Wood, P.J. (1993) The place of plantations in tropical forestry. *ITTO Tropical Forest Update* 3, 3–5.

Expert Panel 'Social Balance of Forestry' (1975) *Eine erste Sozialbilanz des Waldes.* [*A First Attempt at a Social Balance of Costs and Benefits of Forests and Forestry in Germany*]. Fed. Min. Food, Agric. and Forestry.

Facelli, J.M. and Pickett, S.T.A. (1991) Plant litter: its dynamics and effects on plant community structure. *The Botany Review* 57, 1–32.

Fahey, T.J. and Hughes, J.W. (1994) Fine root dynamics in a northern hardwood forest ecosystem, Hubbard Brook Experimental Forest, NH. *Journal of Ecology* 82, 533–548.

FAO (1985a) *Intensive Multiple-use Forest Management in the Tropics. Analysis of Case Studies from India, Africa, Latin-America and the Caribbean.* FAO, Rome.

FAO, Committee on Forest Development in the Tropics (1985b) *The Tropical Forestry Action Plan.* FAO, Rome.

FAO (1988) *FAO/UNESCO Soil Map of the World; Revised Legend.* World Soil Resources Report 60. FAO, Rome.

FAO (1989a) *Management of Tropical Moist Forests in Africa.* Forestry Paper no. 88. FAO, Rome.

FAO (1989b) *Review of Forest Management Systems in Tropical Asia.* Forestry Paper no. 89. FAO, Rome.

FAO (1989c). *Household Food Security and Forestry – an Analysis of Socio-economic Issues.* FAO, Rome.

FAO (1993) *Forest Resources Assessment 1990, Tropical Countries.* Forestry Paper no. 112. FAO, Rome.

FAO (1994) *FAO Model Code of Forest Harvesting Practice.* FAO, Rome.

FAO–UNESCO (1992) *Forests and the Cultures in Asia.* Paris–Bangkok UNESCO.

Federal Environment Ministry (1994) *Environmental Policy, Climate Protection in Germany.* Federal Ministry for the Environment, Bonn.

Fickinger, H. (1992) [*On the Regeneration of some Commercial Tree Species in*

Selectively Utilized Humid Forests in the Republic of Congo]. Gött. Beitr. Land-Forstwirtsch. Trop. and Subtrop., vol. 75.

Figueroa, C., Wadsworth, J.C. and Branham, S. (eds) (1987) *Management of the Forests of Tropical America: Prospects and Technologies*. Institute of Tropical Forestry, Southern Forest Experiment Station and University of Puerto Rico.

Finegan, B. (1992) *El potencial de manejo de los bosques humedos secundarios neotropicales de las tierras bajas*. Turrialba, CATIE, Informe Tecnico 188.

Flenley, J.R. (1979) *The Equatorial Rainforest: a Geological History*. Butterworth, London.

Flenley, J.R. (1992) Palynological evidence relating to disturbance and other ecological phenomena of rainforests. In: Goldammer, J.G. (ed.) *Tropical Forests in Transition*. Birkhäuser Verlag, Basel, pp. 17–24.

Forest Department Sabah (1972) *Manual of silviculture in Sabah for Use in the Productive Forest Estate*. Kota Kinabalu, Sabah.

Forest Department Sarawak (1959) *The Sarawak Forest Manual*. Forest Department, Kuching.

Forest Department Sarawak (1961) (a) *Forest Management Code*. Government Printer, Kuching; (b) *Forest Inventory Code*. Forest Department Sarawak, Kuching.

Forest Department Sarawak (1971) Silvicultural Research Programme 1971–1975 with a summary of work done in 1966–70 and supporting papers. Kuching, Sarawak.

Forest Department Sarawak (1982) *Instructions for the Inspection of Logging Areas*. Forest Department Sarawak, Kuching.

Forest Department Sarawak (1990a) *Forestry in Sarawak, Malaysia*. Forest Department Sarawak, Kuching.

Forest Department Sarawak (1990b) *Longterm Research and Development Programme for Forestry in Sarawak*. Forest Department Sarawak, Kuching.

Forest Department Sarawak (1995) *Topography, Soil, Hydrology and Forest Reservation*. Proceedings Workshop Kuching, 9 May 1995, Sarawak Forest Department/ITTO Project Model Forest Management Area-MFMA. Forest Department Sarawak, Kuching.

Forest Research Institute Malaysia (1993) *The Economic Case for Natural Forest Management*. ITTO PCV (VI)/13.

Fox, J.E.D. (1968) Logging damage and the influence of climber cutting prior to logging in the lowland dipterocarp forest of Sabah. *Malay Forester* 31, 326–347.

Fox, J.E.D. (1969) The soil damage factor in present day logging in Sabah. *Sabah Society Journal*, 5, 43–52.

Francke, A. (1941) Aus der Waldwirtschaft Britisch Malaysias, ein Beispiel zum Waldbau im tropischen Regenwald. *Kolonialforstlidel Mitteilungen* IV, 95–140.

Francke, A. (1942) *Zur Kenntnis bisheriger Verjüngungssysteme im tropischen Regenwald unter besonderer Berücksichtigung ihrer Grundlagen im afrikanischen Äquatorialwald*. Habilitationsmanuscript, Bundesforschungsanstalt für Forst- und Holzwirtschaft, Hamburg.

Freeman, J.D. (1955) *Iban Agriculture. A Report on the Shifting Cultivation of Hill Rice by the Iban in Sarawak*. Colonial Research Studies no.18. HMSO, London.

Freeman, J.D. (1970) *Report on the Iban*. Athlone Press, London.

Frühwald, A., Wegener, G., Krüger, S. and Beudert, M. (1994) Holz-ein Rostoff der zukunft, nachhaltig verfügbar und umweltgerecht. *Informationsdienst Holz. Munich*, 23 pp.

Fundacion Natura (1991) *The Environment of the Amazon Region: Prospects for the Year 2000.* WWF, Gland.

Furtado, J.I. and Ruddle, K. (1986) The future of tropical forests. In: Polunin, N. (ed.) *Ecosystem Theory and Application.* John Wiley, New York, pp. 145–171.

Gane, M. (1992) Sustainable forestry. *Commonwealth Forestry Review,* 71, 83–90.

Gardner, M. (1992) Forscher mit der Axt im Walde. *DUZ* (German University Newspaper) 22, 31–33.

Garnier-Sillam, E., Kabala, M., Senechal, J. and Valerie, M. (eds) (1988) *Facteurs et Conditions du Maintien de la Fertilite du Milieu Tropical Humide.* UNESCO–MAB, Paris.

Gartlan, J.S. Newbery, D. McC., Thomas, D.W. and Waterman, P.G. (1986) The influence of topography and soil phosphorus on the vegetation of Korup Forests Reserve, Cameroun. *Vegetatio* 65, 131–148.

Gartlan, S. (1990) Practical constraints on sustainable logging in Cameroon. Paper to Conf. Cons. Util. Rat. For. D. Afr. Cent. et de L'Ouest, 5–9 November 1990, African Dev. Bank, IUCN, World Bank, Abidjan.

Gates, D.M. (1965) Energy, plants and ecology. *Ecology* 46, 1–13.

Gates, D.M. (1966) Transpiration and energy exchange. *Quarterly Review of Biology* 41, 353–364.

Gates, D.M. (1968) Energy exchange and ecology. *Bio-Science* 18, 90–95.

Geddes, W.R. (1954) *The Land Dayaks of Sarawak.* Colonial Research Study No. 14. HMSO, London.

German Federal Government (1993) *Protection and Management of the Tropical Forests.* Third tropical forest report of the German Federal Government. Fed. Min. Food, Agriculture Forestry and Fed. Min. Economic Cooperation and Development, Bonn.

German Forestry Association (Deutscher Forstverein, Kommittee für Internationale usammenarbeit) (1987) *Conservation and Sustainable Use of Tropical Rainforest.* BMZ – Bundesministerium für Wirtschaftliche Zusammenarbeit, Bonn.

Gist, C.S. (1973) *Some Tropical Modelling Efforts.* Utah State University, Logan/Utah, Ecology Centre, Mimeograph. Report.

Glover, N. and Adams, N. (eds) (1994) *Tree Improvement of Multipurpose Species.* Faculty Forestry, Kasetsart University Multipurpose Tree Species Network Technical Series, vol. 2, Bangkok.

Godoy, R.A. and Bawa, K.S. (1993) The economic value and sustainable harvest of plants and animals from the tropical forest: assumptions, hypotheses and methods. *Economic Botany* 47, 215–219.

Godoy, R.A., Laubowski, R. and Markandaya, A. (1993) A method for the economic valuation of non-timber tropical forest products. *Economic Botany* 47, 220–233.

Goenner, C. (1992) *Bird Density in the Sepilok Virgin Jungle Reserve – a Comparison of Census Techniques.* Library of the Forest Department Sandakan.

Goldammer, J.G. and Seibert, B. (1989) Natural rainforest fires in eastern Borneo during the pleistocene and holocene. *Naturwissenschaften* 76, 518–519.

Golley, F.B. (1978) 10. Gross and net primary production and growth parameters. 11. Secondary production. In: UNESCO (UNEP/FAO (eds). *Tropical Forest Ecosystems.* UNESCO–UNEP, Paris, pp. 233–255.

Golley, F.B. (Hrsg.) (1983) *Tropical Rainforest Ecosystems – Structure and Function. Ecosystems of the World, vol. 14A.* Elsevier Scientific Publishing, Amsterdam.

Gomez-Pompa, A., Whitmore, T.C. and Hadley, M. (eds) (1991) *Rainforest Regeneration and Management*. UNESCO, Paris and Parthenon, Carnforth.

Gonggryp, J.W. (1942) Die Holzzufuhr aus den Tropen nach Europea. *Intersylva* 2, 232–247.

Goodland, R. and Ledec, G. (1987) Neoclassical economics and principles of sustainable development. In: *Ecological Modelling*. Elsevier, Amsterdam, pp. 19–45.

Goudberg, N., Bonell, M. and Benzaken, D. (eds) (1991) *Tropical Rainforest Research in Australia. Present Status and Future Directions for the Institute for Tropical Rainforest Studies*. Proceedings of a workshop held in Townsville, Australia, 4–6 May 1990. Institute Tropical Rainforest Studies, Townsville.

Graaf, N.R. de (1982) Sustained timber production in the tropical rainforest of Surinam. In: Wienk, J.F. and de Wit, H.A. (eds) *Management of Low Fertility Acid Soils of the American Humid Tropics*. IICA, San José, Costa Rica, pp. 179–189.

Graaf, N.R. de (1986) A silvicultural system for natural regeneration of tropical rainforest in Surinam. PhD Thesis, Wageningen Agricultural University.

Graaf, N.R. de (1991) Managing natural regeneration for sustained timber production in Surinam: the CELOS silvicultural and harvesting system. In: Gomez-Pompa A., Whitmore, T.C. and Hadley, M. (eds) *Rainforest Regeneration and Management*. UNESCO, Paris and Parthenon, Carnforth.

Grainger, A. (1985) *A Model of World Trade in Tropical Hardwoods*. Oxford Forestry Institute, Oxford.

Grainger, A. (1986) The future role of the tropical rainforests in the world forest economy. Dissertation, Department of Plant Sciences, University of Oxford.

Grainger, A. (1989) Forests and rangelands chapter – forests section. In: *World Resources Report 1988–89*. World Resources Institute, Washington.

Grammel, R. (1995) Entwicklung und Erprobung eines pfleglichen Holzernteverfahrens als Mittel einer nachhaltigen Tropenwaldbewirtschaftung. Project report, Monte Dourado/Jari, Para, Brazil. Freiburg, University of Freiburg, Inst. Forest Utilisation and Forestry Work Science, D-7800 Freiburg, 26 pp.

Grayson, A.J. (ed.) (1995) *The World's Forests: International Initiatives Since Rio*. Commonwealth Forestry Association, Oxford.

Greenpeace (1994) *Principles and Guidelines Towards Ecologically Responsible Forest Use*. Greenpeace International.

Grimes, A., Loomis, S., Jahnige, P. *et al.* (1994) Valuing the rainforest: the economic value of non-timber forest products in Ecuador. *Ambio* 23, 405–410.

Grosmann, W.D. (1979) A model for the interactions of forest and environment. In: Adisoemarto, S. and Bruenig, E.F. (eds) *Transactions of the 2nd International MAB–IUFRO Workshop on Tropical Rainforest Ecosystem Research, Jakarta, 21–25 October 1978*. Chair for World Forestry, Hamburg, pp. 166–185.

Grossmann, W.D. and Watt, K.E.F. (1992) Viability and sustainability of civilizations, corporations, institutions and ecological systems. *Systems Research* 1, 3–41.

Grut, M. (1989) Economics of managing the African rainforest. Paper presented to the 13th Commonwealth Forestry Conference, September 1989, Rotorua, New Zealand.

Guerreiro, A.J. (1988) Swidden agriculturists and planned change. *Borneo Research Bulletin* 20, 3–14.

Hadisuparto, H. (1993) The effects of timber harvesting and conversion on peatswamp forest dynamics and environment in West Kalimantan. In: *Conference on Tropical Rainforest Research, Current Issues, Brunei, 9–17 April 1993* (in press).

Hadley, M. (ed.) (1988) *Rainforest Regeneration and Management.* Report of a workshop, Guri, Venezuela, 1986. Biology International, special issue 18. IUCN, Paris.

Hadley, M. and Schreckenberg, K. (1989) *Contribution to Sustained Resource Use in the Humid and Sub-humid Tropics: some Research Approaches and Insights.* UNESCO, Paris.

Haeckel, E. (1866) *Generelle Morphologie der Organismen.* Reimer Verlag, Berlin.

Hahn-Schilling, B. (1988) [*Dryobalanops aromatica* Gaertn.f. and *Shorea parvifolia* Dyer. A silvicultural analysis of planted dipterocarp stands on the Malaysian Peninsula]. Thesis, University of Goettingen.

Hahn-Schilling, B. (1994) Struktur, sukzessionale Entwicklung und Bewirtschaftung selektiv genutzter Moorwälder in Malaysia. PhD Thesis, University of Göttingen.

Hahn-Schilling, B., Heuveldop, J. and Palmer, J. (1994) *A Comparative Study of Evaluation Systems for Sustainable Forest Management (including Principles, Criteria and Indicators).* BFH, Hamburg, Institute Silviculture, University Goettingen, OFI, Oxford.

Haig, I.T., Huberman, M.A. and U Aung Din (1958) *Tropical Silviculture,* vols I and II. FAO, Rome.

Halenda, C. (1985) Site protection and rehabilitation by shifting cultivators in the Julau, Pakan and Kapit areas. Report on a field survey. Forest Department Sarawak, R.O. (047) 62 (VIII) – 1, 2 August 1985. pp. 4.

Halenda, C. (1988a) *The Ecology of an* Acacia mangium *Plantation Established after Shifting Cultivation in Niah Forest Reserve.* Forest Department Sarawak, Kuching.

Halenda, C. (1988b) *The Ecology of a* Leucaena leucocephala *Plantation Established after Shifting Cultivation in Niah Forest Reserve.* Forest Department Sarawak, Kuching.

Halenda, C. (1988c) *The Ecology of a* Gruelina arborea *Plantation Established after Shifting Cultivation in Niah Forest Reserve.* Forest Department Sarawak, Kuching.

Halenda, C. (1989) *The Ecology of Fallow Forest after Shifting Cultivation in Niah Forest Reserve.* Forest Department Sarawak, Kuching.

Hall, P. (1994) Structure stand dynamics and species compositional change in three mixed dipterocarp forests of northwest Borneo. PhD thesis, Boston University.

Hall, P. and Bawa, K. (1993) Methods to assess the impact of extraction of non-timber tropical forest products on plant populations. *Economic Botany* 47, 234–247.

Haller, K.E. (1969) The determination of the variable round-wood potential from measurable inventory data as a new way of quality assessment in tropical forest inventories, demonstrated on the species *Shorea albida.* In: Panzer, K.F. (ed.) [*Problems of the Measurement, Estimation and Evaluation Methods of Tropical-forests Inventories*]. IUFRO Section 25. Mitt. Bundesforsch. anst. Forst-Holzwirtsch. no. 74, Hamburg, pp. 95–141.

Hamid Bugo (1984) *The Economic Development of Sarawak. The Effects of Export Instability.* Summer Times, Singapore.

Hamilton, A.C. (1982) *Environmental History of East Africa*. Academic Press, London.

Hamilton, A.C. and Taylor, D. (1991) History of climate and forests in tropical Africa during the last 8 million years. *Climate Change* 19, 65–70.

Hamilton, L.S. (1983) Removing some of the myths and misses about the soil and the water impacts of tropical forest land uses. International Conf. Soil Erosion and Conservation, East–West Centre, Honolulu.

Hamrick, J.L. and Murawski, D.A. (1991) Levels of allozyme diversity in populations of uncommon neotropical tree species. *Journal of Tropical Ecology* 7, 395–399.

Hartig, G.L. (1795) Anweisung zur Taxavion der Forste, oder zur Bestirumung des Holzertrages der Wälder. Giessen.

Hartig, G.L. (1820) *Lehrbuch für Förster und die es werden wollen*, 6th edn. J.G. Cotta'sche Buchhandlung, Stuttgart and Tübingen.

Hasel, K. (1985) *Forstgeschichte*. Paul Parey, Hamburg.

He, S.Y. and Yu, Z. (1984) The studies on the reconstruction of vegetation in tropical coastal eroded land in Guandong. *Tropical and Subtropical Forest Ecosystem* 2, 87–90.

Heany, L.H. (1991) A synopsis of climatic and vegetational change in Southeast Asia. *Climate Change* 19, 53–61.

Hendersen-Sellers, A. and Robinson, P.J. (1988) *Contemporary Climatology*. Longman, Harlow.

Heske, F. (1931a) Problems of forestry in less developed countries as a subject of teaching and research. Lecture at the opening of the Institute of Foreign and Colonial Forestry, Tharandt, September 1931. Translated, edited and revised by Bruenig, E.F. (1992). In: *60 Jahre Weltforstwirtschaft*. Mitt. Bundesforsch.Anst. Forst-Holzwirtsch., no. 170, Hamburg.

Heske, F. (1931b) [The establishment of sustainable forest estate in less developed countries, the major forestry problem of the 20th century]. Translated, edited and revised by Bruenig, E.F. (1992). In: *60 Jahre Weltforstwirtschaft*. Mitt. Bundesforsch.Anst. Forst-Holzwirtsch., no. 170, Hamburg.

Heske, F. (1932) Aus den britischen Besitzungen in Asien, Malaiische Staaten. *Zielschrift für Weltforstwirtschaft* 1, 756–757.

Heske, F. (1937/38) Ziele und Wege der tropischen Waldwirtschaft. *Zeilschrift für Weltforstwirtschaft* V, 133–148.

Heske, F. (1942) Ziele, Aufgaben und Organisation moderner kolonialforstlicher Forschung. *Kolonialforstliche Mitteilungen* IV, 345–383.

Hesmer, H. (1966, 1970) *Der kombinierte land- und forstwirtschaftliche Anbau. I. Tropisches Afrika. II. Tropisches und subtropisches Amerika*. Schriftenreihe BM Wirtsch, Stuttgart.

Hesmer, H. (1975) *Leben und Werk von Dietrich Brandis, 1824–1907*. Westdeutscher Verlag, Opladen.

Hesmer, H. (1986) *Einwirkungen der Menschen auf die Wälder der Tropen*. Westdeutscher Verlag, Opladen.

Heuveldop, J. (1978) The International Amazon Rainforest Ecosystem MAB pilot project at the San Carlos de Rio Negro: micrometeorological studies. In: Adisoemarto, S. and Bruenig, E.F. (eds) *Transactions of the 2nd International MAB–IUFRO Workshop on Tropical Rainforest Ecosystem Research*. Chair of World Forestry, Hamburg-Reinbek, p. 18.

Heuveldop, J. (1980) Bioklima von San Carlos de Rio Negro, Venezuela. *Amazoniana*, VII, 7–17.

Heuveldop, J. (1994) *Assessment of Sustainable Tropical Forest Management.* Mitt. Bundesforsch.Anst. Forst-Holzwirtsch., no. 178, Hamburg.

Heuveldop, J. and Neumann, M. (1980) Structure and function of a rainforest in the International Amazon Ecosystem Project: preliminary data on growth rates and natural regeneration from a pilot study. *Turrialba* 33, 25–38.

Heuveldop, J., Kriebitzch, W.U. and Schneider, T.W. (1993) *Potentiale der Kohlenstoffixierung durch Ausweitung der Waldflächen als Manahme zur Klimastabilisierung.* Report for the Enquête Commission of the Federal German Parliament. Bundesforschungsanstaltfür Forst-und Holzwirtschaft, Institut für Weltforstwirtschaft, Hamburg.

Heyligers, P.C. (1963) *Vegetation and Soil of a White-sand Savanna in Surinam.* North Holland Publisher, Amsterdam.

Heywood, V.H. and Stuart, S.N. (1992) Species extinction in tropical forests. In: Whitmore, T.C. and Sayer, J.A. (eds) *Tropical Deforestation and Species Extinction.* Chapman & Hall, London, pp. 91–117.

Hilton, G. (1985) Nutrients in rainwater, do they make a significant contribution? *Klinkii,* 3, 100–109.

Hodnett, M.G., Silva, L.P. da, Senna, R.C. and Rocha, H. (1992) A comparison of dry season soil water depletion beneath Amazonian pasture and rainforest. Cited in Shuttleworth and Nobre (1992).

Högberg, P. and Alexander, I.J. (1995) Roles of root symbioses in African woodland and forest: evidence from ^{15}N abundance and foliar analysis. *Ecology* 83, 217–224.

Holm-Nielsen, L.B., Nielsen, I.C. and Balslev, H. (eds) (1989) *Tropical Forests. Botanical Dynamics, Speciation and Diversity.* Academic Press, London.

Hong, E. (1987) *Natives of Sarawak – Survival in Borneo's Vanishing Forests.* Inst. Masyarakat, Penang.

Hornung, M. and Skeffington, R.A. (eds) (1993) *Critical Loads, Concepts and Application.* HMSO, London.

Hose, C. (1926) *Natural Man. A Record from Borneo.* Macmillan, London.

Hose, C. (1927) *Fifty Years of Romance and Research in Borneo.* Hutchinson, London.

Houghton, R.A. and Hackler, J.L. (1993) The net flux of carbon from deforestation in South and Southeast Asia. In: Dale, V. (ed.) *Effects of Land Use Change in Atmospheric CO_2 Concentrations: Southeast Asia as a Case Study.* Springer-Verlag, Berlin.

Houghton, R.A. and Woodwell, G.M. (1989) Global climate change. *Scientific American* 260, 36–44.

Houghton, R.A., Lefkowitz, D.S. and Skole, D.L. (1991) Changes in the landscape of Latin America between 1850 and 1985. I. Progressive loss of forests. II. Net release of CO_2 to the atmosphere. *Forest Ecology and Management* 38, 143–172; 173–199.

Howe, H.F. (1990) Seed dispersal by birds and mammals: implications for seedling demography. In: Bawa, K.S. and Hadley, M. (eds) *Reproductive Ecology of Tropical Forest Plants.* Parthenon, Carnforth and UNESCO, Paris.

Howlett, D. and Sargent, C. (1991) *Technical Workshop to Explore Options for Global Forestry Management, Bangkok, 1991.* IIED, London.

Huang, Y.W. and Bruenig, E.F. (1987) Species richness, diversity, structure and pattern changes along ecological gradients. I. Preliminary results of the Borneo:Hainan comparison. 14th International Botanical Congress, Abstracts, Symposia 6–11, Tropical Forests, p. 351.

Hubbel, S.P. and Foster, R.B. (1986) Canopy gaps and the dynamics of a neotropical forest. In: Crawley M.J. (ed.) *Plant Ecology*. Blackwell, Oxford, pp. 77–96.

Huggett, R. (1980) *System Analysis in Geography*. Clarendon Press, Oxford.

Humboldt, A.V. (1847) *Kosmos*. [*Design of a physical description of the world*], vols 1–5. Cotta's Publ., Stuttgart and Tübingen.

Hundeshagen, J. Ch. (1828) *Encyclopädie der Forstwissenschaften*, 2nd edn. H. Laupp, Tübingen.

Hutchinson, I.D. (1977) *Study to Establish Interim Guidelines for Silviculture and Forest Management of the Mixed Dipterocarp Forest in Sarawak*. FAO, Rome.

Hutchinson, I.D. (1981) *Liberation Thinnings: a Tool in the Management of Mixed Dipterocarp Forest in Sarawak*. FO:MAL/76/008, Field Document no. 5.

Hutchinson, I.D. (1991) Diagnostic sampling to orient silviculture and management in natural tropical forest. *Commonwealth Forestry Review* 70, 113–132.

Hyde, W.F., Newman, D.H. and Sedjo, R.A. (1991) *Forest Economics and Policy Analysis. An Overview*. World Bank Discussion Papers, 134.

IIED (1993) *The Economic Linkage between the International Trade in Tropical Timber and the Sustainable Management of Tropical Forests*. International Institute for Environment and Development, London.

IMBAS (1988) *Indonesien: Irrweg Transmigrasi*. Albatros Verlag, Frankfurt.

Inoue, E. (1963) On the turbulent structure of airflow within crop canopies. *Journal of the Meteorological Society of Japan* 46, 817–325.

Inoue, T. and Abg. Abdul Hamid (eds) (1994) *Plant Reproductive Systems and Animal Seasonal Dynamics. Long-term Study of Dipterocarp Forests in Sarawak*. Center for Ecological Research, Kyoto University.

Insam, H. (1990) *Regeneration of a Degraded Forest Soil in Tropical China: Organic Matter and Soil Microbial Processes*. Project report TS2 0174–D(AM). LFA, Institut für Bodenbiologie, Braunschweig.

INTECOL Newsletter (1991–1995) Vols 21–25. Savannah River Ecology Laboratory, University of Georgia, Aiken, SC 29802, USA.

International Trade Centre UNCTAD/GATT (1990) *Wooden Household Furniture, a Study of Major Markets*. ITTO, Yokohama, 449 pp.

IPCC (1992) *Prediction of the Regional Distributions of Climate Change and Associated Impact Studies, Including Model Validation Studies*. IPCC Secretariat, Geneva.

Ismail Bin Haji Ali (1966) A critical review of Malayan silviculture in the light of changing demand and form of timber utilisation. *Malay Forester* 29, 228–233.

ITTO (1990a) *The Promotion of Sustainable Forest Management: a Case Study in Sarawak, Malaysia*. Report by ITTO Mission, Earl of Cranbrook (ed.) ITTC, Denpasar, Indonesia.

ITTO (1990b) *ITTO Guidelines for the Sustainable Management of Natural Tropical Forests*. ITTO, Yokohama.

ITTO (1990c) *Criteria and Priority Areas for Programme Development and Project Work*. ITTO, Yokohama.

ITTO (1990d) Project report PD 74/90. ITTO, Yokohama, unpublished.

ITTO (1990–1995) *Tropical Forest Update*, vols 1–5. ITTO, Yokohama.

ITTO (1991) *Beyond the Guidelines – an Action Program for Sustainable Management of Tropical Forests*. ITTO, Yokohama.

ITTO (1992a) *ITTO Guidelines for the Establishment and Sustainable Management of Planted Tropical Forests*. ITTO, Yokohama.

ITTO (1992b) *Criteria for the Measurement of Sustainable Tropical Forest Management*. ITTO, Yokohama.

ITTO (1993a) *ITTO Guidelines on the Conservation of Biological Diversity in Tropical Production Forests.* ITTO, Yokohama.

ITTO (1993b) *Rehabilitation of Logged-over Forests in Asia/Pacific Region.* ITTO, Yokohama.

ITTO (1993c) *Analysis of Macro-economic Trends in the Supply and Demand of Sustainably Produced Timber from the Asia-Pacific Region. Report by Reid, Collins Forest Consultants,* Bradshaw, K.P. and McLellan, J. (eds). ITTO, Yokohama.

ITTO (1993d) *The Economic and Environmental Values of Mangrove Forests and their Present State of Conservation.* ITTO, Yokohama.

ITW (1994a) *Criteria and Indicators for Sustainable Management of Tropical Forests.* Initiative Tropenwald, Berlin.

ITW (1994b) *Sustainable Management of Forests, a Challenge and Chance for the Timber Industry.* Initiative Tropenwald, Berlin.

ITW (1994c) *Timber Certification Proposals and Propositions.* Initiative Tropenwald, Berlin.

ITW (1994d) *The Need for European Harmonization.* Initiative Tropenwald, Berlin.

IUCN (1988/1995) (1) *IUCN Tropical Forest Programme Newsletter* 1, 1988 and following issues. (2) *IUCN Forest Conservation Programme Newsletter* no. 1–21 (April 1995). World Conservation Union, CH 1196 Gland.

Jacobs, M. (1987) *The Tropical Rainforest, a First Encounter.* Springer, Berlin.

Jayasuriya, A.H.M. (1995) National conservation review: the discovery of extinct plants in Sri Lanka. *Ambio* 24(5), 313–316.

Jentsch, F. (1933/1934) Boden und Wald in warmen Ländern. *Zeitschrift für Weltforstwirtschaft* I, 529–535.

Jermy, A.C. and Kavanach, K.P. (1982) *Gunung Mulu National Park, Sarawak. An Account of its Environment and Biota being the Results of the Royal Geographical Society/Sarawak Government Expedition and Survey 1977–1978.* part I. Sarawak Museum Journal, vol. XXX no. 51.

Jetten, V.G. (1994) *Modelling the Effects of Logging on the Water Balance of a Tropical Rainforest. A Study in Guyana.* Tropenbos Foundation, Wageningen.

Jia, C.Y. and Insam, H. (1991) Microbial biomass and relative contributions of bacteria and fungi in soil beneath tropical rainforest, Hainan Island, China. *Journal of Tropical Ecology* 7, 385–393.

JOFCA (1993) *Rehabilitation of Logged-over Forests in Asia/Pacific region.* Japanese Overseas Forestry Consultants Association for ITTO, Yokohama.

Johansson, P.-O. (1993) *Cost–Benefit Analysis of Environmental Change.* Cambridge University Press, Cambridge.

John, S. St (1862) *Life in the Forests of the Far East,* vols I and II. Smith, Elder and Co. London.

John, S. St (1879) *The Life of Sir James Brooke, Rajah of Sarawak.* W. Blackwood, London.

John V. Ward Associates (1990) *The Japanese Market for Tropical Timber: an Assessment for The International Tropical Timber Organisation.* ITTO, Yokohama.

Johns, A.D. (1988) Effects of selective timber extraction on rainforest structure and composition and some consequences for frugivores and folivores. *Biotropica* 20, 31–37.

Johns, A.D. (1989) *Timber, the Environment and Wildlife in Malaysian Rain Forests: Final Report.* Institute of South East Asian Biology, University of Aberdeen, Scotland.

Johns, A.D. (1992) Species conservation in managed tropical forests. In: Whitmore, T.C. and Sayer, J.A. (eds) *Tropical Deforestation and Species Extinction.* Chapman & Hall, London, pp. 15–53.

Johns, R.J. (1992) The influence of deforestation and selective logging operations on plant diversity in Papua New Guinea. In: Whitmore, T.C. and Sayer, J.A. (eds) *Tropical Deforestation and Species Extinction.* Chapman & Hall, London, pp. 143–147.

Johnson, N. and Cabarle, B. (1993) *Surviving the Cut: Natural Forest Management in the Humid Tropics.* World Resources Institute, Washington D.C.

Jonkers, W.B.J. (1982) *Options for Silviculture and Management of Mixed Dipterocarp Forest.* Forest Development Project Sarawak, FO: MAL/76/008, working paper no. 11.

Jonkers, W.B.J. (1987) *Vegetation Structure, Logging Damage and Silviculture in a Tropical Rain Forest in Surinam.* Agricultural University, Wageningen, The Netherlands.

Jordan, C.F. (1985) *Nutrient Cycling in Tropical Forest Ecosystems.* John Wiley, Chichester.

Jordan, C.F. (ed.) (1989) *An Amazonian Rainforest. The Structure and Function of a Nutrient-stressed Ecosystem and the Impact of Slash and Burn Agriculture.* UNESCO, Paris and Parthenon, Carnforth.

Jordan, C., Golley, F., Hall, J. and Hall, P. (1980) Nutrient scavenging of rainfall by the canopy of an Amazonian rainforest. *Biotropica* 12, 61–66.

Kallio, M., Dykstra, D.P. and Binkley, C.S. (1987) *The Global Forest Sector.* Wiley-Interscience Publication, London.

Kamis Awang, Lee, S.S., Lai, F.S., Abdul Rhaman, M.D. and Ali Abod, S. (eds) (1982) *Ecological Basis for Rational Resource Utilization in the Humid Tropics of Southeast Asia.* University Pertanian Malaysia, Serdang.

Kato, M. (1994) Plant–pollinator interactions at the forest floor of a lowland mixed dipterocarp forest in Sarawak. In: Abdul Hamid and Inoue, T. (eds) *Plant Reproductive Systems and Animal Seasonal Dynamics.* Kyoto University, pp. 151–157.

Kato, R., Tadaki, Y. and Ogawa, H. (1978) Plant biomass and growth increment studies in the Pasoh forest. *Malayan Natural Journal* 30, 211–224.

Kedit, P.M. (1982) An ecological survey of the Penan. *Sarawak Museum Journal* 51 (NS), 226–279.

Kehlenbeck, U. (1993) Kennzeichnung des Wurzelraumes tropischer Waldökosysteme in Südchina. (Characterization of the rooting sphere of tropical forest ecosystems in South China.) DSc Thesis, Faculty Geosciences, Hamburg University.

Kehlenbeck, U. (1994) [*Soil Analyses and Mapping in Sabal F.R., Sarawak, Malaysia*]. Chair of World Forestry, Hamburg University.

Kemp, R.H. (1992) *Commonwealth Forestry. A Proposed International Initiative.* Commonwealth Secretariat, Marlborough House, Pall Mall, London SW1Y 5HX.

Kendawang, J.J. (1992) Foresters' perception of indigenous species for forest plantations in Sarawak. In: Ahmad Said Sajap *et al.* (eds) *Indigenous Species for Forest Plantations.* University Pertanian Malaysia, Serdang, pp. 14–23.

Kenya Indigenous Forest Conservation Programme (1993) *Phase 1 Report.* KIF-CON, Nairobi.

Kerr, B. (1993) Iwokrama: the Commonwealth Rainforest programme in Guyana.

Commonwealth Forestry Review 72, 303–309.

Khen, C.V., Speight, M.R. and Holloway, J.D. (1992) Comparison of biodiversity between rainforest and plantations in Sabah, using insects as indicators. In: Miller, F.R. and Adam, K.L. (eds) *Wise Management of Tropical Forests 1992.* Oxford Forestry Institute and University of Oxford, Oxford, pp. 221–224.

Kiew, R. (ed.) (1991) *The State of Nature Conservation in Malaysia.* Malayan Nature Society, Kuala Lumpur.

Kikkawa, J. (1982) Ecological association of birds and vegetation structure in wet tropical forests of Queensland. *Australian Journal of Ecology* 7, 325–345.

Killham, K. (1994) *Soil Ecology,* 2nd edn. Cambridge University Press, Cambridge, 1995.

Killmann, W., Sickinger, T. and Hong, L.T. (1994) *Restoration and Reconstruction of Traditional Malay Kampong Houses.* Forest Research Institute of Malaysia (FRIM), Kepong, Kuala Lumpur, 103 pp.

Kimmins, J.P. (1988) Community organization: methods of study and prediction of the productivity and yield of forest ecosystems. *Canadian Journal of Botany* 66, 2654–2672.

Kimmins, J.P. (1990) A strategy for research on the maintenance of long-term site productivity. In: Proceedings XIX IUFRO World Congress, Div. 1, vol. 1. IUFRO, Montreal.

Kio, P.R.O. (1983) Management potentials of the tropical high forest with special reference to Nigeria. In: Sutton, S.K., Whitmore, T.C. and Chadwick, A.C. (eds) *Tropical Rainforest: Ecology and Management.* Blackwell Scientific Publications, Oxford, pp. 445–463.

Kio, P.R.O. and Ekwebelan, S.A. (1987) Plantations versus natural forests for meeting Nigeria's wood needs. In: Mergen, F. and Vincent, J.R. (eds) *Natural Management of Tropical Moist Forests.* Yale University, New Haven.

Kira, T. (1978) Primary productivity of Pasoh forest – a synthesis. *Malayan Nature Journal* 30, 291–297.

Kira, T. (1995) Forest ecosystems of East and Southeast Asia in a global perspective. In: Box, E.O. *et al.* (eds) *Vegetation Science in Forestry.* Kluwer Academic Publ., Dordrecht, pp. 1–21.

Kira, T., Ogawa, H., Yoda, K. and Ogino, K. (1967) Comparative ecological studies on three main types of forest vegetation in Thailand. IV. Dry matter production, with special reference to the Khao Chong rainforest. *Nature and Life in Southeast Asia* 5, 149–174.

Kleine, M. and Heuveldop, J. (1993) A management planning concept for sustained yield of tropical forests in Sabah, Malaysia. *Forest Ecology and Management* 61, 277–297.

Kleine, M. and Weinland, G. (1994) Natural forest management in Malaysia. *Tropical Forest Update* 4, 22.

Klinge, H. (1969) *Report on Tropical Podzols.* FAO, Rome.

Klinge, H. and Herrera, R. (1978) Biomass studies in Amazon Caatinga forest in Southern Venezuela. I. Standing crop of composite root mass in selected stands. *Tropical Ecology* 19, 93–110.

Klinge, H., Rodrigues, W.E., Fittkau, E.J. and Bruenig, E.F. (1974) Biomass and structure in a Central Amazonian Forest. In: Medina, E. and Golley, F. (eds) *Ecological Studies, vol. II. Tropical Ecological Systems.* Springer, New York, pp. 115–122.

Knoerr, K.R. and Gay, L.W. (1965) Tree heat energy balance. *Ecology* 46, 17–24.

Kochummen, K.M., Lafrankie, J.V. and Manokaran, N. (1991) Floristic composition of Pasoh Forest Reserve, a lowland rainforest in Peninsular Malaysia. *Journal of Tropical Forest Science* 3, 1–13.

Kochummen, K.M., Lafrankie, J.V. and Manokaran, N. (1992) Floristic composition of Pasoh Forest Reserve, a lowland rainforest in Peninsular Malaysia. In: Yap S.K. and Lee, S.W. *In Harmony with Nature*. Malayan Nature Society, Kuala Lumpur, pp. 545–554.

Koerner (1993) Discussion intervention. 14th Commonwealth Forestry Conference, Kuala Lumpur, September 1993.

Kohyama, T. (1993) Size-structured tree populations in gap-dynamic forest – the forest architecture hypothesis for the stable coexistence of species. *Journal of Ecology* 81, 131–143.

Kollert, W., Norini Haron and Ismariah Ahmad (1993) A conceptual approach for the evaluation of forest plantations of high quality species established on degraded logged-over forest areas. Kepong, FRIM, CFFPR, abstracts p. 12.

Korsgaard, S. (1985) *Guidelines for Sustained Yield Management of Mixed Dipterocarp Forests of South East Asia*. GCP/PAS/106/JPN, Field Document no. 8, Bangkok.

Korsgaard, S. (1988a) *An Analysis of the Potential for Timber Production under Conservation Management in the Tropical Rainforest of South East Asia*. Research Council for Development Research, Copenhagen.

Korsgaard, S. (1988b) *A Manual for the Stand Table Simulation Model*. IUFRO Conference Growth and Yield in Tropical Mixed Moist Forests, Kuala Lumpur and Research Council for Development Research, Copenhagen.

Korsgaard, S. (1992) *An Analysis of Growth Parameters and Timber Yield Prediction, based on Research Plots in the Permanent Forest Estate of Sarawak, Malaysia*. Council for Development Research, Copenhagen.

Korsgaard, S. (1993a) *Methodology for Forest Management Site Productivity Classification as Applied to Individual Compartments*. Field Document no. 5, Forestry Department HQ, Kuala Lumpur.

Korsgaard, S. (1993b) *Recommendation for Management and Silviculture of the Permanent Forest Estate of Peninsular Malaysia*. Field Document no. 7, Forestry Department HQ, Kuala Lumpur.

Korsgaard, S. (1993c) *Recommendations for Forest Management Data Collection*. Forest Inventory and Management Systems as Part of Forest Resources Conservation Programme, Technical Note no. 11.2, Forest Department HQ, Kuala Lumpur, in prep.

Korsgaard, S. (1993d) Briefing notes, visit to ITTO project study area compartment 50, Sungai Lalang Forest Reserve, 17 May 1993.

Kotschwar, A. (1949) Ursprung und internationale Bedeutung des deutschen forstlichen Nachhaltsgedankens. PhD thesis, Faculty Biology, University of Hamburg.

Kriebitzsch, W.V., Schneider, T.W. and Neuveldop, J. (1993) Aufforstungen sind keine Lösung zur Klimastabilisierung. *Holz-Zentralblatt*. 119, 1541–1542.

Kummer, D.M. (1992) *Deforestation in the Postwar Philippines*. Ateneo de Manila University Press and University of Chicago Press.

Kung, E. (1961) Derivations of roughness parameter from wind profile data. Annual Report, DA-36-039-SC-80282, University of Wisconsin, pp. 27–61.

Kuntz, S., Siegert, F. and Streck, C. (1995) Tropical rainforest and use of land investigation (TRULI). *TREES ERS – 1, 1994 study final workshop*, IRC, ISPRA, Italy, 23–24 February 1995.

Kurz, W.A. (1983) Biomasse eines amazonischen immergrünen Feuchtwaldes: Entwicklung von Biomasseregressionen. Thesis, Hamburg University.

Kurz, W. and Kimmins, J.P. (1987) Analysis of some sources of error in methods used to determine fine root production in forest ecosystems: a simulation approach. *Canadian Journal of Forestry Research* 17, 909–912.

Lagan, P.M. and Glauner, R. (1994) *A Guide to the Soils of Sabah.* Library of the Forest Department Sabah, Forest Research Centre Sepilok, Sandakan.

Lai, F.S. (1993) *Sediment and Solute Yield of a Hill Forest Catchment in Selangor, Peninsular Malaysia.* UBD/RGS Conference Tropical Rainforest Research, Current Issues, Brunei, UBD.

Lai, K.K. (1976) *Performance of planted* Shorea albida. Forest Research Report 14. Forest Department Sarawak, Kuching.

Lai, K.K. (1978) *Line planting trials of* Shorea albida. Forest Research Report SR21, Forest Department Sarawak, Kuching.

Langub, J. (1991) Penan responses to change and development. Conference Interactions of People and Forests in Kalimantan, New York Botanical Garden, 21–23 June 1991.

Lamb, D. (1990) *Exploiting the Tropical Rain Forest: An Account of Pulpwood Logging in Papua New Guinea.* Parthenon, Carnforth and UNESCO, Paris.

Lambert, F. (1990) Long-term effects of selective logging operations on the Malaysian avifauna. *University of Aberdeen Tropical Biological Newsletter* no. 59, 1.

Lamprecht, H. (1986) *Waldbau in den Tropen.* Paul Parey, Hamburg.

Lamprecht, H. (1989) *Silviculture in the Tropics.* GTZ, Eschborn.

Lanly, J.P. (1982) *Tropical Forest Resources.* FAO Forestry Paper no. 30. FAO, Rome.

Lawton, J.H. and May, R.M. (eds) (1995) *Extinction Rates.* Oxford University Press, Oxford.

Leaky, R.B. and Newton, A.C. (1994) *Domestication of Tropical Trees for Timber and Non-timber Products.* UNESCO, Paris.

Lee, H.S. (1979) Natural regeneration and reforestation in the peatswamp forests of Sarawak. *Tropical Agricultural Series* no. 12. Tropical Agriculture Research Centre, Ministry of Agriculture, Forestry and Fisheries, Tokyo, pp. 51–60.

Lee, H.S. (1981) *The Ecological Effects of Shifting Cultivation on Tropical Forest Ecosystems and their Significance on Reforestation and Rehabilitation Efforts in Sarawak.* Forest Research Report SR 22. Forest Department Sarawak, Kuching.

Lee, H.S. (1982) The development of silvicultural systems in the hill forests of Malaysia. *Malay Forestry* 45, 1–9.

Lee, H.S. (1991) Utilization or conservation of peatswamp forests in Sarawak. In: Aminuddin, B.Y. *et al.* (eds) *Tropical Peat.* MARDI, Kuala Lumpur, pp. 286–293.

Lee, H.S. and Lai, K.K. (1977) *A Manual of Silviculture for the Permanent Forest Estate of Sarawak.* Forest Department, Sarawak.

Lee, R. and Gates, D.M. (1964) Diffusion resistance in leaves as related to their stomatal anatomy and microstructure. *American Journal of Botany* 51, 963–975.

Lee, Y.F., Chung, A.Y.C., Swaine, M.D. and Jong, K. (1993) Insect visitors to flowers of an indigenous Bornean rattan, *Calamus subinermis. Tropical Biology Newsletter, University Aberdeen,* 65, 2–3.

Lemon, E.L., Allen, H. and Mueller, L. (1969) Photosynthesis in a tropical rainforest. In: *Vertical Carbon Dioxide Fluxes.* USDA, Agric. Res. Service, Soil and Water Conserv. Res. Div., Northeast Branch, Ann. Progr. Rep., 1968, Ithaca.

Leslie, A.J. (1977) Where contradictory theory and practice co-exist. *Unasylva* 29, 115.

Leslie, A.J. (1987) A second look at the economics of natural management systems in tropical mixed forests. *Unasylva* 39, 46–58.

Levi-Strauss, C. (1978) *Traurige Tropen.* Suhrkamp, Frankfurt.

Lieth, H. (1984) Biomass pools and primary productivity of natural and managed ecosystem types in a global perspective. Options. Institute Agron. Mediterranean de Zaragoza, pp. 7–14.

Lieth, H. and Whittaker, R.H. (1975) *Primary Productivity of the Biosphere.* Ecological Studies no. 14, Springer Verlag, Berlin.

Liew, T.F. and Ong, C.C. (1986) Logging damage in the Mixed Dipterocarp Forests of Sarawak. Forest Department Sarawak, Family Tree, 30 June 1986, 10, 24–30.

Livingstone, D.A. (1975) The late quarternary climatic changes in Africa. *Annual Review Ecology and Systematics* 6, 249.

Livingstone, D.A. and Hammen, T. van der (1978) Paleogeography and paleoclimatology. In: *Tropical Forest Ecosystems, a State-of-knowledge-report Prepared by UNESCO/UNEP/FAO.* UNESCO Natural Resources Research, Paris, pp. 61–90.

Loetsch, F. and Haller, K.E. (1964) *Forest Inventory,* vol. I. BLV, München.

Loetsch, F., Zoehrer, F. and Haller, K.E. (1973) *Forest Inventory,* vol. II. BLV, München.

Logan, W.E.M. (1947) Timber exploitation in the Gold Coast. *Empire Forestry Review* 26, 30–53.

Longman, K.A. and Jenik, J. (1987) *Tropical Forest and its Environment,* 2nd edn. Longman, Harlow.

Loomis, W.E. (1965) Absorption of radiant energy by leaves. *Ecology* 46, 14–17.

Lovejoy, T.E., Bierregard, R.O., Rankin, J. and Schubart, H.O.R. (1983) Ecological dynamics of tropical forest fragments. In: Sutton, S.L. Whitmore T.C. and Chadwick, A.C. (eds) *Tropical Rainforest: Ecology and Management.* Blackwell, Oxford, pp. 377–384.

Low, H. (1848) *Sarawak, its Inhabitants and Productions.* Pustaka Delta Pelajaran, Selangor.

Lowman, M.D. (1992) Leaf growth dynamics and herbivory in five species of Australian rainforest canopy trees. *Journal of Ecology* 80, 433–447.

Lubchenco, J., Olson, A.M., Brubaker, L.B., Carpenter, S.R., Holland, M.M., Hubbell, S.P., Levin, S.A., MacMahon, J.A., Matson, P.A., Melillo, J.M., Mooney, H.A., Peterson, C.H., Pulliam, H.R., Real, L.A., Regal, P.J. and Risser, P.G. (1991) The Sustainable Biosphere Initiative: an ecological research agenda. *Ecology* 72, 371–412.

Lugo, A.E., Parrotta, J.A. and Brown, S. (1993) Loss in species caused by tropical deforestation and their recovery through management. *Ambio* 22, 106–109.

Lundgren, B. (1979) Research strategy for soils in agroforestry. In: Mongi, H.O. and Huxley, D.A. (eds) *Soils Research in Agroforestry.* ICRAF, Nairobi, pp. 523–538.

MacArthur, R.H. (1972) *Geographical Ecology.* Harper & Row, New York.

McIntosh, R.P. (1967) An index of diversity and the relation of certain concepts of diversity. *Ecology* 48, 392–403.

Mackie, C. (1984) The lessons behind East Kalimantan's forest fires. *Borneo Research Bulletin* 16, 62–75.

MacKinnon, D.S. and Freedman, B. (1993) Effects of silvicultural use of the herbicide glyphosate on breeding birds of regenerating clearcutts in Nova

Scotia, Canada. *Journal of Applied Ecology* 30, 395–406.

Makey, B.G. (1992) A spatial analysis of the environmental relations of rainforest structural types. In: Stott, P. (ed.) *Global Ecology and Biogeography Letters* Vol. 3, 1993.

Malmer, A. and Grip, H.P. (1992) Soil disturbance and loss of infiltrability caused by media, mixed and manual extraction of tropical rainforest in Sabah, Malaysia. *Forest Ecology and Management* 38, 1–12.

Manila, A.C. (1989) Growth responses and economic production of residual dipterocarp stands to timber stand improvement (TSI) treatments. PhD thesis, University of the Philippines at Los Baños.

Manokaran, N. and Kochummen, K.M. (1987) Recruitment, growth and mortality of tree species in a lowland dipterocarp forest. *Journal of Tropical Ecology* 3, 315–330.

Manokaran, N. and Kochummen, K.M. (1991) A re-examination of data on structure and floristic composition of hill and lowland dipterocarp forest in Peninsular Malaysia. *Malayan Nature Journal* 44, 61–76.

Manokaran, N. and Lafrankie, J.V. (1991) Stand structure of Pasoh Forest Reserve, a lowland rainforest in Peninsular Malaysia. *Journal of Tropical Forest Science* 3, 14–24.

Manokaran, N., Lafrankie, J.V. and Roslan Ismail (1991) Structure and composition of the Dipterocarpaceae in a lowland rainforest in Peninsular Malaysia. In: Soerianegara, I., Tjitrosomo, S.S., Umali, R.C. and Umboh, I. (eds) *Proceedings 4th Roundtable Conference on Dipterocarps.* Seamco/Biotrop, Bogor, pp. 317–331.

Manser, B. (1992) *Voices from the Rainforest. Evidences of an Endangered People.* WWF and Zytglogge Publ., Berlin (original in German).

Markl, H. (1986) *Evolution, Genetik und menschliches Verhalten. Zur Frage der wissenschaftlichen Verantwortung. [Evolution, Genetics and Human Behaviour. On the Question of Responsibility of Science.]*. München, Zürich, Serie Piper.

Marsh, C. (1991) *Report on Forest Management Study Visit to East Kalimantan, 1991.* Innoprise, Kota Kinabalu, Forestry Information Paper no. 36.

Marsh, S. and Gait, B. (1988) Effects of logging on rural communities. *Sabah Society Journal* XIII, 394–430.

Marshall, A.G. and Swaine, M.D. (1992) Tropical rainforest: disturbance and recovery. *Philosophical Transactions of the Royal Society B* 335, 135.

Marss R.H., Thomson, J., Scott, D. and Proctor, J. (1991) Nitrogen mineralization and nitrification in Tierra firma forest and savanna soils in Ilha de Maraca, Roraima, Brazil. *Journal of Tropical Ecology* 7, 123–137.

Maskayu (1993) Bulletin of the Malaysian Timber Industries Development Board, Kuala Lumpur, no. 11/93 and following issues.

Mattson Marn, H. and Jonkers, W.B.J. (1981) *Logging Damage in Tropical High Forest.* FO: MAL/76/008, Working paper no. 5, Kuching (paper presented at the International Forestry Seminar, 1980, Kuala Lumpur).

Maurer, B.A. (1994) *Geographic Population Analysis: Tools for the Analysis of Biodiversity.* Blackwell Scientific Publications, Oxford.

Maury-Lechon, G. (1991) Comparative dynamics of tropical rainforest regeneration in French Guyana. In: Gomez-Pompa, A., Whitmore, T.C. and Hadley, M. (eds) *Rainforest Regeneration and Management.* UNESCO, Paris and Parthenon, Carnforth, pp. 285–294.

Mawdsley, N. (1993) The theory and practice of extrapolating regional and global species richness. Tropical Rainforest Research: Current Issues. Univ. Brunei Darussalam and Roy. Geogr. Soc. Joint Conf., Brunei 1993 (in press).

May, R.M. (1986) Thoughts on ecology. *Nature* 324, 514–519.

Medina, E., Sobrado, M.A. and Herrera, R. (1978) Significance of leaf orientation for temperature in Amazonian sclerophyll vegetation. *Journal of Radiation Environment and Biophysics* 15, 131–140.

Michon, G. (1993) The damar gardens: existing buffer zones adjacent to Barisan Selatan National Park. *ITTO Tropical Forest Management Update* 3, 7–8.

Mikola, P. (ed.) (1980) *Tropical Mycorrhiza Research.* Clarendon Press, Oxford.

Milliman, J.D. and Meade, R.H. (1983) World-wide delivery of river sediments to the oceans. *Journal of Geology* 91, 1, 1–21.

Ministry of Forestry Indonesia (1993) *National Masterplan for Forest Plantations. Vol. 1, Synopsis.* Ministry of Forestry, Directorate General of Reforestation and Land Rehabilitation, Jakarta.

Mitchell, B.A. (1963) Possibilities for forest plantations. *Malayan Forester* 26, 259–289.

Moeller, A. (1922) *Der Dauerwaldgedanke. Sein Sinn und seine Bedeutung. [The Sustainable (Permanent) Forest Concept. Meaning and Relevance].* Springer, Berlin.

Moeller, A. (1929) *Der Waldbau (Silviculture).* Springer, Berlin.

MOF–CIRAD (1994) *The Development of Silvicultural Techniques for the Regeneration of Logged-over Forest in East Kalimantan (STREK-project).* Agency for Forestry Research and Development, Ministry of Forestry, Jakarta and CIRAD-Forêt de France.

Mohr, E.C.L. and Baren, F.A. van (1954) *Tropical Soils.* Netherlands Roy. Trop. Inst., Amsterdam.

Mohren, G.M.J., Burg, J. van den and Burger, F.W. (1986) Phosphorus deficiency induced by nitrogen input in Douglas fir in the Netherlands. *Plant and Soil* 95, 191–200.

Mok, S.T. (1980) A conservational approach to sustained yield. In: *Proceedings ASEAN Seminar on Management of Tropical Forests, Chiangmai, Thailand.* ASEAN Institute Forest Management, Kuala Lumpur.

Monteith, J.L. (1963) Gas exchange in plant communities. In: Evans, L.T. (ed.) *Environmental Control of Plant Growth.* Academic Press, London.

Morley, R.J. (1982) A palaeoecological interpretation of a 10,000 year pollen record from Danau Padang, Central Sumatra, Indonesia. *Journal of Biogeography* 9, 151–190.

Morley, R.J. and Flenley, J.R. (1987) Late Ceinozoic vegetational and environmental changes in the Malay Archipelago. In: Whitmore, T.C. (ed.) *Biogeographical Evolution of the Malay Archipelago.* Cambridge University Press, Cambridge, pp. 50–59.

Moura-Costa, P.H. (1993) Large scale enrichment planting with dipterocarps, methods and preliminary results. In: Suzuki, K. (ed.) *Proceedings of the BIO-REFOR/IUFRO/SPDC Workshop: Bio-reforestation in Asia-Pacific Region* (in press).

Moura-Costa, P.H., Lundoh, L. and Uren, S.C. (1992) Effects of nitrogen and light on growth, development and nitrogen metabolism of *Dryobalanops lanceolata* Burck (Dipterocarpaceae) wildings grown in nursery. *Journal of Tropical Forest Science* (submitted).

Moura-Costa, P.H., Yap, S.W., Ong, C.L., Ganing, A. and Lundoh, L. (1994) Large scale enrichment planting with dipterocarps as an alternative for carbon offset-methods and preliminary results. Paper prepared for the 15th Round-Table Conference on Dipterocarps, Chiang Mai, Thailand, November 1994.

Mueller, D. and Nielson, J. (1965) Production brute, pertes par respiration et production nette dans la forêt ombrophile tropicale. *Forstliges Forsvaes Danmark* 29, 69–160.

Muller, J. (1972) Polynological evidence for change in geomorphology, climate and vegetation the Mio-Pliocene of Malesia. In: Ashton, P.S. and Ashton, M. (eds) *The Quarternary Era in Malesia*. Department of Geography, University of Hull, Misc. Ser., no. 13, pp. 6–16.

Munn, R.E. (ed.) (1981) *Carbon Cycle Modelling*. SCOPE Series no. 16. John Wiley, Chichester.

Munn, R.E. (ed.) (1984) *The Role of Terrestrial Vegetation in the Global Carbon Cycle*. SCOPE Series no. 23. John Wiley, Chichester.

Myint Zan (1994) Review of J.C. Garnett: states and state-centred perspectives. In: Baylis, J. and Rengger, N.J. (eds) *Dilemmas of World Politics. International Issues in a Changing World*. Oxford University Press, Oxford.

National Parks Office (1989) *Bako Briefing Pack*. Forest Department Sarawak, Kuching.

National Research Council (1980) *Research Priorities in Tropical Biology*. National Research Council, Washington, DC.

Nectoux, F. and Kuroda, Y. (1989) *Timber from the South Seas: An Analysis of Japan's Tropical Timber Trade and Environmental Impact*. WWF, Gland, 134 pp.

Neil, P.E. (1981) *Problems and Opportunities in Tropical Rainforest Management*. CFI Occasional Paper no. 16. University of Oxford, Oxford.

Nepstad, D.C. and Schwartzman, S. (eds) (1992) *Non-timber Products from Tropical Forests, Evaluation of a Conservation and Development Strategy*. Advances in Economic Botany, Vol. 9. The New York Botanical Garden, 164 + xii pp.

Newbery, D. McC. (1985) Analysis of spatial pattern of trees in caatinga forest at San Carlos de Rio Negro. Unpublished research report, University of Stirling.

Newbery, D. McC. (1991) Floristic variation within Kerangas (heath) forest: re-evaluation of data from Sarawak and Brunei. *Vegetatio* 96, 43–86.

Newbery, D. McC. and Proctor, J. (1984) Ecological studies in four contrasting lowland rainforests in Gunung Mulu National Park, Sarawak. IV. Associations between tree distribution and soil factors. *Journal of Ecology* 72, 475–493.

Newbery, D. McC., Renshaw, E. and Bruenig, E.F. (1986) Spatial pattern of trees in Kerangas forest, Sarawak. *Vegetatio* 65, 77–89.

Newbery, D.M., Alexander, I.J., Thomas, D.W. and Gartlan, J.S. (1988) Ectomycorrhizal rain-forest legumes and soil phosphorus in Korup National Park, Cameroon. *New Phytologist* 109, 433–450.

Newsome, J. and Flenley, J.R. (1988) Late quarternary vegetational history of the central highlands of Sumatra. II. Palaeopalynology and vegetational history. *Journal of Biogeography* 15, 555–578.

Ng, F.S.P. (1983) Ecological principles of tropical lowland rain forest conservation. In: Sutton, S.L., Whitmore, T.C. and Chadwick, A.C. (eds) *Tropical Rain Forest Ecology and Management*. Blackwell, Oxford.

Ngui, S.K. (1986) Regeneration survey of 'Ulat bulu' areas in the Baram peatswamp forests. Unpublished report, Forest Department, Sarawak.

Nichol, J.E. (1993) Remote sensing of tropical blackwater rivers: a method for environmental water quality analysis. *Applied Geography* 13, 169–188.

Nicholson, D.I. (1958) An analysis of logging damage in tropical rainforest, North Borneo. *Malay Forestry* XXI, 235–245.

Nicholson, D.I. (1979) *The Effects of Logging and Treatment on the Mixed Dipterocarp Forests of Southeast Asia.* FAO, Rome.

Niekisch, M. (1991) [ITTO, facts, critics and requests to the Federal Government.] *Newsletter, German Association of Tropical Ecology,* December 1991, 5–9.

Nik, A.R. and Yusop, Z. (1986) Stream water quality of undisturbed forest catchments in Peninsular Malaysia. Proc. Workshop on Impact of Man's Activities on Tropical Upland Forest Ecosystems. Faculty Forestry, University Pertanian Malaysia.

Noij, I.G.A.M., Jansse, B.H., Wesselink, L.G. and Van Grinsven, J.J.M. (1993) *Modelling Nutrient and Moisture Cycling in Tropical Forests.* Tropenbos Foundation, Wageningen.

North, R.D. (1995) End of the green crusade. *New Scientist* 145, 38–41.

Nussbaum, R., Anderson, J. and Spencer, T. (1993) Investigation of treatments to promote regeneration of log landings and skid trails after selective logging of rainforest in Sabah, Malaysia. In: Suzuki, K. (ed.) *Proceedings of the BIO-REFOR/IUFRO/SPDC Workshop: Bio-reforestation in Asia-Pacific Region* (in press).

Nye, J.S. (1990) The transformation of world power. *Dialogue* 4, 2–7.

Nykvist, N., Grip, H., Boon, L.S., Malmer, A. and Fui, K.W. (1994) Nutrient losses in forest plantations in Sabah, Malaysia. *Ambio* 23, 210–215.

Odum, H.T. and Pigeon, R.F. (eds) (1970) *A Tropical Rainforest.* US Atomic Energy Commission (USAEC), Washington.

OEEC (1951) *American Forest Operations and Tropical Timber Production.* OEEC, Paris.

OEEC (1952) *American Silviculture and Tropical Forest Problems.* OEEC, Paris.

Oesten, G. (1993) Anmerkungen zur Nachhaltigkeit als Leitbild fur naturverträgliches Wirtschaften. *Forstwissenschaftliches* 112, 313–319.

Ogawa, H. (1978) Litter production and carbon cycling in Pasoh forest. *Centralblatt Malayan Nature Journal* 30, 367–373.

Ogawa, H., Ogino, K., Shidei, T., Ratanawongse, D. and Apasuti, L. (1965) Comparative ecological study on three main types of forest vegetation in Thailand. I. Structure and floristic composition. II. Plant biomass. In: Kira, T. and Iwata, K. (eds) *Nature and Life in Southeast Asia* 4, 13–80.

Oldeman, R.A.A. (1989) Dynamics in tropical rainforests. In: Holm-Nielsen, L.B., Nielsen, I.C. and Balslev, H. (eds) *Tropical Forests. Botanical Dynamics, Speciation and Diversity.* Academic Press, London, pp. 3–22.

Oldeman, R.A.A. (1990) *Forests: Element of Silvology.* Springer, Heidelberg.

Oldfield, S. (1988) *Buffer Zone Management in Tropical Moist Forests: Case Studies and Guidelines.* IUCN Tropical Forest Programme.

Olivieri, S.T. and Backus, E.H. (1992) Geographic information systems (GIS) applications in biological conservation. *Biology International.,* no. 25, 10–16.

Ong, R.C. and Kleine, M. (1995) DIPSIM: Dipterocarp forest growth simulation model. In: *Dipterocarp Ecosystem Ecology, Sustainable Management and Products.* Indonesian German Forestry Project, Samarinda (in press).

Orgle, T.K., Swaine, M.D. and Thompson, J. (1994) History of drought and forest fire in Ghana. *University of Aberdeen, Tropical Biology Newsletter* 66; 67, 5.

Osho, J.S.A. (1995) Optimal sustainable harvest models for a Nigerian Tropical Rainforest. *Journal of Environmental Management* 45, 101–108.

Oxford Forestry Institute (1988) *The Future of the Tropical Rainforest.* Oxford Forestry Institute, Oxford.

Palmer, J. (1991) Jari: Lesson for land managers in the tropics. In: Gomez-Pompa, A., Whitmore, T.C. and Hadley, M. (eds) *Rainforest Regeneration and Management.* UNESCO, Paris and Parthenon, Carnforth, pp. 419–430.

Palo, M. and Salmi, J. (1987–1990) *Deforestation or Development in the Third World?,* vols I–III. Finnish Forest Research Institute, Helsinki.

Panzer, K.F. (1976) [*Quantification of Heart Rot in Hollow Trees in the Mixed Dipterocarp Forest of Sarawak (Borneo)*]. Mitt. Bundesforschanst. Forst-Holzwirtsch., no. 111, Hamburg.

Paoletti, M.G., Taylor, R.A.J., Stinner, B.R., Stinner, D.H. and Benzing, D.H. (1991) Diversity of soil fauna in the canopy and forest floor of a Venezuelan cloud forest. *Journal of Tropical Ecology* 7, 373–383.

Parnwell, M.J.G. (1993) Environmental change and Iban communities in Sarawak, East Malaysia: impact and response. 15th International Botanical Congress (in press).

Parnwell, M.J.G. and Taylor, D.M. (1995) Environmental degradation, non-timber forest products and Iban communities in Sarawak: impact, response and future prospects. In: Parnwell, M.J.G. and Bryant, R. (eds) *Environmental Change in Southeast Asia: Rendering the Human Impact Sustainable.* Routledge, London (in press).

Paterson, S.S. (1956) *The Forest Area of the World and its Potential Productivity.* Université Royal, Göteborg.

Perring, C. (1995) Ecology, economics and ecological economics. *Ambio* 24, 60–64.

Petch, B. (1985) *The Nature and Rate of Soil Erosion in Sarawak Forests: a Review.* Forest Department Sarawak, Kuching.

Peters, C.M., Gentry, A.H. and Mendelsohn, R. (1989) Valuation of an Amazonian rain forest. *Nature* 339, 655–656.

Philips, C. (1993) Preliminary observations on the effects of logging on a hill forest in Sabah. Cited in Putz (1993).

Pinard, M.A. (1994) The reduced-impact logging project. A pilot carbon offset project in Sabah, Malaysia. *Tropical Forest Update* 4, 11–12.

Pitt, C.J.W. (1961) *Application of Silvicultural Methods to Some Forests of the Amazon.* FAO, Rome.

Poker, J. (1989) *Struktur und Wachstum in selektiv genutzten Beständen im Grebo National Forest von Liberia.* GTZ, Eschborn.

Poker, J. (1992) [Structure and dynamics of the forest mosaic of tropical rainforests]. Thesis, Faculty of Biology, Hamburg University.

Poker, J. (1995) Vegetation pattern, structure and dynamics in different types of humid tropical evergreen forests of Southeast Asia. In: Box, E.O. *et al.* (eds) *Vegetation Science in Forestry.* Kluwer Academic Publ., Dordrecht, pp. 557–571.

Poore, D. and Sayer, J. (1988) The management of tropical moist forest lands: ecological guidelines. IUCN Tropical Forest Programme, Gland, pp. 72. See also: 'Overview', natural forest management for sustainable timber production. A report by Poore *et al.* to IUCN, Gland.

Poore, D., Burgess, P., Palmer, J., Rietbergen, S. and Synnott, T. (1989) *No Timber Without Trees.* Earthscan Publications, London.

Porrit, V. (1993) Economic development: Forestry, a review of forestry in Sarawak. Thesis, Murdoch University, Australia.

Posey, D.A. and Balee, W. (eds) (1989) *Resource Management in Amazonia, Indigenous and Folk Strategies.* Advances in Economic Botany 7.

Prabhu, B.R., Weidelt, H.J. and Leinert, S. (1993) *Sustainable Management of Tropical Rainforests: Experiences, Risks, and Opportunities. An Investigation Based on Four Case Studies.* Federal Ministry for Economic Cooperation and Development, Bonn.

Prause, G. and Randow, Th.v. (1985) *Der Teufel in der Wissenschaft. Wehe, wenn Gelehrte irren: vom Hexenwahn bis zum Waldsterben. [The Devil in Science. Beware When Scientists Err: from Witch Hunt to Forest Decline].* Rasch u. Röhrig, Hamburg–Zürich.

Pretzsch, H. (1993) *Analyse und Reproduktion räumlicher Bestandesstrukturen. Versuche mit dem Struktur-generator STRUGEN.* Sauerländer Verlag, Frankfurt.

Pretzsch, J. (1987) *Die Entwicklungsbeiträge von Holzexploitation und Holzindustrie in Ländern der feuchten Tropen, dargestellt am Beispiel der Elfenbeinküste.* Schriftenr. Inst. Landespflege, University Freiburg, no. 11.

Prevost, M.-F. and Sabatier, D. (1993) *Variations Spatials de la Richesse et de la Diversite du Peuplement Arbore en Foret Guyanaise.* Comm. Coll. Phytographic Tropicale, Realites et perspectives, Paris.

Primack, R.B. (1991) Logging, conservation and native rights in Sarawak forests from different viewpoints. *Borneo Research Bulletin* 23, 3–13.

Primack, R.B. and Hall, P. (1991) Species diversity, research in Bornean forests with implications for conservation biology and silviculture. *Tropics* 1, 91–111.

Primack, R.B., Chai, E.O.K., Tan, S.S. and Lee, H.S. (1987) The silviculture of dipterocarp trees in Sarawak, Malaysia I. Introduction; II. Improvement fellings in primary forest stands. III. Plantation forests. IV. Seedling establishment in logged forest. *Malay Forestry* 50, 29–42, 43–61, 148–160, 162–178.

Proctor, J. (ed.) (1989) *Mineral Nutrients in Tropical Forest and Savanna Ecosystems.* Blackwell Scientific Publications, Oxford.

Proctor, J. (1992) Soils and mineral nutrients: what do we know, and what do we kneed to know, for wise rainforest management. In: Miller, F.R. and Adam, K.L. (eds) *Wise Management of Tropical Forests 1992.* OFI, Oxford, pp. 27–36.

Proctor, J., Anderson, J.M., Chai, P. and Vallack, H.W. (1983a) Ecological studies in four contrasting lowland rainforests in Gunung Mulu National Park, Sarawak. I. Forest environment, structure and floristics. *Journal of Ecology* 71, 237–260.

Proctor, J., Anderson, J.M., Fogden, S.C.L. and Vallack, H.W. (1983b) Ecological studies in four contrasting lowland rainforests in Gunung Mulu National Park, Sarawak. II. Literfall, litter standing crop and preliminary observation in herbivory. *Journal of Ecology* 71, 261–283.

Putz, F.E. (1993) Studies in silviculture and forest ecology in Sabah. Proceedings of a course on research methods in silviculture and ecology, 11–24 July 1993.

Putz, I.E. and Pinard, M. (1991) *Natural Forest Management in the American Tropics, an Annotated Bibliography.* Tropical Conservation and Development Program, University of Florida, Gainesville.

Putz, F.E. and Pinard, M.A. (1993) *Reducing the Impacts of Logging as a Carbon Offset Method.* Department of Botany, University of Florida, Gainesville.

Queensland Department of Forestry (undated) *Guidelines for the Selective Logging of Rainforest Areas in North Queensland State Forests and Timber Reserves. (2) Tree Marking for Mill Logs.* Queensland Department of Forestry, Brisbane, Australia.

Queensland Department of Forestry (1983) *50 Years of Rainforest Research in North Queensland.* Queensland Department of Forestry, Brisbane, Government Printer.

Rai, S.N. (1989) Rate of diameter growth of tree species in humid tropics of Western Ghats, India. In: Wan Razali, W.M., Chan, H.T. and Appanah, S. (eds) *Growth and Yield in Tropical Mixed-moist Forests*, FRIM, Kepong, pp. 106–116.

Rama Krishna, K. and Woodwell, G. (eds) (1993) *World Forests for the Future. Their Use and Conservation.* Yale University, New Haven.

Reifsnyder, W.F. (1967) Forest meteorology: the forest energy balance. In: *International Review of Forest Research*, vol. 2. Academic Press, London.

Remigo, A.A. (1992) Philippine forest resource policy in the Marcos and Aquino governments: a comparative assessment. Proceedings Workshop on The Political Ecology of South East Asia's Forests: 23–24 March 1992, SOAS, London University. In: Stott, P. (ed.) *Global Ecology and Biogeography Letters*, Vol. 3, 1993.

Reyes, M.R. (1958) Useful derivation and application of diameter growth of commercial dipterocarps in the Basilian Working Circle. *Philippine Journal of Forestry* 2, 16–21.

Richards, P.W. (1936) Ecological observations on the rainforest of Mt. Dulit, Sarawak. *Journal of Ecology* 24, 1–60.

Richards, P.W. (1952) *The Tropical Rainforest.* Cambridge University Press, Cambridge.

Richards, P.W. (1965) Soil conditions in some Bornean lowland plant communities. In: *Proceedings UNESCO Symposium Ecol. Res. in Humid Trop. Veg., Kuching, 1963.* UNESCO, Tokyo, pp. 198–204.

Richter, D.D., Saplaco, S.R. and Nowak (1985) Water management problems in humid tropical uplands. *Nature and Resources* 21, 10–21.

Riswan, S. (1991) Kerangas forest at Gunung Pasir, Samboja, East Kalimantan: status of nutrients in the leaves. In: *Fourth Round-Table Conference on Dipterocarps.* Biotropical Special Publication 41, pp. 279–294.

Riswan, S. and Kartaniwata, K. (1988) A lowland dipterocarp forest 35 years after pepper plantation in East Kalimantan, Indonesia. In: Soemodihardjo, S. (ed.) *Some Ecological Aspects of Tropical Forests of East Kalimantan.* Indonesian Institute of Science and Indonesian National MAB Committee, Jakarta, pp. 1–39.

Riswan, S. and Kartaniwata, K. (1989) Regeneration after disturbance in a lowland mixed dipterocarp forest in East Kalimantan, Indonesia. *Ecology Indonesia* 1, 9–28.

Riswan, S. and Kartaniwata, K. (1991) Species strategy in early stage of secondary succession associated with soil properties status in a lowland mixed dipterocarp forest and kerangas forest in East Kalimantan. *Tropics* 1, 13–34.

Riswan, S., Kenworthy, J.B. and Katarniwata, K. (1985) The estimation of temporal processes in tropical rainforest: A study of primary mixed dipterocarp forest in Indonesia. *Journal of Tropical Ecology* 1, 171–182.

Roberts, J. and Cabral, O.M.R. (1993) ABRACOS: a comparison of climate, soil moisture and physiological properties of forests and pastures in the Amazon basin. *Commonwealth Forestry Review* 72, 310–315.

Rodgers, W.A. (1991) Forest preservation plots in India. I. Management status and value. *Indian Forester* 117, 425–433.

Roos, M.C. (1993) State of affairs regarding Flora Malesiana: progress in revision work and publication schedule. *Flora Malesiana Bulletin* 11, 133–142.

Roth, H.L. (1896) *The Natives of Sarawak and British North Borneo.* Truslove and Hanson, London.

Routledge, R.D. (1980) The effect of potential catastrophic mortality and other un-predictable events on optimal forest rotation policy. *Forest Science* 26, 389–399.

Rukuba, M.L.S.B. (1992) Linkages between wise forest management for forest production and local markets and industries: Example from Uganda. In: Miller, F.R. and Adam, K.L. (eds) *The Wise Management of Tropical Forests 1992.* Oxford Forestry Institute, Oxford, pp. 161–169.

Rule, A. (1947) Exploitation of tropical forests and the problem of secondary species. *Empire Forestry Journal,* Rev. 26, 83–86.

Sagal, A.P.S. (1991) Gap cutting in the hill forests. Asia Pacific Forest Industries, May 1991, pp. 23–25.

Salati, E. (1987) The forest and the hydrological cycle. In: Dickinson, R.E. (ed.) *The Geophysiology of Amazonia.* John Wiley, New York.

Salleh Mohd. Nor (1992) *Forestry Research in the Asia-Pacific.* FORSPA, publ. 1.

Salmon, G. (1993) Conservation and environmental management: an NGO's view. *Commonwealth Forestry Review* 72, 233–241.

Sanchez, P.A. (1989) Soils. In: Lieth and Werger, M.J.A. (eds) *Tropical Rainforest Ecosystems. Ecosystems of the World, vol. 14B.* Elsevier, Amsterdam, p. 714.

Sanchez, P.A. and Cochrane, T.T. (1980) Soil constraints in relation to major farming systems of tropical America. In: *Priorities for Alleviating Soil-related Constraints to Food Production in the Tropics.* International Rice Research Institute, Los Baños, pp. 107–139.

Sandin, B. (1956) The westward migration of the Sea Dayak. *Sarawak Museum Journal* 7 (NS), 54–81.

Sargent, C., Husain, T., Kotey, N.A., Mayers, J., Prah, E. , Richards, M. and Treve, T. (1994) Incentives for the sustainable management of the tropical high forest in Ghana. *Commonwealth Forestry Review* 73, 155–163.

Sarre, A. (1992) Cable logging: increasing or reducing damage to the forest? *ITTO Tropical Forest Management Update* 2, 3–5.

Saulei, S.M. and Aruga, J.A. (1993) *The Status and Prospects of Non-timber Forest Products Development in Papua New Guinea.* ITTO, PD 34/93 (F), annex pp. 11–52.

Sawyer, J.A. (1993) *Plantations in the Tropics – Environmental Concerns.* IUCN Forest Conservation Programme.

Sayer, J.A., McNeely, J.A. and Stuart, S.N. (1989) The conservation of tropical forest vertebrates. In: *International Symposium on Vertebrate Biogeography and Systematics in the Tropics, Bonn, 6–8 June 1989.* IUCN, Gland.

Sayer, J.A., Harcourt, C.S. and Collins, N.M. (1992) The Conservation Atlas of Tropical Forests: Africa. Simon and Schuster, New York.

Scharpenseel, H.W. and Pfeiffer, E.M. (1990) [Soil organic matter and nutrient availability in the tropics]. *Giessener Beitrräge zur Entw. forschung, ser. I* 18, 53–66.

Schlich, Sir W. (1889) *Schlich's Manual of Forestry. Vol. I. The Utility of Forests and Principles of Silviculture. Vol. II. (1891) Silviculture. Vol. III (1895) Forest Management. Vol. IV (1895) Forest utilization* (by W.R. Fisher). *Vol. V (1908) Forest protection* (by W.R. Fisher). Bradbury & Agnew, London.

Schmidt, R. (1987) Tropical rainforest management: A status report. *Unasylva* 156, 2–17.

Schmidt-Lorenz, R. (1986) Die Böden der Tropen und Subtropen. [The soils of the tropics and subtropics] In: Blanckenburg, P.v. and Cremer, H.D. (eds) *Handbuch der Landwirtschaft und Ernährung in des Entwicklungsländern. Vol. 3 Grundlagen des Pflanzenbaus in den Tropen und Subtropen.* Ulmer, Stuttgart, pp. 46–92.

Scholz, I. and Wiemann, J. (1993) *Ecological Requirements to be Satisfied by Consumer Goods – a New Challenge for Developing Countries' Exports to Germany.* German Development Institute, Berlin.

Schönwiese, C.D. (1988) Der Einfluß des Menschen auf das Klima. *Naturwissenschaftliche Rundschau* 41, 387–390.

Schulz, J.P. (1960) *Ecological Studies on Rainforest in Northern Surinam.* North Holland Publ., Amsterdam.

Schulze, E.D. and Mooney, H.A. (1993) *Biodiversity and Ecosystem Function.* Springer, Berlin.

Seibold, E. (1988) Vorhersagen in den Geowissenschaften. *Naturwissenschaftliche Rundschau* 41, 351–357.

Senada, D. (1977) The role of forestry in the socio-economic development of rural Sarawak. *Malay Forestry* 2–13.

Seward, P.D. and Woomer, P.L. (1992) *The Biology and Fertility of Tropical Soils.* TSBF, c/o UNESCO–ROSTA, Nairobi.

Sharma, N. (ed.) (1992) *Managing the World's Forests.* Kendall and Hunt Publ., Falls Church, VA.

Shearer, W. (1985) Bioprecipitation, a tropical forest interaction. In *Frontiers, the Science Column.* United Nations University Bulletin, December 1985.

Sheil, D. (1994) Evaluating the nature of longterm change in tropical forests: lessons from half a century of Ugandan experience. *Journal of Tropical Ecology* (in press).

Sheil, D., Burslem, D.F.R.D. and Alder, D. (1995) The interpretation and misinterpretation of mortality rate measures. *Journal of Ecology* 83, 331–333.

Shepherd, G. (1993) Local and national level forest management strategies – competing priorities at the forest boundary. *Commonwealth Forestry Review* 72, 316–318.

Shepherd, K.R. and Richter, H.V. (eds) (1983) *Managing Tropical Forest.* Australian National University, Canberra.

Shuttleworth, W.J. and Nobre, C.A. (1992) Wise forest management and its linkages to climate change. In: Miller, F.R. and Adam, K.L. (eds) *Wise Management of Tropical Forests 1992.* Oxford Forestry Institute, Oxford, pp. 77–90.

Shyamsunder, S. and Parameswarappa, S. (1987) Forestry in India – the forester's view. *Ambio* 16, 332–337.

Silva, J.N.M., Carvalho, J.O.P. de and Lopes C.A. do (1989) Growth of a logged-over tropical rainforest of the Brazilian Amazon. In: Wan Razali, W.M., Chan, H.T. and Appanah, S. (eds) *Growth and Yield in Tropical Mixed-moist Forests.* FRIM, Kepong, pp. 117–136.

Silver, W.L. and Vogt, K.A. (1993) Fine root dynamics following single and multiple disturbances in a subtropical wet forest ecosystem. *Journal of Ecology* 81, 729–738.

Smith, C.R. (1978) *Diagnostic Sampling and Forest Class Survey in Remnant Mixed Dipterocarp Forest Ten Years after Logging.* Forest Research Report SR20, Forest Department Sarawak, Kuching.

Smythies, B.E. (1963) History of forestry in Sarawak. *Malay Forestry* 26, 232–253.

Soedjito, H. (1988) Spatial patterns, biomass, and nutrient concentrations of root systems in primary and secondary forest trees of a tropical rainforest in Kalimantan, Indonesia. In: Soemodihardjo, S. (ed.) *Some Ecological Aspects of Tropical Forests of East Kalimantan.* Indonesian Institute of Science and Indonesian National MAB Committee, Jakarta, pp. 41–59.

Sokolov, V.E., Shilova, S.A. and Shchipanov, N.A. (1994) Peculiarities of small mammal populations as criteria for estimating anthropogenic impacts on tropical ecosystems. *International Journal of Ecology and Environmental Science* 20, 375–386.

Solbrig, O.T. (ed.) (1991a) *From Genes to Ecosystems: A Research Agenda for Biodiversity.* Report of an IUBS–SCOPE–UNESCO Workshop. IUBS, Paris.

Solbrig, O.T. (1991b) *Biodiversity, Scientific Issues and Collaborative Research Proposals.* MAB Digest 9, Paris.

Solbrig, O.T., Emden, H.M. van, Oordt, P.G.W.J. van (eds) (1992) *Biodiversity and Global Change.* IUBS, Paris.

Sombroek, W.C. (1966) *Amazon Soils.* Center for Agric. Public and Doc., Wageningen.

Sombroek, W.G. (1986) The need for common methodologies. EC-Tropenbos Consultation, October 1986. International Soil Reference and Information Center, NL 6700 AJ, Wageningen.

Sombroek, W.G. (1990) Soils on a warmer earth: tropical regions. In: Scharpenseel, H.W., Schomaker, M. and Ayoub, A. (eds) *Soils on a Warmer Earth.* Elsevier, Amsterdam, pp. 157–174.

Sombroek, W.G., Nachtergaele, F.O. and Hebel. A. (1993) Amounts dynamics and sequestering carbon in tropical and subtropical soils. *Ambio* 22, 417–426.

Songan, P. (1993) A naturalistic inquiry into participation of the Iban peasants in the land development project in the Kalaka and Saribas districts, Sarawak, Malaysia. *Borneo Research Bulletin* 25, 101–122.

Speidel, G. (1984) *Forstliche Betriebswirtschaftslehre,* 2nd edn. Paul Parey, Hamburg, pp. 43–49.

Spurway, B.I.C. (1993) Jelutong as a plantation crop. *Malay Forestry* 2, 178–179.

Steenis, C.G.G.I. van (1958) *Vegetation Map of Malesia, 1,5,000,000 with Commentary.* UNESCO Humid Tropics Research Project, Paris.

Steinlin, H. (1985) Waldschäden weltweit. *Forstarchiv* 52, 61–65.

Steinlin, H. (1987) Kommerzielle Nutzung und Export von Holz aus tropischen Feuchtwäldern. *Allgemeine Forst- und Jagdztgeitung* 158, 50–55.

Stocker, G.C. (1985) Aspects of gap regeneration theory and the management of tropical rainforests. In: Shepherd, K.R. and Richter, H.V. (eds) *Managing the Tropical Forests.* Development Studies Center, Canberra.

Stoorvogel, J.J. (1993) *Gross Inputs and Outputs of Nutrients in Undisturbed Forest, Tai area, Côte d'Ivoire.* Tropenbos Series 5, Wageningen.

Sutisna, M. (1994) Cost effectiveness of TPTI system. Paper 2, app. 12, in MOF–CIRAD (1994).

Sutlive, V.H. (1992) *Tun Jugah of Sarawak. Colonialism and the Iban response.* Penerbit Fajar Bakti, Kuala Lumpur.

Sutton, S.L., Whitmore, T.C. and Chadwick, A.C. (eds) (1983) *Tropical Rainforest Ecology and Management.* Blackwell Scientific Publications, Oxford.

Swaine, M.D., Hawthorne, W.D., Orgle, T.K. and Agyemang, V.K. (1990) Fifteen years of forest succession in Ghana. *University of Aberdeen Tropical Biology Newsletter* no. 58, 1–2.

Swift, M.J. (1986) *Tropical Soil Biology and Fertility: Inter-regional Research Planning Workshop*. IUBS and UNESCO–MAB, Biology International Special Issue 13.

Synnott, T.J. (1979) *A Manual of Permanent Plot Procedures for Tropical Rainforests*. Tropical Forest Paper no. 14, Commonwealth Forestry Institute, Oxford.

Tans, S.S. and Lee, B.M.H. (1986) Engkabang plantations in Sarawak. 9th Mal. For. Conf. 13–20 October 1986, Kuching.

Tang, H.T. and Watley, H.E. (1976) *Report on the Survival and Development Survey of Areas Reforested by Line-planting in Selangor*. Research Pamphlet no. 67, Forest Department HQ, Peninsular Malaysia.

Tang, T. (1994) Conclusions of recent workshop on use of tropical hardwoods in Hong Kong. Seminar on trade of timber from sustainably managed forest. 5–6 April 1994, Kuala Lumpur, Malaysia, pp. 17.

Tansley, A. (1923) *Introduction to Plant Ecology*. Allen & Unwin, London.

Taylor, D.M., (1992) Pollen evidence from Muchoya Swamp, Rukiga Highlands (Uganda), for abrupt changes in vegetation during the last 21,000 years. *Bulletin de Société Geologie de France* 163, 77–82.

Taylor, D.M., Hortin, D., Parnwell, M.J.G., King, V.T. and Marsden, T.K. (1994) The degradation of rainforests in Sarawak, East Malaysia, and its implications for future management policies. *Geoforum* 25, 351–369.

Taylor, D.M., Hamilton, A.C., Whyatt, J.D., Mucunguzi, P. and Ziraba, R.B. (1995) A 25 year record of tree population dynamics in mid-altitude, semi-deciduous tropical forest. 1. Preliminary analysis of data from Mpanga Research Forest, 1968–1993. *Journal of Tropical Ecology* (in press).

Thang, H.C. (1987) *Selective Management System Concept and Practice (Peninsular Malaysia)*. Forest Department, Kuala Lumpur.

Thang, H.C. and Yong, T.K. (1989) Status of growth and yield studies in Peninsular Malaysia. In: Wan Razali, W.M., Chan, H.T. and Appanah, S. (eds) *Growth and Yield in Tropical Mixed-moist Forests*. FRIM, Kepong, pp. 137–148.

Thieme, F.K. (1988) [Roughness of the canopy and bird-species richness in northwestern German coniferous forests in Forest District Sellhorn]. Msc thesis, Faculty Biology, University of Hamburg.

Tierney, I. (1991) A wager on the world's resources. *Dialogue* 4, 60–65.

Tie Yiu Liong (1982) *Soil Classification in Sarawak*. Soils Division, Research Branch, Department of Agriculture Sarawak, Kuching.

Titin, J. and Glauner, R. (1993) Residual arsenic in soils in forest stands treated with sodium arsenite under the 'Modified' Malaysian Uniform System. Soil Science Conference of Malaysia, Penang, 1993. Forest Research Centre, Sepilok, Sandakan, Library No. 136.

Toman, M.A. (1992) The difficulty in defining sustainability. *Resources* 106, 3–6.

Touche, D. (1993) *Integrated Land-use Planning for Sarawak. Vol. II. Spatial Planning Framework*. Sarawak Agricultural Development Project, State Planning Unit, Sarawak. Kuching.

Trefil, I. (1991) Earth's future climate. *Dialogue* 3, 25–31.

Trevor, C.G. and Smythies, E.A. (1923) *Practical Forest Management*. Government Press, Allahabad.

Tropenbos (1986–1987) Information Series 1–4; Technical Series 1. The Tropenbos Foundation, Ede, Niederland.

Troup, R.S. (1921) *The Silviculture of Indian Trees*. Clarendon Press, Oxford.

Troup, R.S. (1928) *Silvicultural Systems*. Clarendon Press, Oxford.

Udarbe, M.P., Uebelhör, K., Klemp, C.D., Kleine, M., von der Heide, B., Glauner, R.

and Benneckendorf, W. (1993) *Standards for Sustainable Management of Natural Forests in Sabah.* Sabah Forestry Department, Malaysian–German Sustainable Forest Management Project, Sandaken.

Udarbe, M.P., Glauner, R., Kleine, M. and Uebelhör, K. (1994) Sustainability criteria for Sabah. *ITTO Update* 4, 13–17.

Uebelhör, K. and Abalus, R. (1991) Silvicultural considerations for the management of logged-over Dipterocarp forest in the Philippines. Sustainable management of natural forests in the Philippines: possibility or illusion? Philippine–German Dipterocarp Forest Management Project PN 88.2047.4 Proceedings, project seminar 12–16 November 1991, pp. 111–127. GTZ, Eschborn, Manila, pp. 111–127.

Uebelhör, K. and Heyde, B. von der (1993) An approach towards multiple-use forest management in logged-over forests in Sabah. *Zeitschrift Wirschaftsgeographie Geogr.* 37, 102–116.

Uebelhör, K., Lagundino, B. and Abalus, R. (1990) *Appraisal of the Philippine Selective Logging System. Philippine–German Dipterocarp Forest Management Project Technical Report 7.* DENR, GTZ, DFS, Manila–Eschborn, 47 pp.

Uganda Protectorate Forest Policy. Entebbe, 9 June 1948. *Empire Forestry Review* 27 (1948), 272–275.

UNCED (1992) *Report of the United Nations Conference on Environment and Development*, vols I–III. United Nations.

UNESCO (1978) *Tropical Forest Ecosystems, a State-of-knowledge Report Prepared by UNESCO/UNEP/FAO.* UNESCO, Natural Resources Research, Paris.

UNESCO (1993) *Cooperative Ecological Research Project (CERP) – Phase 1.* FIT/509/CPR/40, Terminal Report, FMR/SC/ECO/93/229 (FIT). UNESCO, Paris.

Valkenburg, J.L.C.H. van and Ketner, P. (1994) Non-timber forest products of East Kalimantan. *Tropenbos Newsletter* 8, 2–4.

Vanclay, J.K. (1989) A growth model for North Queensland rainforests. *Forest Ecology and Management* 27, 245–271.

Vanclay, J.K. (1990) Effects of selection logging on rainforest productivity. *Australian Forestry* 53, 200–214.

Vanclay, J.K. (1991a) Modelling the growth and yield of tropical forests. DSc thesis, University of Queensland, Brisbane.

Vanclay, J.K. (1991b) Review: data requirements for developing growth models for tropical moist forests. *Commonwealth Forestry Review* 70, 248–271.

Vanclay, J.K. (1991c). Modelling changes in the merchantability, of individual trees in tropical rainforest. *Commonwealth Forestry Review* 70, 105–112.

Vanclay, J.K. (1994) Sustainable timber harvesting: simulation studies in the tropical rainforests of North Queensland. *Forest Ecology and Management* 69, 299–320.

Varangis, P., Primo Braga, C.A. and Takeuchi, K. (1993) Tropical timber trade policies: what impact will eco-labelling have? Environment and Trade Symposium, Federal Economic Chamber, Vienna, 22–23 March 1993.

Vavoulidou-Theodorou, E. and Babel, U. (1987) Zur Wurzelproduction in Wäldern – Methoden und Ergebnisse. *Allgemeine Forst- und Jagdzgeitung* 157, 232–238.

Veevers-Carter, W. (1991) *Riches of the Rainforest.* Oxford University Press, Singapore.

Vernhes, J.R. and Younés, T. (1993) Inventorying and monitoring biodiversity under the Diversitas programme. *Biology International* no. 27, 3–14.

Verstappen, H. Th. (1994) Climate change and geomorphology in South and

Southeast Asia. *Geo-Ecology in the Tropics* 16, 101–147.

Vester, F. (1980) *Neuland des Denkens. [New realms of thought].* DVA, Stuttgart.

Vigus, T. (ed.) (1986) *The Future of Forestry in Papua New Guinea.* Lae University of Technology, Papua New Guinea.

Vitousek, P.M. (1984) Litterfall, nutrient cycling and nutrient limitation in tropical forests. *Ecology* 65, 285–298.

Vogt, H.H. (1991) Frühe Kulturen in Amazonien (citing Gibbons, A. (1990) *Science* 248, 1488). *Naturwissenschaftliche Rundschau* 44, 111–112.

Vorhoeve, A.G. (1965) *Liberian High Forest Trees.* Agricultural Research Reports 625, State Agricultural University, Department of Plant Taxonomy and Geography, Wageningen.

Wadsworth, F.H. (1951/1952) Forest management in the Luquillo Mountains. I. The setting. II. Planning for multiple land use. III. Selection of products and silvicultural policies. *Caribbean Forestry* 12, 93–114; 13, 49–61; 13, 93–119.

Wadsworth, F.M. (1987) Applicability of Asian and African silviculture systems to naturally regenerated forests of the neotropics. In: Mergen, F. and Vincent, J.R. (eds) *Natural Management of Tropical Moist Forests.* Yale University, School of Forestry and Environment Studies, New Haven.

Waidi Sinun, Wong, W.M., Douglas, I. and Spencer, T. (1992) Throughfall, stemflow, overland flow and throughflow in the Ulu Segama rainforest, Sabah, Malaysia. *Philosophical Transactions of the Royal Society B* 335, 389–395.

Walter, H. and Breckle, S.W. (1984) *Ökologie der Erde [Ecology of the Earth], vol. 2, Spezielle Ökologie der Tropischen und Subtropischen Zonen.* Fischer Verlag, Stuttgart.

Walton, A.B. (1932) Artificial regeneration. *Malay Forestry* 1, 107–110.

Walton, A.B., Barnard, R.C. and Wyatt-Smith, J. (1952) The silviculture of Lowland Dipterocarp forest in Malaya. *Malay Forestry* 15, 181–197.

Wan Razali, W.M., Chan, H.T. and Appanah, S. (eds) (1989) *Growth and Yield in Tropical Mixed-moist Forest.* FRIM, Kuala Lumpur.

Wang, C.W. (1961) *The Forests of China.* Harvard, M.M. Cabot Foundation Publ. no. 5, Cambridge, Mass.

Watson, J.G. (1928) *Mangrove Forests of the Malay Peninsula.* Malayan Forest Record, no. 6, Kepong.

Watson, J.G. (1934) Jelutong, distribution and silviculture. *Malay Forester* 3, 57–61.

Webb, L.J. (1959) A physiognomic classification of Australian rainforest. *Journal of Ecology* 47, 551–570.

Webb, L.J. (1982) The human face in forest management. In: Hallsworth, E.G. (ed.) *Socio-economic Effects and Constraints in Tropical Forest Management.* John Wiley, Chichester and Academic Press, London.

Weck, J. (1960) Klima Index und Forstliches Produktionspotential. *Forstarchiv* 31, 101–104.

Weck, J. (1970) An improved CVP-Index for the delimitation of the potential productivity zones of the forest lands of India. *Indian Forestry* 96, 565–572.

Weidelt, H.J. (1986) *Die Auswirkungen waldbaulicher Pflegemaßnahmen auf die Entwicklung exploitierter Dipterocarpaceen-Wälder.* Göttinger Beiträge zur Land- und Forstwirtschaft der Tropen und Subtropen, 19, Gottingen.

Weidelt, H.J. and Banaag, V.S. (1982) *Aspects of Management and Silviculture of Philippine Dipterocarp Forests.* GTZ-SR vol. 132, Eschborn.

Weinland, G., Killmann, W. and Heuveldop, J. (1993) *Compensatory Plantation*

Projects. Malaysian Forest Research Institute, GTZ–FRIM, Kepong.

Weiscke, A.S. (1982) [Structure and functions in forest ecosystems: Structure comparison between Kerangas and Caatinga]. Diploma Thesis, Faculty of Biology, Hamburg University.

Whitmore, T.C. (1974) *Change with Time and the Role of Cyclones in Tropical Rain Forest on Kolombangara, Solomon Islands.* Commonwealth Forestry Institute, Oxford.

Whitmore, T.C. (1975a) *Tropical Rainforests of the Far East.* Clarendon Press, Oxford.

Whitmore, T.C. (1975b) *Palms of Malaya.* Oxford University Press, Kuala Lumpur.

Whitmore, T.C. (1984) A new vegetation map of Malesia at a scale of 1.5 million. *Journal of Biogeography*, 11, 461–471.

Whitmore, T.C. (1989) Tropical forest nutrients, where do we stand? In: Proctor, J. (ed.) *Mineral Nutrients in Tropical Forest and Savanna Ecosystems.* Blackwell Scientific Publications, Oxford.

Whitmore, T.C. (1990) *An Introduction to Tropical Rain Forest.* Clarendon Press, Oxford.

Whitmore, T.C. (1991) Tropical rain forest dynamics and its implications for management. In: Gomes-Pompa *et al.* (eds) *Rain Forest Regeneration and Management.* UNESCO, Paris and Parthenon, Carnforth.

Whitmore, T.C. and Sayer, J.A. (eds) (1992) *Tropical Deforestation and Species Extinction.* Chapman & Hall, London.

Wiebecke, C. and Bruenig, E.F. (1974) Education for World Forestry. University of Toronto, Forestry Lecture Series, 1974, 165.

Wigley, T.M.L. (1993) Climate change and forestry. *Commonwealth Forestry Review* 72, 256–264.

Wilhelmy, H. (1989) *Welt und Umwelt der Maya.* Piper, Muenchen.

Wilson, E.O. and Peter, F.M. (1988) *Biodiversity.* National Academy Press, Washington.

Winkler, H. (1914) *Die Pflanzendecke Südostborneos. [The Vegetation of South East Borneo].* Botanisches Jahrbuch 50, Suppl., pp. 188–208.

Wirawan, N. (1987) Good forest within the burned forest area in East Kalimantan. In: Kostermans, A.J.G.H. (ed.) *Proceedings Third Roundtable Conference on Dipterocarps.* UNESCO Southeast Asia, Jakarta, pp. 413–425.

Woell, H.J. (1989) *Struktur und Wachstum von kommerziell genutzten Dipterocarpaceenmischwäldern und die Auswirkungen von waldbaulicher Behandlung auf deren Entwicklung, dargestellt am Beispiel von Dauerversuchsflächen auf den Philippinen. [Structure and growth of commercially used Mixed Dipterocarp forests and the results of silvicultural treatment, examples from permanent research plots in the Philippines.]* Mitt. Bundesforsch. Anst. Forst-Holzwirtsch. Hamburg, no. 161.

Woldegabriel, G., White, T.D., Suwa, G., Renne, P., Heinzelin, J. de, Hart, W.K., Heiken, G. (1994) Ecological and temporal placement of early Pliocene hominids at Aramis, Ethiopia. *Nature* 371, 330–336.

Wood, T.W.W. (1965) A study of the correlation between some soil factors and distribution of four tree species and their regeneration in the Sungai Dalam Forest Reserve, Sarawak. In: *Proceedings, Symposium Ecology Research Humid Tropical Vegetation, Kuching, July 1963.* Government of Sarawak, Kuching Office and UNESCO Southeast Asia, Tokyo, pp. 206–216.

Woodley, E. (ed.) (1991) *Medicinal Plants of Papua New Guinea. I: Morobe Province.* Markgraf, Weikersheim and Wau Ecological Institute, PNG.

Woods, P. (1987) Drought and fire in tropical forests in Sabah – an analysis of rainfall patterns and some ecological effects. In: Kostermans, A.J.G.H. (ed.) *Proceedings Third Roundtable Conference on Dipterocarps.* UNESCO Southeast Asia Jakarta, pp. 367–383.

Woomer, P.L. and Swift, M.J. (eds) (1994) *Biological Management of Tropical Soil Fertility.* John Wiley, Chichester and Sayce Publ., Exeter.

Workshop on Funding Priorities for Ecological Research into More Effective Conservation and Management of Biologically Diverse Ecosystems in Tropical Asia, US National Science Foundation, USAID and National Research Council of Thailand, Bangkok, 27–30 March 1989, unpublished report.

World Bank (1991a) *Malaysia Forestry Subsector Study.* World Bank, Washington.

World Bank (1991b) *Forest Policy Paper.* World Bank, Washington.

World Commission on Environment and Development (1987) *Our Common Future.* Oxford University Press, Oxford.

World Resources Institute (annual series). *World Resources 1986–1994.* Basic Books, New York.

World Resources Institute (1991) *Tropical Forests: A Call for Action.* WRI, Washington.

World Resources Institute (WRI), World Conservation Union (IUCN) and United Nations Environment Programme (UNEP) (1992) *Global Biodiversity Strategy.* Washington.

Wrobel, S. (1977) Holzanatomische Untersuchungen zur Wachstumsrhythmik von drei Laubbaurmarten aus der amazonischen Caatinga. MA Thesis, Universität Hamburg.

Wyatt-Smith, J. (1963) *Manual of Malayan Silviculture for Inland Forest,* Parts I–III. Malayan Forest Record no. 23, Forest Research Institute, Kepong.

Wyatt-Smith, J. (1988) *The Management of Tropical Moist forest for the Sustained Production of Timber: Some Issues.* IUCN/IIED Tropical Forest Policy Paper no. 4.

Yamada, I. (1995) Stratification of several peatswamp forest types in Brunei Darussalam. In: Box, E.O. *et al.* (eds) *Vegetation Science in Forestry.* Kluwer Academic Publ., Dordrecht, pp. 529–544.

Yang, J.C. and Insam, H. (1991) Soil microbial biomass and relative contributions of bacteria and fungi in a tropical rain forest, Hainan Island, China. *Journal of Tropical Ecology* 7, 385–393.

Yap, S.W. (1993a) A preliminary study of seedling damage by crawler tractors in a selectively logged forest in central Sabah. In: Putz, F.E. (ed.) *Proceedings of a Course on Research Methods for Silviculture and Ecology.* Forest Research Centre, Sabah, pp. 30–37.

Yap, S.W. (1993b) Climbing habits of two *Dinochlea* species in Deramakot Forest Reserve, Sabah. In: Putz, F.E. (ed.) *Proceedings of a Course on Research Methods for Silviculture and Ecology.* Forest Research Centre, Sabah, pp. 140–152.

Yap, S.W. and Majuakim, L. (1993c) Dipterocarp seedling density in relation to canopy cover in Deramakot Forest Reserve, Sabah. In: Putz, F.E. (ed.) *Proceedings of a Course on Research Methods for Silviculture and Ecology.* Forest Research Centre, Sabah, pp. 220–228.

Yap, S.W., Tay, J., Ong, R.C. and Tangah, J. (1993d) Logging damage by reduced impact and conventional logging. In: Putz, F.E. (ed.) *Proceedings of a Course on Research Methods for Silviculture and Ecology.* Forest Research Centre, Sabah, pp. 54–60.

Yap, S.K. and Lee, S.W. (eds) (1992) *In Harmony with Nature*. Proceedings International Conference Conservation of Tropical Biodiversity. Malayan Nature Society, Kuala Lumpur.

Yeo, N.T. (1987) *Result of a Skid Trail Density Study in Hill Mixed Dipterocarp Forest*. Forest Research Report FE/1987, Forest Department Sarawak, Kuching.

Yoneda, T. and Tamin, R. (1990) Dynamics of above-ground big woody organs in a foothill Dipterocarp forest, West Sumatra, Indonesia. *Ecological Research (Japan)* 5, 111–130.

Yu, Z.Y. and Pi, Y.P. (1984) *The Studies on the Reconstruction of Forest Vegetation in Coastal Eroded Land in Guangdong*. South China Institute of Botany, A.C., Guangzhou.

Zoehrer, F., Forster, H. and Schindele, W. (1987) *Forsteinrichtungswerke für Entwicklungsländer*. Feldkirchen, München, DFS-Mitt. 6.

Reference to Major Newsletters

ESFRN-Newsletter and ATSAF Circular, Council for Tropical and Subtropical Agricultural Research and Secretariat, European Tropical Forest Research Network, Hans-Boeckler-Strasse 5, D-53225, Bonn, Germany. Mixture of newsletter and short professional articles, very good value.

INTECOL Newsletter (six times a year), International Association for Ecology, Savannah River Ecology Laboratory, University of Georgia, Aiken, SC 29802, USA; focus on ecology.

ISTF News (four times a year), International Society of Tropical Foresters, 5400 Grosvenor Lane, Bethesda, MD 20814, USA.

ITTO Tropical Forest Management Update. ITTO, Yokohama. Very useful mixture of newsletter and short professional notes and articles on management.

IUCN Forest Conservation Programme, The World Conservation Union, Tropical Forest Programme, Avenue du Mont Blanc, CH-1196 Gland, Switzerland.

LTER Network News. Long-Term Ecological Research Network, c/o Department of Forest Science, Oregon State University, Corvallis, OR 97331, USA.

Prosea Newsletter, Plant Resources of South-East Asia, Herbarium Bogorienses CRDB, PO Box 234, Bogor 16122, Indonesia.

South-East Asian Rainforest Research Programme Newsletter, The Royal Society, c/o Institute of South-east Asian Biology, Department of Zoology, University of Aberdeen AB9 2TN, Scotland, UK. General information and specific reports on research progress of the programme.

Tropinet, Biotropica Supplement (four times a year), Association for Tropical Biology and the Organization for Tropical Studies, PO Box DM, Duke Station, Durham, NC 27706, USA.

CIFOR News, Center of International Forestry Research, Bogor, Jalan Gunung Batu 5, Bogor 16001; POB 6596 JKPWB, Jakarta 10065.

Flora Malesiana Bulletin. Veldkamp, J.F. and Rifai, M.A. (eds), annual periodical since 1947 published by Rijksherbarium/Hortus Bobanicus, PO Box 9514, NL 2300 RA Leiden. The Netherlands.

The PROSEA Programme Plant Resources of Southeast Asia. Jansen, P.C.M., POB 341, NL 6700 RH Wageningen, The Netherlands.

UNESCO programme on Man and the Biosphere (MAB) with MAB report series, issues as the case arises; Info MAB 'Man and the Biosphere (MAB) programme', continued series; MAB Digest, a series of publications on current issues; Man and the Biosphere Book series, UNESCO Paris and Parthenon Publishing, Carnforth. Address for information: UNESCO, 7, place de Fontenoy, F 75352 Paris 07 SP France.

Tropenbos Newsletter, ed. The Tropenbos Foundation, Lawickse Allee 11, PO Box 232, NL 6700 AE Wageningen, The Netherlands, series, published several times a year. Also very informative and good value are the annual reports, latest available issue for 1993.

Index of Species and Major Non-timber Forest Products

321

Subject Index

Accident rate 89–93

Acquisition motif 73, 75, 76, 79, 84, 88, 158–9, 244, 252–4

Acrisol, orthic acrisol 6, 12–24, 39–48, 223–4

Adaptability *see* Robustness

Adat, fuzzy complex of lore and norms 79–87

Advancement, cultural 73–84, 131–2, 229–32, 255–61

Aerodynamic roughness 8, 31, 41, 44, 56, 125–7, 179, 180, 183, 191–2, 196, 201, 203, 264

Afforestation 219–21, 223–4, 265, 267
of lallang areas 219–21
multiple matching of planting stock, site and environ 219–21

African rainforest region 94–8, 134–44, 183–8, 229–32, 235–9, 242–6

African Timber Organization (ATO) 139–44

Age of trees 171–2

Agricultural improvement to stop shag 159–61

Agriculture, irrigated 79–84

Agroforestry 79–84, 203–5, 207–18

AIDS 89, 91, 92

Air-borne timber extraction 113–16

Airships for logging 113–16

Albedo 30–32, 125–7

Allelopathic compounds 12–20, 194–201

Amazonia, soils 1–6, 12–20
species richness 39–48
silviculture 144, 153, 240
protection 161

Amenity (beauty) value 87–9, 179–83, 194, 254–61

America, tropical 134–55

Anduki Forest Reserve, Brunei Darussalam 188–94

Andulau Forest Reserve, Brunei Darussalam 179–83

Annual cut reduction, Sarawak 154–8, 186, 187

Anti-tropical timber campaign 94–8, 229–32, 251–2

APARAI, NGO in Rondonia, Brazil 84–7, 242–6

Apo Kayan 152–3

Aquic spodosol *see* Podzol soils

Aramis, Ethopia 73–9

Arena Forest Reserve, Trinidad 139–44

Arenosol 12–20

Asian region 94–8, 134–9, 229–32, 242–6

Assessments of forest resource productivity 65–72, 171–2
global and regional 37, 38, 65–72
individual stands 65–72, 171–2
sustainable yield of merchantable timber 65–72, 171–2

Atmospheric pollution from PSF 194–201